T0212536

Communications in Computer and Information Science 687

Commenced Publication in 2007
Founding and Former Series Editors:
Alfredo Cuzzocrea, Dominik Ślęzak, and Xiaokang Yang

Editorial Board

More information about this series at http://www.springer.com/series/7899

Vladimir Voevodin · Sergey Sobolev (Eds.)

Supercomputing

Second Russian Supercomputing Days, RuSCDays 2016
Moscow, Russia, September 26–27, 2016
Revised Selected Papers

 Springer

Editors
Vladimir Voevodin
Research Computing Center (RCC)
Moscow State University
Moscow
Russia

Sergey Sobolev
Research Computing Center (RCC)
Moscow State University
Moscow
Russia

ISSN 1865-0929   ISSN 1865-0937 (electronic)
Communications in Computer and Information Science
ISBN 978-3-319-55668-0   ISBN 978-3-319-55669-7 (eBook)
DOI 10.1007/978-3-319-55669-7

Library of Congress Control Number: 2017934455

Printed on acid-free paper

This Springer imprint is published by Springer Nature
The registered company is Springer International Publishing AG
The registered company address is: Gewerbestrasse 11, 6330 Cham, Switzerland

Preface

The Second Russian Supercomputing Days Conference (RuSCDays 2016) was held during September 26–27, 2016, in Moscow, Russia. It was organized by the Supercomputing Consortium of Russian Universities and the Federal Agency for Scientific Organizations. The conference was supported by the Russian Foundation for Basic Research and our respected platinum sponsors (T-Platforms, RSC, Intel, NVIDIA), gold sponsors (IBM, Mellanox, Dell EMC, Hewlett Packard Enterprise), and silver sponsor (NICEVT). The conference was organized in a partnership with the ISC High-Performance conference series and NESUS project.

The conference was born in 2015 as a union of several supercomputing events in Russia and quickly became one of the most notable Russian supercomputing meetings. The conference caters to the interests of a wide range of representatives from science, industry, business, education, government, and students – anyone connected to the development or the use of supercomputing technologies. The conference topics cover all aspects of supercomputing technologies: software and hardware design, solving large tasks, application of supercomputing technologies in industry, exaflops computing issues, supercomputing co-design technologies, supercomputing education, and others.

All papers submitted to the conference were reviewed by three referees and evaluated for the relevance to the conference topics, scientific contribution, presentation, approbation and related works description. For this proceedings volume, a second review round was performed. The 28 best works were carefully selected to be included in this volume.

The proceedings editors would like to thank all the conference committee members, especially the Organizing and Program Committee members as well as other reviewers for their contributions. We also thank Springer for producing these high-quality proceedings of RuSCDays 2016.

January 2017

Vladimir Voevodin
Sergey Sobolev

Organization

The Second Russian Supercomputing Days Conference (RuSCDays 2016) was organized by the Supercomputing Consortium of Russian Universities and the Federal Agency for Scientific Organizations, Russia. The conference organization coordinator was Moscow State University Research Computing Center.

Steering Committee

V.A. Sadovnichiy (Chair)	Moscow State University, Russia
V.B. Betelin (Co-chair)	Russian Academy of Sciences, Moscow, Russia
A.V. Tikhonravov (Co-chair)	Moscow State University, Russia
J. Dongarra (Co-chair)	University of Tennessee, Knoxville, USA
A.I. Borovkov	Peter the Great Saint-Petersburg Polytechnic University, Russia
Vl.V. Voevodin	Moscow State University, Russia
V.P. Gergel	Lobachevsky State University of Nizhni Novgorod, Russia
G.S. Elizarov	NII Kvant, Moscow, Russia
V.V. Elagin	Hewlett Packard Enterprise, Moscow, Russia
A.K. Kim	MCST, Moscow, Russia
E.V. Kudryashova	Northern (Arctic) Federal University, Arkhangelsk, Russia
N.S. Mester	Intel, Moscow, Russia
E.I. Moiseev	Moscow State University, Russia
A.A. Moskovskiy	RSC Group, Moscow, Russia
V.Yu. Opanasenko	T-Platforms, Moscow, Russia
G.I. Savin	Joint Supercomputer Center, Russian Academy of Sciences, Moscow, Russia
V.A. Soyfer	Samara University, Russia
L.B. Sokolinskiy	South Ural State University, Chelyabinsk, Russia
I.A. Sokolov	Russian Academy of Sciences, Moscow, Russia
R.G. Strongin	Lobachevsky State University of Nizhni Novgorod, Russia
A.N. Tomilin	Institute for System Programming of the Russian Academy of Sciences, Moscow, Russia
A.R. Khokhlov	Moscow State University, Russia
B.N. Chetverushkin	Keldysh Institutes of Applied Mathematics, Russian Academy of Sciences, Moscow, Russia

| E.V. Chuprunov | Lobachevsky State University of Nizhni Novgorod, Russia |
| A.L. Shestakov | South Ural State University, Chelyabinsk, Russia |

Program Committee

Vl.V. Voevodin (Chair)	Moscow State University, Russia
R.M. Shagaliev (Co-chair)	Russian Federal Nuclear Center, Sarov, Russia
M.V. Yakobovskiy (Co-chair)	Keldysh Institutes of Applied Mathematics, Russian Academy of Sciences, Moscow, Russia
T. Sterling (Co-chair)	Indiana University, Bloomington, USA
A.I. Avetisyan	Institute for System Programming of the Russian Academy of Sciences, Moscow, Russia
D. Bader	Georgia Institute of Technology, Atlanta, USA
P. Balaji	Argonne National Laboratory, USA
M.R. Biktimirov	Russian Academy of Sciences, Moscow, Russia
A.V. Bukhanovskiy	ITMO University, Saint Petersburg, Russia
J. Carretero	University Carlos III of Madrid, Spain
V.E. Velikhov	National Research Center "Kurchatov Institute", Moscow, Russia
V.Yu. Volkonskiy	MCST, Moscow, Russia
V.M. Volokhov	Institute of Problems of Chemical Physics of Russian Academy of Sciences, Chernogolovka, Russia
R.K. Gazizov	Ufa State Aviation Technical University, Russia
B.M. Glinskiy	Institute of Computational Mathematics and Mathematical Geophysics, Siberian Branch of Russian Academy of Sciences, Novosibirsk, Russia
V.M. Goloviznin	Moscow State University, Russia
V.A. Ilyin	National Research Center "Kurchatov Institute", Moscow, Russia
I.A. Kalyaev	NII MVS, South Federal University, Taganrog, Russia
H. Kobayashi	Tohoku University, Japan
V.V. Korenkov	Joint Institute for Nuclear Research, Dubna, Russia
V.A. Kryukov	Keldysh Institutes of Applied Mathematics, Russian Academy of Sciences, Moscow, Russia
J. Kunkel	University of Hamburg, Germany
J. Labarta	Barcelona Supercomputing Center, Spain
A. Lastovetsky	University College Dublin, Ireland
M.P. Lobachev	Krylov State Research Centre, Saint Petersburg, Russia
Y. Lu	National University of Defense Technology, Changsha, Hunan, China
T. Ludwig	German Climate Computing Center, Hamburg, Germany
V.N. Lykosov	Institute of Numerical Mathematics, Russian Academy of Sciences, Moscow, Russia

M. Michalewicz	University of Warsaw, Poland
A.V. Nemukhin	Moscow State University, Russia
G.V. Osipov	Lobachevsky State University of Nizhni Novgorod, Russia
A.V. Semyanov	Lobachevsky State University of Nizhni Novgorod, Russia
Ya.D. Sergeev	Lobachevsky State University of Nizhni Novgorod, Russia
H. Sithole	Centre for High Performance Computing, Cape Town, South Africa
A.V. Smirnov	Moscow State University, Russia
R.G. Strongin	Lobachevsky State University of Nizhni Novgorod, Russia
H. Takizawa	Tohoku University, Japan
M. Taufer	University of Delaware, Newark, USA
H. Torsten	ETH Zurich, Switzerland
V.E. Turlapov	Lobachevsky State University of Nizhni Novgorod, Russia
E.E. Tyrtyshnikov	Institute of Numerical Mathematics, Russian Academy of Sciences, Moscow, Russia
V.A. Fursov	Samara University, Russia
L.E. Khaymina	Northern (Arctic) Federal University, Arkhangelsk, Russia
B.M. Shabanov	Joint Supercomputer Center, Russian Academy of Sciences, Moscow, Russia
N.N. Shabrov	Peter the Great Saint-Petersburg Polytechnic University, Russia
L.N. Shchur	Higher School of Economics, Moscow, Russia
R. Wyrzykowski	Czestochowa University of Technology, Poland
M. Yokokawa	Kobe University, Japan

Industrial Committee

A.A. Aksenov (Co-chair)	Tesis, Moscow, Russia
V.E. Velikhov (Co-chair)	National Research Center "Kurchatov Institute", Moscow, Russia
V.Yu. Opanasenko (Co-chair)	T-Platforms, Moscow, Russia
Yu.Ya. Boldyrev	Peter the Great Saint-Petersburg Polytechnic University, Russia
M.A. Bolshukhin	Afrikantov Experimental Design Bureau for Mechanical Engineering, Nizhny Novgorod, Russia
R.K. Gazizov	Ufa State Aviation Technical University, Russia
M.P. Lobachev	Krylov State Research Centre, Saint Petersburg, Russia
V.Ya. Modorskiy	Perm National Research Polytechnic University, Russia

A.P. Skibin	Gidropress, Podolsk, Russia
S. Stoyanov	T-Services, Moscow, Russia
N.N. Shabrov	Peter the Great Saint-Petersburg Polytechnic University, Russia
A.B. Shmelev	RSC Group, Moscow, Russia
S.V. Strizhak	Hewlett-Packard, Moscow, Russia

Educational Committee

V.P. Gergel (Co-chair)	Lobachevsky State University of Nizhni Novgorod, Russia
Vl.V. Voevodin (Co-chair)	Moscow State University, Russia
L.B. Sokolinskiy (Co-chair)	South Ural State University, Chelyabinsk, Russia
Yu.Ya. Boldyrev	Peter the Great Saint-Petersburg Polytechnic University, Russia
A.V. Bukhanovskiy	ITMO University, Saint Petersburg, Russia
R.K. Gazizov	Ufa State Aviation Technical University, Russia
S.A. Ivanov	Hewlett-Packard, Moscow, Russia
I.B. Meerov	Lobachevsky State University of Nizhni Novgorod, Russia
V.Ya. Modorskiy	Perm National Research Polytechnic University, Russia
I.O. Odintsov	RSC Group, Saint Petersburg, Russia
N.N. Popova	Moscow State University, Russia
O.A. Yufryakova	Northern (Arctic) Federal University, Arkhangelsk, Russia

Organizing Committee

Vl.V. Voevodin (Chair)	Moscow State University, Russia
V.P. Gergel (Co-chair)	Lobachevsky State University of Nizhni Novgorod, Russia
V.Yu. Opanasenko (Co-chair)	T-Platforms, Moscow, Russia
S.I. Sobolev (Scientific Secretary)	Moscow State University, Russia
A.A. Aksenov	Tesis, Moscow, Russia
A.P. Antonova	Moscow State University, Russia
K.A. Barkalov	Lobachevsky State University of Nizhni Novgorod, Russia
M.R. Biktimirov	Russian Academy of Sciences, Moscow, Russia
O.A. Gorbachev	RSC Group, Moscow, Russia
V.A. Grishagin	Lobachevsky State University of Nizhni Novgorod, Russia
V.V. Korenkov	Joint Institute for Nuclear Research, Dubna, Russia
D.A. Kronberg	Moscow State University, Russia

I.B. Meerov	Lobachevsky State University of Nizhni Novgorod, Russia
I.M. Nikolskiy	Moscow State University, Russia
N.N. Popova	Moscow State University, Russia
N.M. Rudenko	Moscow State University, Russia
L.B. Sokolinskiy	South Ural State University, Chelyabinsk, Russia
V.M. Stepanenko	Moscow State University, Russia
A.V. Tikhonravov	Moscow State University, Russia
A.Yu. Chernyavskiy	Moscow State University, Russia
B.M. Shabanov	Joint Supercomputer Center, Russian Academy of Sciences, Moscow, Russia
M.V. Yakobovskiy	Keldysh Institutes of Applied Mathematics, Russian Academy of Sciences, Moscow, Russia

Russian
Supercomputing
Days 2016

Contents

The Future of Supercomputing: New Technologies

The Present of Supercomputing: Large Tasks Solving Experience

Accelerating Assembly Operation in Element-by-Element FEM on Multicore Platforms

Sergey Kopysov$^{(\boxtimes)}$, Alexander Novikov, Nikita Nedozhogin, and Vladimir Rychkov

Institute of Mechanics, Ural Branch of the Russian Academy of Sciences, 34 ul. T. Baramzinoy, Izhevsk 426067, Russia
s.kopysov@gmail.com,sc_work@mail.ru,Nedozhogin@inbox.ru,bob.r@mail.ru

Abstract. The speedup of element-by-element FEM algorithms depends not only on the peak processor performance but also on the access time to shared mesh data. Eliminating memory boundness would significantly speedup unstructured mesh computations on hybrid multicore architectures, where the gap between processor and memory performance continues to grow. The speedup can be achieved by ordering unknowns so that only those elements are processed in parallel which do not have common nodes. If vectors are composed with respect to the ordering, memory conflicts will be minimized. Mesh was partitioned into layers by using neighborhood relationship. We evaluated several partitioning schemes (block, odd-even parity, and their modifications) on multicore platforms, using Gunther's Universal Law of Computational Scalability. We performed numerical experiments with element-by-element matrix-vector multiplication on unstructured meshes on multicore processors accelerated by MIC and GPU. We achieved 5-times speedup on CPU, 40-times — on MIC, and 200-times — on GPU.

Keywords: Finite element · Element-by-element matrix-vector multiplication · Mesh partitioning · Multicore processors · Universal law of computational scalability

1 Introduction

Performance and scalability modeling of computational algorithms on modern multicore/manycore platforms is a crucial problem. On multicore platforms, it is important to take into account different types of delays, which can be caused by: exchange of shared rewritable data between the processor caches and between the processors and main memory (i), synchronization locks (serialization) shared data available for recording (ii), waiting for memory access to complete operations (iii), etc. Evolution of parallel speedup models derived from Amdahl's law is described in [1,8]. In particular, they refer to Hill-Marty model [5], which includes additional parameters defining a number of computing hardware resources. However, estimation of these parameters for real platforms is not trivial.

© Springer International Publishing AG 2016
V. Voevodin and S. Sobolev (Eds.): RuSCDays 2016, CCIS 687, pp. 3–14, 2016.
DOI: 10.1007/978-3-319-55669-7_1

For over a decade, the Universal Scalability Law [3] has been successfully applied to model diverse software applications on modern hardware platforms [4] but it has been largely ignored by the parallel simulation community. This model takes into account contention for shared resources, retrograde scaling (latency due to exchange of data between caches), saturation resulting from resource limitations. In this work, we apply this model to FEM simulations. To the best of our knowledge, this is the first application of this model to parallel numerical algorithms.

Predictive models based on the results of real computational experiments allow for detecting and eliminating the sources of delays. Many FEM algorithms demonstrate irregular memory accesses, which significantly reduces the efficiency of parallel processing. This irregularity is caused by unstructured meshes. As a result, the global assembly phase becomes the main performance bottleneck in many FEM algorithms [2, 7]. This operation is characterized by low data locality and therefore low potential for parallelization. Global assembly inherently results in memory conflicts, which can be avoided, for example, with help of OpenMP critical sections. Unfortunately, the latter significantly slow down computations.

In this paper, we design new parallel algorithms on unstructured meshes, with improved data locality and reduced delays in memory access. We propose a new ordering of FEM unknowns, which does not allow for simultaneous summations over the cells that contain common vertices, and therefore does not require resolution of memory conflicts. Using different neighborhood relationship schemes, we partition mesh into nonadjacent layers, which are then grouped into subdomains and assigned to different processes. We use and extend the Universal Scalability Law to evaluate the efficiency of these algorithms in terms of data locality.

This paper is structured as follows. In Sect. 2, we show how element-by-element FEM can be parallelized on shared memory architecture. In Sect. 3, we present new partitioning schemes, which significantly improve data locality and performance of element-by-element FEM algorithms. In Sect. 4, we describe experimental platforms and their memory features. In Sect. 5, we present an extension of the Universal Scalability Law for multi-processor/multi-socket node. In Sect. 6, we present experimental results and scalability analysis. Section 7 concludes the paper.

2   Element-by-Element FEM on Shared Memory Platforms

In this section, we show how element-by-element FEM can be parallelized for shared memory architecture.

The global assembly is a performance-critical stage of parallel FEM algorithms. It is based on some data partitioning, which targets a certain level of data locality and load balancing. The assembly stage is a part of a step of the algorithm while the partitioning stage precedes this step. In FEM, assembling usually means adding local element stiffness matrices into global stiffness matrix.

In element-by-element schemes [2], assembling is a part of solving the finite element system of equations. Assembly operator is applied not to the matrices but to the vectors resulting from the matrix-vector multiplication:

$$q = Kp = \sum_{e=1}^{m} C_e^T \tilde{K}_e C_e p = \sum_{e=1}^{m} C_e^T \tilde{q}_e = \sum_{e=1}^{m} q_e, \qquad (1)$$

where q, p, q_e are the vectors of size N; K is the global matrix $N \times N$; $\tilde{K}_e$ is the local stiffness matrix $N_e \times N_e$ of a finite element e; C_e is the incidence matrix $N_e \times N$ that maps the local space of numbers of unknowns (degrees of freedom) $[1, 2, \ldots, N_e]$ into the global space $[1, 2, \ldots, N]$; m is the number of finite elements. This mapping can be done either by indirect indexing of mesh nodes to unknowns (i) or by multiplying by the incidence matrix, which can be efficiently implemented on GPUs (ii).

Let us consider a parallel version of (1) for shared memory. Here vectors p and q are in the shared memory of parallel processes/threads, and matrices $\tilde{K}_e$, C_e, C_e^T and sparse vector q_e are in the local memory of processes. Then summation of the vectors q_e that belong to different processes but have nonzero components with the same indices may result in conflicts and cause errors.

The product $q = Kp$ can be replaced by two operations: element-by-element multiplication $\tilde{q}_e = \tilde{K}_e C_e p$, $e = 1, 2, \ldots, m$ and assembling $q_e = \sum_{e=1}^{m} C_e^T \tilde{q}_e$. These operations have different computation and communication costs because of different data locality, especially in case of unstructured meshes. They have different memory access patterns, determined by local and global node indexing, and therefore have different potential for parallelization.

Assume that each process computes over $m_i \approx m/n$ finite elements, where i is the process number, and n is the number of processes. Then the product can be expressed in the matrix form as follows:

$$q = Kp = \sum_{i=1}^{n} C^{T(i)} \tilde{K}^{(i)} C^{(i)} p = \sum_{i=1}^{n} C^{T(i)} \tilde{q}^{(i)} = \mathcal{A}(\tilde{q}), \qquad (2)$$

where $C^{(i)}$ is the incidence matrix assembled from m_i matrices C_e; $\tilde{K}^{(i)}$ is the block-diagonal matrix assembled from m_i matrices $\tilde{K}^{(i)}$ as blocks; $\tilde{q}^{(i)}$ is the vector assembled from m_i vectors $\tilde{q}_e$; and $\mathcal{A}$ is the assembly operator for the vector q. In the case of indirect indexing, the assembly operator will be denoted as $\mathcal{A}^{ind}$; in the case of the product of the incidence matrix – $\mathcal{A}^{inc}$.

To make this algorithm efficient on shared memory platforms, it is necessary to minimize concurrent memory accesses. This can be achieved by splitting finite elements into sets that do not share nodes with each other. Then, these sets are assigned to different processes/threads, and the assembly operation (2) is performed on nonadjacent elements.

Operator $\mathcal{A}^{inc}$ uses matrix C^T, which is formed by aggregating matrices C_e^T. This sparse matrix is stored in a modified CSR format so that the non-zero elements of matrix C^T are not stored.

This operator can be efficiently executed on GPU, in parallel by the rows of the matrix C^T. The matrix C^T will be formed again after the mesh is adaptive refined.

3   FEM Algorithms with Layer-by-Layer Partitioning

In this section, we present new partitioning schemes, which significantly improve data locality and performance of element-by-element FEM algorithms on shared memory platforms.

We assume that two subsets of mesh cells are nonadjacent at some moment of time if the cells simultaneously taken from these subsets do not contain common vertices. The moment of time is taken when the subsets are accessed for computations. This means that the cells in a subset and the subsets themselves can be connected topologically.

We partition the mesh Ω into nonoverlapping layers of cells $s_j, j = 1, 2, \ldots, n_s$ (Fig. 1a), then combine the layers to get the subsets of cells (subdomains) Ω_i, $i = 1, 2, \ldots, n_\Omega$. We use different schemes to construct subdomains Ω_i from layers. Our target is to reorder cells so that the cells with common vertices are not accessed simultaneously from parallel processes (threads).

In the **block scheme**, the layers are combined consecutively, then these combined layers are divided into subdomains with approximately the same number of finite elements (Fig. 2a). In the **layer index parity scheme**, the layers are enumerated (Fig. 2c). The dashed line in Fig. 2b represents the end of one parallel OpenMP section and the beginning of another.

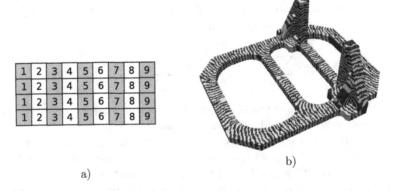

a)

b)

Fig. 1. Layer-by-layer mesh partitioning: (a) scheme; (b) result: unstructured mesh with 137 layers, 485843 tetrahedrons.

We compared these schemes by measuring the performance and load imbalance of the element-by-element matrix-vector product without assembling the vectors. Therefore, we compared fully parallel operations. We performed experiments with the unstructured mesh shown in Fig. 1 on eight-core processor Xeon

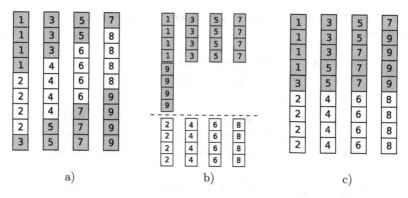

Fig. 2. Combining layers into blocks for four processes ($n_\Omega = 4$): (a) block scheme (bl); (b) layer index parity scheme (ev); (c) balanced layer index parity scheme (ev + bal).

E5-2690. Figure 3 shows the speedup obtained with four types of the mesh partitioning: block (bl), odd-parity (ev), balanced odd-parity (ev+bal), and multilevel graph partitioning by METIS). The load imbalance (in percent to uniform data distribution) is shown above the bars. The load imbalance of (bl) and (ev+bal) was negligible in all experiments.

These results show that the load imbalance can affect the speedup and indeed is a limiting factor for achieving the linear speedup in the fully parallel computational operation. The load imbalance caused by the block and balanced odd-parity mesh partitioning schemes proved to be less than the one caused by the multilevel partitioning of the mesh dual graph [6].

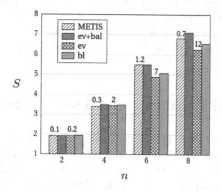

Fig. 3. The speedup of the matrix-vector product on Xeon E5-2690 for different mesh partitioning schemes and numbers of processes ($n = 2, 4, 6, 8$)

Table 1 shows the execution time of assembly operators $\mathcal{A}^{ind}$ and $\mathcal{A}^{inc}$ on mesh partitions (bl) and (ev+bal) measured on different multi-core processors.

Table 1. Execution time, t in seconds, of the assembly schemes with different numbers of threads, n_p.

Processors	Assembly schemes					
	$\mathcal{A}^{ind}(\tilde{q})$ bl		$\mathcal{A}^{ind}(\tilde{q})$ ev+bal		$\mathcal{A}^{inc}(\tilde{q})$	
Unstructured mesh						
Xeon E5-2690	8	**0.002**	8	**0.002**	8	0.007
Xeon Phi 7110X	48	**0.004**	60	0.007	240	0.007
GTX980	32	0.037	60	0.052	45356	**0.002**
Quasi-structured mesh						
Xeon E5-2690	8	**0.003**	8	**0.003**	8	0.004
Xeon Phi 7110X	60	**0.005**	60	0.007	240	**0.005**
GTX980		—		—	65536	**0.002**
	n_p	t	n_p	t	n_p	t

The results are presented for the unstructured mesh shown in Fig. 1(b) and for a quasi-structured mesh, which contains $m = 507904$ hexahedrons. The best results are highlighted in bold.

With layer-by-layer partitioning, assembly operator $\mathcal{A}^{ind}$ performed best on universal processor Xeon E5-2690 and accelerator Xeon Phi 7110X. The scalability of this operator is limited by indirect indexing and the number of layers in the mesh partitioning. The 44-times speedup was achieved for the block partitioning on the Xeon Phi 7110X with 60 OpenMP threads.

Assembly operator $\mathcal{A}^{inc}$ on GPU performed 18–26 times better than $\mathcal{A}^{ind}$. For this operator, the speedup of 94 times was achieved on the quasi-structured mesh and 65536 CUDA threads.

The rest of the paper will be focused on how layer-by-layer partitioning can be used to detect the sources of delays in parallel FEM algorithms on multi-core/manycore platforms.

4   Experimental Platforms and Their Memory Features

In this section, we describe experimental platforms and their memory features.

Parallel execution of FEM algorithms on hybrid multi-core platforms significantly increases the frequency of memory access so that memory bandwidth becomes a performance bottleneck. Multiple processes/threads compete for memory, which results in memory contention. This situation is further complicated by multi-level caches, which allow for reusing frequently accessed data but require maintaining data consistency (cache coherence).

Memory can be utilized more efficiently if multiple writes to the same memory location issued by different processes are executed sequentially (i), and there is a gap in time between write and read operations issued from different processes to the same memory location (ii).

We performed experiments with parallel element-by-element FEM algorithms on several platforms, which represent most popular multi-core architectures:

- **Xeon E5-2690**: 2 × eight-core CPUs Intel Xeon E5-2690, 64 GB RAM NUMA, L1 cache 512 KB (8 × 32 KB instructions, 8 × 32 KB data), L2 cache 8 × 256 KB, L3 cache 20 MB. The processors are connected by two QuickPath Interconnect (QPI) links, which utilize 8.0 GT/s.
- **Xeon E5-2609**: 2 × quad-core CPUs Intel Xeon E5-2609, 64 GB RAM NUMA, L1 cache 256 KB (4 × 32 KB instructions, 4 × 32 KB data), L2 cache 1 MB (4 × 256 KB), L3 cache 10 MB. Memory bandwidth 51,2 GB/s.
- **Opteron 8435**: 4 × six-core CPUs AMD Opteron 8435, HyperTransport with data rates up to 4,8 GT/s on a link, 64 GB RAM, L1cache 384 KB (6 × 64 KB), L2 cache 3 MB (6 × 512 KB), L3 cache 6 MB.
- **Xeon Phi 7110X**: single Intel Xeon Phi 7110X coprocessor, 61 cores, up to 4 simultaneous threads per core at 1.1 GHz, memory controllers, PCEe interface, bi-directional bus, L1 cache 32 KB + 32 KB, L2 cache 512 KB with 16 thread hardware prefetching, inclusive L1, L2 caches (data duplicated in L1 and L2), quick access to caches of neighbor cores (distributed catalog), RAM 8 GB. Memory bandwidth 320 GB/s.
- **NVIDIA GTX 980**: single NVIDIA GTX 980/GM204 GPU, 4 graphics processing clusters (GPC), with 4 streaming multiprocessors (SMM) each, with 32 ALU each, 2048 stream processors (SP). Global graphics memory 4096 MB (L2 cache 2 MB, universal L1 24 K); shared memory 96 KB; texture memory L1 24 KB. Memory bandwidth 224.3 GB/s. Core frequency 1126 MHz. Each SMM has four warp-schedulers.

All these platforms have different memory structure. Not all CPUs have an integrated memory controller, whereas accelerators typically have several controllers. Accelerators' memory is faster. Memory access patterns on GPUs are more regular, so that the access time is more predictable.

CPUs use cache to reduce memory latency, while GPUs use cache to increase memory bandwidth. In case of CPU, memory latency is reduced by increasing the cache size and predicting the code forking. In case of GPU, memory latency is hidden by overlapping data transfers and computations across thousands of simultaneously executed threads.

Unlike GPUs, MIC coprocessors have coherent cache, and, along with x86, use specific instruction set. Execution becomes more predictable, so that instruction scheduling can be performed by compiler. Each core supports 4-way simultaneous multithreading, with 4 copies of each processor's register. MICs provide explicit cache control instructions and prefetching in L1, L2 caches.

For accurate scalability analysis, it is important to take into account hardware and software (compiler optimizations) aspects of memory access. In the following section, we apply the Universal Scalability Law to model their combined effect.

5   Scalability Model for Multi-processor Nodes

In this section, we briefly describe the Universal Scalability Law, estimate its parameters for the chosen architectures, and propose an extended model for homogeneous multi-processor/multi-socket nodes.

The Universal Scalability Law [4] is a generalization of Amdahl's law that takes into account overheads related to synchronizations between cores/processors and maintaining data coherence. The model is considered to be universal in the sense that it does not require a prior information about algorithm (such as the parallel part in Amdahl's law), middleware software, hardware architecture. The model assumes that speedup is a function that has the global maximum value and depends on the number of cores/processors, the degree of contention, and the lack of coherency. The parameters of the USL model are estimated by regression over the results of speedup measurements: $S(n) = t_1/t_n$, where t_1, t_n is the execution time of one or n threads/processes.

The Universal Scalability Law represents the speedup by the following function:

$$S_U(n) = \frac{n}{1 + \sigma(n-1) + \lambda n(n-1)}, \tag{3}$$

where $0 \leqslant \sigma, \lambda < 1$; σ defines the degree of contention; λ defines the lack of coherency. The denominator is summed up from three terms. The second term depends linearly on the number of processes, while the third one – quadratically. They can be interpreted as follows:

– Linear scalability, ideal parallelism: $\sigma = 0, \lambda = 0$.
– Scalability is limited by contention for shared rewritable data, which results in serialization and queuing: $\sigma > 0, \lambda = 0$.
– Scalability is limited by maintaining data coherency at different levels of memory hierarchy: $\sigma > 0, \lambda > 0$.

When $\lambda = 0$, the ratio (3) is reduced to Amdahl's law:

$$S_A(n) = \frac{1}{(1-f) + \frac{f}{n}}, \tag{4}$$

where f is the parallel part of the execution, $(1-f)$ is the serial part. If $\lambda = 0$, contention represents the serial part: $\sigma = (1-f)$.

Parallel algorithms for multi-core architectures can be characterized by the values of σ and λ. These parameters allow us to assess and predict the costs of maintaining data coherence and synchronizing data access from parallel threads. The USL model provides a method of estimation of these parameters.

In Table 2, we present the parameters of the USL model estimated for the assembly operation (2) on the experimental platforms described in Sect. 4. Numerical experiments were carried out for unstructured meshes (one of them shown in Fig. 1). The balanced layer index parity partitioning scheme was used as the most efficient in terms of data movement and memory access.

Table 2. USL parameters of the assembly operation on different platforms.

	Xeon E5-2609	Opteron 8435	Xeon E5-2690	Xeon Phi 7110X	GeForce GTX 980
n	4	6	8	61	2048
σ	0.1	0.021	0.08	0.006	0.0018
λ	0.0001	0.02	0.0001	0.00006	0.00000044
R^2	0.99	0.982	0.9965	0.994	0.999
$S_{U_{\max}}/n_{U_{\max}}$	8.41/94	3.56/7	10.09/96	46.76/129	318.96/1511
$S_{\max}/n_{\max}$	3.1/4	3.48/5	5.19/8	39.43/60	219/384

Non-linear regression analysis was used to estimate the model parameters. The quality of estimation was evaluated with the R^2 measure. We found the number of processor cores at which the maximum scalability is achieved:

$$n_{U_{\max}} = \sqrt{(1-\sigma)/\lambda}. \tag{5}$$

These numbers along with the maximum speedup for our experimental platforms are also presented in Table 2.

The Universal Scalability Law can be extended to model other homogeneous platforms, such as multi-processor (multi-socket) computing nodes, which consists of identical processors. Let us define the parameters of a multi-processor computing node as follows: p is the number of multi-core processors in a single compute node; $n_p n/p$ is the number of cores per one processor; σ_p defines the degree of contention between processors; Λ_p defines the lack of coherency in the node. We generalize (3) for a multiprocessor as follows:

$$S_U(n,p) = \frac{pS(n)}{1 + \sigma_p(p-1)S(n) + \lambda_p(p-1)pS^2(n)}, \tag{6}$$

where the speedup function of the multiprocessor $S_U(n,p)$ has an extra argument, the number of processors, and depends on the speedup of individual processors $S(n)$. If $p = 1$ and $\sigma_p = 0$ (a single processor), this formula is reduced to (3). Similarly to (6), USL can be generalized for homogeneous clusters.

6   Experimental Results

In this section, we present the results of experiments with the parallel element-by-element FEM algorithms on the selected platforms. In performance and scalability analysis, we use both the original and extended USL models.

Using the Universal Scalability Law, we can find the numbers of processors for which scalability deteriorates. For example, we detected scalability problems in the parallel vector assembly on shared memory. The mesh was partitioned by a graph partitioning scheme. The parallel algorithm used critical sections, which become the main source of delays (Figs. 4 and 5).

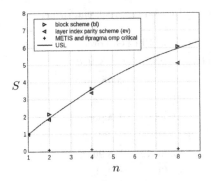

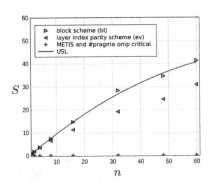

Fig. 4. Scalability of assembly operation on Xeon E5-2690

Fig. 5. Scalability of assembly operation on Xeon Phi 7110X

With the layer-by-layer partitioning algorithms proposed in this paper (Figs. 4 and 5), we can avoid using critical sections in the parallel assembly algorithm and as a result achieve better scalability on both CPUs and accelerators. Figures 6, 7, 8 and 9 show the USL models and the experimentally measured speedup of the parallel vector assembly for different platforms.

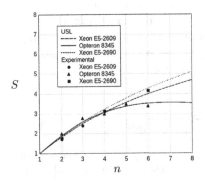

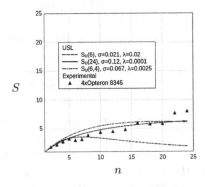

Fig. 6. Scalability of assembly operation on multi-core CPUs

Fig. 7. Scalability of assembly operation on multi-socket node

Significant, three to five-fold, acceleration of parallel assembly was achieved on Xeon E5, which is about a half of the maximum possible acceleration (see Fig. 6). This is a very good result for the computational algorithm based on unstructured meshes. Due to delays incurred by the three-level cache on 6-core Opteron, the maximum speedup was achieved when only five cores were involved in computations.

Figure 8 demonstrates that element-by-element vector assembly can be very efficiently performed on MIC coprocessors. In addition, for MIC we performed experiments with quasi-structured meshes. The assembly operation had almost the same level of scalability (approx. 40 times speedup on 61 cores).

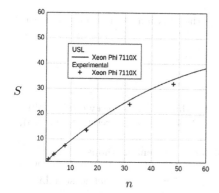

Fig. 8. Scalability of assembly operation on MIC

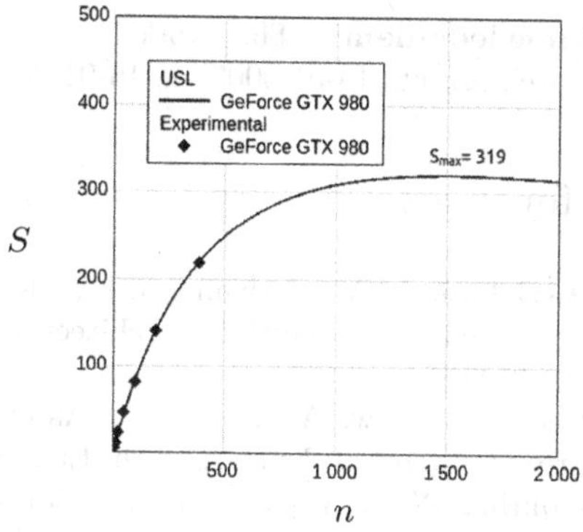

Fig. 9. Scalability of assembly operation on GPU

In the assembly algorithm for GTX 980, memory access is hidden, which is illustrated by the minimum values of σ and λ among all architectures. The maximum speedup was achieved with 1511 cores (see Fig. 9). Experimentally achieved acceleration is limited only by the ability to construct the optimal layers on a given mesh and by the availability of enough RAM.

Figure 7 shows the scalability on the 4-socket Opteron 8345 node with NUMA memory. $S_U(6)$ is the USL model of a single 6-core processor. This model becomes inadequate when the number of cores grows. $S_U(24)$ is the USL model of 24 cores that does not take into account sockets and nonuniform memory access. According to this model, the maximum speedup is 7, which is not realistic. $S_U(6, 4)$, the extended USL model, more accurately estimates the maximum speedup (approx. 6).

7   Conclusion

In this work, we used scalability model for performance analysis of element-by-element FEM operations on modern multi-core platforms. Universal Scalability Law allowed us to estimate and predict the scalability of parallel finite element assembly operators with layer-by-layer partitioning. We estimated the USL parameters for multiple architectures. These parameters were in good agreement with widely accepted mechanisms of reducing delays related to shared memory access and maintaining data consistency. We extended this model for multi-socket architecture, GPU, MIC and demonstrated that it adequately approximates the scalability. We performed numerical experiments with element-by-element matrix-vector multiplication on unstructured meshes on multicore processors accelerated by MIC and GPU. With layer-by-layer partitioning, we achieved 5-times speedup on CPU, 40-times speedup on MIC, and 200-times speedup on GPU. In future, we plan to apply layer-by-layer partitioning to other parallel operations of the finite element, finite volume and domain decomposition methods.

Acknowledgments. This work is supported by the Russian Foundation for Basic Research (grants 14-01-00055-a, 16-01-00129-a).

References

1. Al-Babtain, B.M., Al-Kanderi, F., Al-Fahad, M., Ahmad, I.: A survey on Amdahl's law extension in multicore architectures. Int. J. New Comput. Archit. Appl. **3**(3), 30–46 (2013)
2. Cecka, C., Lew, A., Darve, E.: Assembly of finite element methods on graphics processors. Int. J. Numer. Meth. Eng. **85**(5), 640–669 (2011). doi:10.1002/nme.2989
3. Gunther, N.J.: A new interpretation of Amdahl's law and geometric scalability. CoRR cs.DC/0210017 (2002). http://arxiv.org/abs/cs.DC/0210017
4. Gunther, N.J., Puglia, P., Tomasette, K.: Hadoop superlinear scalability. Commun. ACM **58**(4), 46–55 (2015)
5. Hill, M.D., Marty, M.R.: Amdahl's law in the multicore era. Computer **41**(7), 33–38 (2008)
6. Karypis, G., Kumar, V.: A fast and high quality multilevel scheme for partitioning irregular graphs. SIAM J. Sci. Comput. **20**(1), 359–392 (1999)
7. Markall, G.R., Slemmer, A., Ham, D.A., Kelly, P.H.J., Cantwell, C.D., Sherwin, S.J.: Finite element assembly strategies on multi-core and many-core architectures. Int. J. Numer. Methods Fluids **71**(1), 80–97 (2013)
8. Yavits, L., Morad, A., Ginosar, R.: The effect of communication and synchronization on Amdahl's law in multicore systems. Parallel Comput. **40**(1), 1–16 (2014)

Block Lanczos–Montgomery Method with Reduced Data Exchanges

Nikolai Zamarashkin and Dmitry Zheltkov[✉]

INM RAS, Gubkina 8, Moscow, Russia
nikolai.zamarashkin@gmail.com, dmitry.zheltkov@gmail.com
http://www.inm.ras.ru

Abstract. We propose a new implementation of the block Lanczos–Montgomery method with reduced data exchanges for the solution of large linear systems over finite fields. The theoretical estimates obtained for parallel complexity indicate that the data exchanges in the proposed implementation for record-high matrix sizes and block sizes 50 require less time than those in ideally parallelizable computations with dense blocks. According to numerical results, the acceleration depends almost linearly on the number of cores (up to 2,000 cores). Then, the dependence becomes close to the square root of the number of cores.

Keywords: Large linear systems over finite field · Block Lanczos–Montgomery · Block Wiedemann-Coppersmith · Parallel computation

1 Introduction

The need to solve large linear systems over finite fields arises in a variety of applications, such as large composite number factorization and discrete logarithming in a large prime field. The most efficient methods for such problems are based on algorithms of two types:

1. Lanczos-Montgomery-like algorithms (see [3,5,6,9,10,13,14]);
2. Wiedemann-Coppersmith-like algorithms (see [1,4,7,8]).

The algorithms of both types have much in common. Usually, the most time-consuming calculation is the repeated serial multiplication of a sparse matrix (and/or its transposed matrix) by a vector. The number of such multiplications is of the order of the doubled matrix size. Therefore, in terms of algorithmic complexity, both approaches are considered to be equivalent. However, when solving record-high-complexity problems, the efficient implementations on powerful computing systems with distributed memory plays a key role.

By now, the common opinion seems to be that the parallel versions of Lanczos-Montgomery-like techniques are generally inferior to Wiedemann-Coppersmith-like methods by their performance. All record-high RSA numbers were factorized using some modifications of the block Wiedemann-Coppersmith method for linear systems over $\mathbb{F}_2$ obtained from the GNFS method [7,8].

V. Voevodin and S. Sobolev (Eds.): RuSCDays 2016, CCIS 687, pp. 15–26, 2016.
DOI: 10.1007/978-3-319-55669-7_2

The following simple idea of parallelization makes the Wiedemann-Coppersmith-like methods attractive. The idea is to multiply by a block consisting of K vectors, rather than use a multiplication of the matrix by a single vector. Since each column of the block can be multiplied independently, the algorithm acquires a considerable parallel resource.

Obviously, the similar block modification can be applied to the Lanczos–Montgomery-like algorithms (see, e.g., [14]). However, in this case, global data exchange occurs on each iteration. In the Montgomery method implemented in [14], the time required from this exchange was not scaled with the block-size growth, which led to a significant slowdown in the calculations. We believe that it is the problem of unscaled data exchanges that largely makes the Lanczos–Montgomery-like methods less popular for solving large sparse systems over finite fields.

In this paper, we propose an implementation of the block Lanczos–Montgomery method where the time required for global exchanges is scaled with the growth of the block size K. This implementation is an improvement of the one proposed in [14]. The theoretical estimates obtained for parallel complexity indicate that the data exchanges in the proposed implementation for record-high matrix sizes and block sizes of $K > 50$ require less time than those in ideally parallelizable computations with dense blocks.

It should be noted that the current practice of complexity analysis for methods of solving large sparse systems of linear equations over finite fields usually ignores the computing with dense blocks. It is assumed that the most complex part of the algorithm (the construction of the Krylov subspace basis) is associated with the multiplication of vectors by a large sparse matrix. However, as is shown below, this assumption is valid only for small blocks (with a size not exceeding 10).

The faster solution of the problem requires an increase in the block size of at least up to $K = 100$. In this case, the calculations with dense blocks cannot be disregarded. Now, the computation time even turns out to exceed the time required for exchanges. Thus, the more complex Wiedemann-Coppersmith-like algorithms actually have no significant advantages over the Lanczos-Montgomery methods; this is true at least for systems with a few dozens of thousands of distributed nodes (here, the number of computing cores can reach up to a million).

This study is organized as follows. Section 2 describes the algorithm and the method of data storage in the improved implementation of the Lanczos–Montgomery algorithm. Section 3 provides theoretical estimates for parallel complexity under the assumption that the computer system nodes are connected according to the "point-to-point" topology. This type of communication is currently widespread and can be implemented using the "fat-tree" topology. The numerical examples are presented in Sect. 4.

2 Description of the Improved Lanczos–Montgomery Algorithm

2.1 The Lanczos Algorithm for Linear Systems over Finite Fields

Like standard numerical methods for $\mathbb{R}$ and $\mathbb{C}$ fields, the Lanczos algorithm for the system $AX = B$ of linear equations over finite fields consists of the following steps:

1. transition to a symmetrized system of the form

$$A^T A X = A^T B, \tag{1}$$

2. calculation of the $A^T A$-orthogonal basis $V_0, V_1, \cdots, V_N$ of the Krylov space constructed for the vectors $B, A^T A B, \left(A^T A\right)^2 B, \cdots, \left(A^T A\right)^{\tilde{N}} B$, where $\tilde{N}$ is an integer normally close to the number of rows N of A).
   At the $(k + 1)$-th iteration, the new vector V_{k+1} is obtained from the short recurrence relations

$$V_{k+1} = A^T A V_k + \sum_{i=k}^{k-m} V_k C_{k+1,i}, \tag{2}$$

   with a recursive depth m, where $m = 1$ for large finite fields and $m \geq 2$ for $\mathbb{F}_2$).
3. iterative refinement of the solution

$$X_{k+1} = X_k + V_{k+1} G_{k+1}, \tag{3}$$

with some $K \times K$ matrix G_{k+1}.

However, implementation of the Lanczos algorithm over finite fields has some specific features. The first and most important feature is that the concept of "approximate solution" is meaningless in the case of finite fields. As a consequence, the number of vectors in the resulting $A^T A$-orthogonal basis of the Krylov space is almost close (or equal) to the size of $A^T A$ (with a probability of 1). In addition, if the number of elements in $\mathbb{F}$ is small, such as for $\mathbb{F}_2$, block methods are used to avoid "breaks" [3,5,6,14].

One of the most efficient implementations for $\mathbb{F}_2$ is the Montgomery implementation [5], where the block size is equal to the number of bits in the computer word. Finally, for small fields, to satisfy the condition of $A^T A$-orthogonality, one has to increase the recursion depth up to $m \geq 2$.

In any case, the general structure of calculations in block Lanczos–Montgomery-like methods remains unchanged and can be described as follows:

The structure of Lanczos–Montgomery-like methods

1. **Start of iteration:** before the start of iteration i, the $N \times K$ blocks $V_i, V_{i-1}, \cdots, V_{im}$ over the finite field $\mathbb{F}$ are known;
2. **Matrix by block:** the product $(A^T A) V_i$ is calculated;
3. **Bilinear forms of blocks:** the bilinear forms $X^T Y \in \mathbb{F}^{K \times K}$ for the blocks $X, Y \in \mathbb{F}^{N \times K}$ are calculated;
4. **Linear combinations of blocks:** the linear combinations $XU + YV$ for the blocks $X \in \mathbb{F}^{N \times K}$ and square $K \times K$ matrices U, V are calculated.

2.2 Parallel Computing in the Improved Lanczos–Montgomery Method

The structure of the Lanczos method described in the previous section specifies the set of basic operations of the algorithm (see [13,14]):

1. multiplication of a sparse matrix by block;
2. calculation of $X^T Y$ for $N \times K$ blocks X and Y;
3. calculation of XU for $N \times K$ block X and $K \times K$ matrix U.

Our aim is to implement an efficient parallelization of these operations in the *improved Lanczos–Montgomery method.*

The improved Lanczos–Montgomery method proposed by us is based on unconventional ideas of the data storage on distributed nodes of the computer system and algorithms for computing with these data. This new way of data representation is very close to the one considered in [14] but there are some differences. These changes make it possible to obtain an implementation of the algorithm with a reduced number of data exchanges. In addition, as will be shown by further analysis and computational experiments, when the block size increases, the time required for exchanges becomes insignificant compared to the time of block operations. This occurs despite the fact that the arithmetic calculations are perfectly parallelized.

Now, we describe the method. Let K and S be two positive integers (hereafter, K denotes the block size). We consider a system of $K \times S$ nodes with distributed memory. For convenience, we assume that the nodes $\mathcal{N}_{i,j}$ form a rectangular lattice with K and S nodes along the rectangular grid sides. Despite the regular arrangement, we assume that the nodes are connected according to the "point-to-point" topology.

To fully describe the technique of data representation in memory, we consider:

1. the storage of a sparse matrix A;
2. the storage of $N \times K$ blocks;
3. the storage of $K \times K$ dense matrices.

We begin with the large sparse matrix A. Despite the fact that a symmetrized system is considered in the algorithm, its matrix $A^T A$ is not calculated explicitly. To calculate $(A^T A)V$, we multiply the block V sequentially by the matrices A and A^T.

A special scheme is used to handle the matrix $A \in \mathbb{F}^{M \times N}$. We represent the matrix as a union of blocks of rows A^i and blocks of columns A_j:

$$A = \begin{bmatrix} A_1 \ A_2 \cdots A_S \end{bmatrix} = \begin{bmatrix} A^1 \\ A^2 \\ \cdots \\ A^S \end{bmatrix}, \tag{4}$$

where each block of rows (columns) is assumed to have the same number of rows (columns) and approximately the same number of nonzero elements of A (this can be achieved by using a special preprocessing technique). In the improved scheme, the node $\mathcal{N}_{i,j}$ contains both the block of rows A_j and the block of columns A^j.

Also, it is necessary to distribute the $N \times K$ blocks in the computer system memory. These blocks are stored in the following way: the block V is considered as the union of KS smaller blocks:

$$
V = \begin{bmatrix} V_1 \\ V_2 \\ \cdots \\ V_{KS} \end{bmatrix},
\tag{5}
$$

and each subblock V_l of size $\frac{N}{KS} \times K$ is stored on the node $\mathcal{N}_{i,j}$ such that $l = K * j + i$.

Hereafter, the $K \times K$ matrices are supposed to be stored on each of the KS nodes $\mathcal{N}_{i,j}$.

This structure of data storage largely determines the parallel implementations for different operations in Lanczos–Montgomery-like method.

Remark 1. *The data storage technique considered above describes only the most general principles; some details of the efficient implementation were omitted. For example, some rows and/or columns of the sparse matrix A should be taken to be dense and stored without using special sparse formats. In the case of large fields, the dense rows (columns) can be of two types: (a) "small-module" elements; (b) "typical" elements for the large field. The sparse matrix is stored in a special cache-independent format, etc. However, to analyze the parallel complexity of the improved Lanczos–Montgomery method, it will suffice to use the rough description of data storage presented above.*

Now, we describe the algorithms. The multiplication of the symmetrized matrix of the system by a vector has the following form:

The algorithm of multiplication of the block X by the symmetrized matrix $A^T A$.

1. *collect the vector X_i (the i-th column in the block X) on each node $\mathcal{N}_{i,j}$ for $j = 1, \cdots, S$;*
2. *compute $Y_{i,j} = A^j X_i$ on the node $\mathcal{N}_{i,j}$;*
3. *collect the vector Y_i (the i-th column of $Y = AX$) on each node $\mathcal{N}_{i,j}$ for $j = 1, \cdots, S$;*
4. *compute $W_{i,j} = A_j Y_i$ on the node $\mathcal{N}_{i,j}$ (the blocks of $W_{i,j}$ are the desired result);*
5. *collect $\frac{N}{KS} \times K$ blocks W_t, for $t = 1, \cdots, KS$.*

The data exchanges in **The algorithm of multiplication of the symmetrized matrix $A^T A$ by the block X** occur at steps 1, 3, and 5. The vectors are collected by calling the collective communication procedure **MPI_Allgather**. The amount of exchanged data does not exceed $2 \max(N, M)$.

The number of operations in this algorithm will be estimated in the next section for the case of systems over $\mathbb{F}_2$.

For the block calculation, we propose

The algorithm for calculating bilinear forms $X^T Y$.

1. compute $X_l^T Y_l^T$ on the node $\mathcal{N}_{i,j}$, where $l = Kj + i$ (for all KS nodes);
2. collect the resulting matrix on all the nodes $\mathcal{N}_{i,j}$.

The data exchanges occur at step 2. The data are collected by calling the collective communication procedure **MPI_Allreduce.**

Finally, we consider the multiplication of the $N \times K$ block by the $K \times K$ matrix:

The algorithm of multiplication XU.

1. compute $W_l = X_l U$ on the node $\mathcal{N}_{i,j}$, where $l = Kj + i$ (for all KS nodes); the subblocks W_l are the computing result.

This algorithm has no data exchanges.

Remark 2. *The collective communication procedures such as* **MPI_Allgather** *and* **MPI_Allreduce** *are supposed to be effective if the nodes of the distributed system are connected according to the "point-to-point" topology.*

3   Parallel Complexity Analysis for the Improved Lanczos–Montgomery Method

3.1   Complexity Estimate for the Lanczos Method over Large Fields

Now, we estimate the parallel complexity of the Lanczos method for linear systems over large prime fields. The number of computer words for an element of the field is denoted by W, and the mean value of nonzeros in one column of A is p. For example, if the field element requires 512 bits, then $W = 8$. The parameter p may vary in a wide range (usually, its value is around 100). Our estimates are for the three main operations described in Subsect. 2.2.
Parallel complexity of the Lanczos method.

1. **The complexity of multiplication of a sparse matrix by a block:**

$$\{\textbf{computational complexity}\} = 2W\frac{pN^2}{KS}, \tag{6}$$

$$\{\textbf{complexity of data exchanges}\} = 2W\frac{N^2}{K}. \tag{7}$$

Note that the computational complexity (including the complexity of multiplications of numbers of a large field) depends only on the first degree of W (but not W^2, as it could be expected). Estimate (6) holds for applications where the initial matrix A is obtained by the GNFS method (for example, see [10]). In this case, almost all elements of A are "small-module" numbers (moreover, 90% of them are 1, and -1 in $\mathbb{F}$). As a result, the multiplication complexity is proportional to W.

Another important observation is that the time required for global exchanges is inversely proportional to the block size K. It is very important since K is the main parallelism resource in the algorithm. It should be noted that the term $2WpN^2$ in (6) is divisible by KS. It is this part of the algorithm that usually was most time-consuming. However, as it will be shown below, the situation changes significantly with the growth of K.

2. **Computational complexity of bilinear forms:**

$$\{\text{computational complexity}\} = 3\frac{W^2N^2}{S}, \tag{8}$$

$$\{\text{complexity of data exchanges}\} = WKN\left(\log_2\left(KS\right) + 1\right). \tag{9}$$

The computational complexity of bilinear forms was estimated by us under the following assumptions: (a) the Montgomery algorithm was used for the multiplication of numbers in a large prime field and (b) the Winograd method was used for dense matrix multiplications. The Winograd algorithm allows us to reduce the number of multiplications of the elements of the large field in two times. It is the multiplications of elements that determine the complexity of algorithms of basic linear algebra in the case of large finite fields. (8) does not depend on K; therefore, starting with some K (with a fixed S), the calculation time for block multiplications will surpass the time for the operations with the sparse matrix.

According to (9), the number of data exchanges grows with K; however, the number of these exchanges in practice is rather small. Indeed, estimate (9) is linear over N. If the system size N is around several millions and K is almost 100, we have $KN \ll \frac{N^2}{K}$ and, therefore, the number of exchanges in (7) is much greater than that in (9).

3. **Complexity of calculations of linear combinations of blocks:**

$$\{\text{computational complexity}\} = \frac{9W^2N^2}{2S}, \tag{10}$$

$$\{\text{complexity of data exchanges}\} = 0. \tag{11}$$

Like in (8), the complexity of computations in (10) corresponds to the Winograd method. The complexity value in (10) is independent of K. There are no data exchanges in this operation.

Remark 3.

It follows from the above estimates that the block operations are not scalable. Thus, the fast realizations of basic linear algebra procedures in finite fields are very important for the efficient implementation of the parallel Lanczos method. The more efficient are these calculations, the larger the block size can be taken.

3.2 Complexity Estimate for the Montgomery Method over $\mathbb{F}_2$

Let us estimate the complexity of parallel computations for the Montgomery method in the case of linear systems over $\mathbb{F}_2$. The estimates are the same as in Subsect. 3.1 accurate to scalar coefficients.

Parallel complexity of the Montgomery method

1. **Multiplication of sparse matrix by block:**

$$\{\text{computational complexity}\} = \frac{1}{32} \cdot \frac{pN^2}{KS}, \tag{12}$$

$$\{\text{complexity of data exchanges}\} = \frac{N^2}{32K}. \tag{13}$$

The coefficient $\frac{1}{32}$ arises because the block size in the Montgomery method is divisible by the computer word size (which is supposed to be 64). The actual block size is taken to be $K \cdot 64$. Obviously, both terms are scaled with K.

2. **Complexity of the calculation of bilinear forms:**

$$\{\text{computational complexity}\} = 3\frac{N^2}{8S}, \tag{14}$$

$$\{\text{complexity of data exchanges}\} = KN\left(\log_2\left(KS\right) + 1\right). \tag{15}$$

Estimate (14) was obtained for the case of the Coppersmith algorithm, which allows one to reduce the number of operations with computer words in 8 times compared to the "naive algorithm".

3. **Complexity of the linear combination of blocks:**

$$\{\text{computational complexity}\} = \frac{5N^2}{8S}, \tag{16}$$

$$\{\text{complexity of data exchanges}\} = 0. \tag{17}$$

Here, the "four Russians" method is used for the dense matrix multiplications in $\mathbb{F}_2$. This algorithm makes it possible to reduce the number operations with computer words in 8 times compared to the "naive" algorithm.

3.3 Parallel Complexity Analysis for the Lanczos–Montgomery-like Methods

We use the estimates obtained in Sects. 3.1 and 3.2 to demonstrate that the main obstacle to the use of more powerful computer systems for Lanczos–Montgomery-like methods is the operations with dense matrices and blocks, rather than data exchanges.

First, we determine the values of parameters K and S for which the complexity of multiplications of sparse matrix by blocks and the complexity of calculations of bilinear forms coincide. This can be easily done by equaling the expressions (6) and (8). In this case, we have

$$\frac{p}{K} = 3W. \tag{18}$$

For $p = 200$, and $W = 8$, we have $K \leq 9$. In other words, even for K of around 10 the parallel complexity (8) is no less than (6). This means that the parallel resource of K is bounded.

Remark 4.

In fact, the situation is slightly better: sparse matrices usually have a random structure; therefore, the speed of operations with them is limited due to inefficient memory access. Therefore, it will be more correct to use an estimate for values of K close to 100. In addition, one can significantly speed up the linear algebra operations (BLAS) by using special accelerators. For example, in case of basic algebra operations over large fields, GPU can be used. The resulting acceleration can increase the block size K. Also, the use of fast algorithms leads to a further acceleration with the increase of K. In this case, estimate (8) should be refined.

Now, we compare the times required for exchanges and computations with dense blocks. We assume the following distributed computing system:

1. the communication network of the computer system efficiently links the nodes via the "point-to-point" topology (for example, using the "fat-tree" technology);
2. the baud rate is about $20\,\mathrm{Gb/s}$ (approximately $300 \cdot 10^8 \frac{W}{c}$);
3. we assume that the computer system is equipped with 8-core nodes running at $3 \cdot 10^8$ GHz. Thus, each node is capable of performing $2.4 \cdot 10^{10} \frac{W}{c}$ computer words operations per second.

The most important consequence of the above description is that the calculations performed by a single multicore node are 100 times faster than the its data transmissions.

By equating (7) and (8), we find the values of K and S such that the times for data exchanges and for computations with dense blocks are comparable:

$$\frac{K}{S} = 100 \frac{2}{3W}. \tag{19}$$

Taking into account that $S \geq 10$ (usually), we obtain $K \geq 100$. Thus, the data exchanges in the improved algorithm imposes less strict limitations on K than the computations with dense blocks.

Similarly, we can show that the same conclusions hold for the field $\mathbb{F}_2$. Note that these conclusions are independent of the matrix sizes. However, they depend on the average number of nonzeros in row p. The larger is p, the larger values of K can be used and the higher is the parallel efficiency of the improved method.

4   Numerical Experiments

The following tables give the absolute times obtained in numerical experiments for the improved Lanczos–Montgomery method applied to linear systems over finite fields. The numbers of system cores are indicated in the tables. All the

Table 1. Acceleration and the time for single basis vector calculation in the case of a large finite field. Block Lanczos method (matrix $2 \cdot 10^6$, with 84 nonzero elements in a column)

Number of cores	1	2	4	8	16	32	64	128
Acceleration	1	1.99	3.95	7.86	15.48	30.49	56.04	103.04
Time, s	38.1	19.2	9.6	4.85	2.46	1.25	0.67	0.37
$S \times K$	1×1	1×1	1×1	1×1	2×1	2×2	4×2	8×4

experiments were performed on the Lomonosov and Lomonosov-2 supercomputers. In the experiments, the matrices were equivalent to the matrices obtained by the GNFS method. In the case of large fields, the numbers were encoded using 512 bits. It should be noted that the block size in some experiments was sufficiently large: $K = 32$. According to the Tables 1, 2, 3 and 4, the resulting acceleration depends almost linearly on the number of cores (up to 2,000 cores). Then, the dependence becomes close to the square root of the number of cores.

Table 2. Acceleration and the time for single basis vector calculation in the case of a large finite field. Block Lanczos method (matrix $2 \cdot 10^6$, with 84 nonzero elements in a column)

Number of cores	256	512	1024	2048	4096	8192
Acceleration	159.48	231.01	391.34	608.96	847.13	1211.4
Time, s	0.239	0.165	0.0974	0.0625	0.0451	0.0316
$S \times K$	16×4	16×8	32×8	32×16	64×16	64×32

Table 5 contains the absolute times of computations and data exchanges for different values of S and K in case of the field $\mathbb{F}_2$. The computations include:

- Sparse matrix by vector product ('MatVec');
- All operations with the dense blocks ('Dense').

The data exchanges occure in the following operations:

- Matrix by vector multiplication ('MatVec');
- Collecting vectors and subblocks before and after matrix by vector multiplication ('Block-Vector');
- Bilinear forms for dense blocks ('Dense').

Synchronizations constitute a substantial part of the time, especially for the exchanges in 'Block-Vector' and 'Dense' parts. That is why the numerical results confirm the theoretical estimates imperfectly for the data exchanges times. However, the results are close to the theory.

Table 3. Acceleration and the time for single basis vector calculation in the case of a field $\mathbb{F}_2$. Block Lanczos method (matrix $2 \cdot 10^6$, with 300 nonzero elements in a row)

Number of cores	1	2	4	8	16	32	64	128	256	512	1024
Acceleration	1	1.88	3.6	6.8	12.4	23.8	43.9	82.4	153.54	268.2	342.0
Time, ms	57.1	30.4	15.9	8.39	4.59	2.40	1.30	0.693	0.372	0.213	0.167
$S \times K$	1×1	1×1	1×1	1×1	2×1	4×1	4×2	8×2	8×4	16×4	16×8

Table 4. Acceleration and the time for single basis vector calculation in case of a field $\mathbb{F}_2$. Block Lanczos method (matrix $2 \cdot 10^7$, with 800 nonzero elements in a row)

Number of cores	1024	2048	4096	8192
Acceleration	342.0	605.1	813.7	1140.2
Time, ms	1.38	0.780	0.579	0.414
$S \times K$	16×8	32×8	64×16	64×32

Table 5. Computation and communication time of different operations ($\mathbb{F}_2$ field, matrix 2127498×2127690, 170 nonzero entries per row)

Parameters			Computation time, s		Communication time, s		
Number of cores	S	K	MatVec	Dense	MatVec	Block-Vector	Dense
14	1	1	2396	382	0	0	0
28	2	1	1220	186	193	0	6
28	1	2	1223	382	0	130	1.5
56	4	1	553	105	206	0	18
56	2	2	692	194	102	70	18
56	1	4	690	380	0	114	8
112	8	1	280	49	200	0	15
112	4	2	277	102	142	39	19
112	2	4	351	195	62	56	19
224	16	1	149	28	244	0	6
224	8	2	142	55	148	21	12
224	4	4	137	105	88	41	14

Acknowledgments. The work was supported by the Ministry of Education and Science of the Russian Federation, agreement no. 14.604.21.0034 (identificator RFMEFI60414X0034).

References

1. Wiedemann, D.H.: Solving sparse linear equations over finite fields. IEEE Trans. Inform. Theor. **32**(1), 54–62 (1986)

2. Lanczos, C.: An iteration method for the solution of the eigenvalue problem of linear differential and integral operators. J. Res. Natl. Bur. Stan. **45**, 255–282 (1950)

3. Coppersmith, D.: Solving linear equations over GF(2): block Lanczos algorithm. Linear Algebra Appl. **193**, 33–60 (1993)

4. Coppersmith, D.: Solving homogeneous linear equations over GF(2) via block Wiedemann algorithm. Math. Comput. **62**(205), 333–350 (1994)

5. Montgomery, P.L.: A block Lanczos algorithm for finding dependencies over GF(2). In: Guillou, L.C., Quisquater, J.-J. (eds.) EUROCRYPT 1995. LNCS, vol. 921, pp. 106–120. Springer, Heidelberg (1995). doi:10.1007/3-540-49264-X_9

6. Peterson, M., Monico, C.: $\mathbb{F}_2$ Lanczos revisited. Linear Algebra Appl. **428**, 1135–1150 (2008)

7. Kleinjung, T., et al.: Factorization of a 768-Bit RSA modulus. In: Rabin, T. (ed.) CRYPTO 2010. LNCS, vol. 6223, pp. 333–350. Springer, Heidelberg (2010). doi:10.1007/978-3-642-14623-7_18

8. Thome, E., et al.: Factorization of RSA-704 with CADO-NFS, pp. 1–4. Preprint (2012)

9. Dorofeev A.Y.: Vychiskenie logarifmov v konechnom prostom pole metodom lineinogo resheta. (Computation of logarithms over finite prime fields using number sieving) Trudy po diskretnoi matematike, vol. 5, pp. 29–50 (2002)

10. Dorofeev, A.Y.: Solving systems of linear equations arising in the computation of logarithms in a finite prime field. Math. Aspects Crypt. **3**(1), 5–51 (2012). (In Russian)

11. Cherepnev, M.A.: A block algorithm of Lanczos type for solving sparse systems of linear equations. Discret. Math. Appl. **18**(1), 79–84 (2008)

12. Cherepnev, M.A.: Version of block Lanczos-type algorithm for solving sparse linear systems. Bull. Math. Soc. Sci. Math. Roumanie **53**, 225–230 (2010). Article no. 3. http://rms.unibuc.ro/bulletin

13. Popovyan, I.A., Nestrenko, Y.V., Grechnikov, E.A.: Vychislitelno slozhnye zadachi teorii chisel. Uchebnoe posobie (Computationally hard problems of number theory. Study guide). Publishing of the Lomonosov Moscow State University (2012)

14. Zamarashkin, N.L.: Algoritmy dlya razrezhennykh sistem lineinykh uravneniy v GF(2). Uchebnoe posobie (Algorithms for systems of linear equations over GF(2). Study guide). Publishing of the Lomonosov Moscow State University (2013)

ChronosServer: Fast In Situ Processing of Large Multidimensional Arrays with Command Line Tools

Ramon Antonio Rodriges Zalipynis[✉]

National Research University Higher School of Economics, Moscow, Russia
rodriges@wikience.org

Abstract. Explosive growth of raster data volumes in numerical simulations, remote sensing and other fields stimulate the development of new efficient data processing techniques. For example, in-situ approach queries data in diverse file formats avoiding time-consuming import phase. However, after data are read from file, their further processing always takes place with code developed almost from scratch. Standalone command line tools are one of the most popular ways for in-situ processing of raster files. Decades of development and feedback resulted in numerous feature-rich, elaborate, free and quality-assured tools optimized mostly for a single machine. The paper reports current development state and first results on performance evaluation of ChronosServer – distributed system partially delegating in-situ raster data processing to external tools. The new delegation approach is anticipated to readily provide rich collection of raster operations at scale. ChronosServer already outperforms state-of-the-art array DBMS on single machine up to 193×.

Keywords: Big raster data · Distributed processing · Command line tools · Delegation approach

1 Introduction

Raster is the primary data type in a broad range of subject domains including Earth science, astronomy, geology, remote sensing and other fields experiencing tremendous growth of data volumes. For example, DigitalGlobe – the largest commercial satellite imagery provider, collects 70 terabytes of imagery on an average day with their constellation of six large satellites [1].

Traditionally raster data are stored in files, not in databases. The European Centre for Medium-Range Weather Forecasts (ECMWF) has alone accumulated 137.5 million files sized 52.7 petabytes in total [2]. This file-centric model resulted in a broad set of raster file formats highly optimized for a particular purpose and subject domain. For example, GeoTIFF represents an effort by over 160 different remote sensing, GIS (Geographic Information System), cartographic, and surveying related companies and organizations to establish interchange format for georeferenced raster imagery [3].

The corresponding software has long being developed to process raster data in those file formats. Many tools are free, popular and have large user communities that are very accustomed to them. For example, ImageMagic is under development since

© Springer International Publishing AG 2016
V. Voevodin and S. Sobolev (Eds.): RuSCDays 2016, CCIS 687, pp. 27–40, 2016.
DOI: 10.1007/978-3-319-55669-7_3

1987 [4], NetCDF common operators (NCO), a set of tools for multidimensional arrays, since about 1995 [5]; Orfeo ToolBox – remote sensing imagery processor now represents over 464,000 lines of code made by 43 contributors [6]. Many tools take advantage of multicore CPUs (e.g., OpenMP), but mostly work on a single machine.

In-situ distributed raster data processing has recently gained increased attention due to explosive growth of raster data volumes in diverse file formats. However, already existing stable and multifunctional tools are largely ignored in this research trend. Thus, raster operations are re-implemented almost from scratch delaying emergence of a mature in-situ distributed raster DBMS.

This paper describes the prototype extension of ChronosServer [7, 8] leveraging existing command line tools for in-situ raster data processing on a computer cluster of commodity hardware. Unlike current systems, it is easier and faster to equip ChronosServer with wide variety of raster operations due to new delegation approach. Thus, it is anticipated that it is possible to quickly develop new distributed file-based raster DBMS with rich functionality and exceptional performance.

2 In-Situ Raster Data Processing

In-database data storage (in-db, import-then-query) requires data to be converted (imported) to internal database format before any queries on the data are possible. Out-of-database (out-db, in-situ, file-based, native) approach operates on data in their original (native) file formats residing in a standard filesystem without any prior format conversions.

2.1 State-of-the-Art

PostgreSQL extensions PostGIS [9] and RasDaMan [10] work on single machine and allow registering out-database raster data in file system in their native formats. Enterprise RasDaMan version claims to be in-situ enabled and distributed, but is not freely available [11]. PostGIS has poor performance on multidimensional arrays (e.g. NetCDF, HDF or Grib formats [12]). No performance evaluation has been ever published for enterprise RasDaMan. SAGA [13] executes only distributed aggregation queries over data in HDF format. SWAMP [14] accepts shell scripts with NCO and parallelizes their execution. Hadoop extensions SciHadoop [15] and SciMATE [16] were never released publicly. They implement drivers reading Hadoop DFS chunks as if they are in HDF or NetCDF formats. Galileo [17] indexes geospatial data with distributed geo-hash. SWAMP launches command line tools but focuses on NCO and requires scripts looping over files with explicitly specified file names. The proposed approach is universally applicable to any tool and abstracts from "file" notion at all.

Commercial ArcGIS ImageServer [18] claims in-situ raster processing with custom implementation of raster operations. However, in a clustered deployment scenario all cluster nodes are recommended to hold copies of the same data or fetch data from a centralized storage upon request what negatively impacts scalability. Commercial Oracle Spatial [19] does not provide in-situ raster processing [20]. Open source SciDB is specially designed for distributed processing of multidimensional arrays [21]. However, it does not operate in-situ and imports raster data only converted to CSV format – very time-consuming and complex undertaking. Moreover, SciDB lacks even core raster operations like interpolation which makes it an immature and not widely used product [22]. Intel released open source TileDB on 04 Apr. 2016. It is yet neither distributed nor in-situ enabled [23].

Hadoop [24] and experimental SciDB streaming [25] allow launching a command line tool, feed text or binary data into its standard input and ingesting its standard output. Note two time-consuming data conversion phases in this case: data import into internal database format and their conversion to other representation to be able to feed to external software. The proposed approach directly submits files to external executables without additional data conversion steps.

SciQL was an effort to extend MonetDB with functionality for processing multi-dimensional arrays [26]. However, it has not yet finished nor its active development is seen so far. Also, SciQL does not provide in-situ raster processing.

2.2   In-Situ Approach Benefits

This section collects in one place advantages and challenges of in-situ approach that are quite scattered in the published literature.

- *Avoid inefficient neighborhood.* Traditionally, BLOB (Binary Large OBject) data type served for in-db raster storage (PostGIS, RasDaMan). Physical layouts where raster data are close to other data types are quite inefficient since the former are generally much larger than the latter.
- *Leverage powerful storage capabilities.* Some raster file formats support chunking, compression, multidimensional arrays, bands, diverse data types, hierarchical namespaces and metadata. These techniques are fundamental for raster storage; their implementation for an emerging in-db storage engine results in yet another raster file format.
- *Avoid conversion bottleneck.* In mission-critical applications it is important to be able to analyze the data before their new portion arrives or a certain event happens. In some cases the conversion time may take longer than the analysis itself. The data arrival rate and their large volumes may introduce prohibitively high conversion overheads and, thus, operational failure.
- *Avoid additional space usage.* Most data owners never delete source files after any kind of format conversions including database import. There are numerous reasons for this including unanticipated tasks that may arise in future that are more convenient, faster, easier or possible to perform on the original files rather than their converted counterparts. Storing both source data and their in-db copies requires additional space that may be saved by in-situ approach.
- *Reduce DBMS dependence.* It is easier to migrate to other DBMS keeping data in a widely adopted storage format independent from a DBMS vendor.
- *Leverage other software tools (this paper).* Out-db raster data in their native formats remain accessible by any other software which was inherently designed to process file-based data.

Key difficulties lie in the ability to perform the same set of operations on data in different file formats. Three data models are most widely used that allow abstracting from file format: Unidata CDM, GDAL Data Model, ISO 19123 (not cited due to space constraints). Most existing command line tools use those models and, thus, are capable to handle data in diverse raster file formats.

3 ChronosServer Architecture

3.1 Raster Data Model: Abstracting from Files, Their Locations and Formats

This paper focuses on climate reanalysis and Earth remote sensing global gridded raster data represented as multidimensional arrays and usually stored in NetCDF, Grib and HDF file formats. For example, AMIP/DOE Reanalysis 2 (R2) spans several decades, from 01.01.1979 to current date with 6 h time interval and contains over 80 variables [27]. The grid resolution is usually $2.5° \times 2.5°$. Global grids for each variable are stored in a sequence of separate files partitioned by time. File names contain variable codename, e.g. files with surface pressure are named pres.sfc.1979.nc, pres.sfc.1980. nc, …, pres.sfc.2015.nc. Where "pres.sfc" denotes surface pressure, 1979 is year, ".nc" is NetCDF file extension. Usually file naming is much more complex (e.g., compare to AIRS/AMSU daily file name for CO_2 satellite data: AIRS.2004.08.01.L3.CO2Std001. v5.4.12.70.X09264193058.hdf). Note, that the data are usually already split by files by data providers.

ChronosServer distributes files among cluster nodes without changing their names and formats. Any file is always located as a whole on a machine in contrast to parallel or distributed file systems. It introduces a data model to work with grids, not files to abstract from "file" notion, file naming, their locations, formats and other details that are unique to every dataset and not relevant for data analysis. ChronosServer dataset namespace is hierarchical. For example, "r2.pressure.surface" refers to surface pressure of R2 reanalysis. ChronosServer provides SQL-like syntax for subsetting grids. For example, "SELECT DATA FROM r2.pressure.surface WHERE TIME_INTERVAL = 01.01.2004 00:00–01.01.2006 00:00 AND REGION = (45, 60, 50, 70)" returns time series of R2 surface pressure in the specified time interval and region between 45°S–50°N and 60°W–70°E. The query execution may involve several cluster nodes. Any grid or time series from any dataset may be extracted with the same syntax regardless of original file format, file split policy and other details.

3.2 Cluster Orchestration

ChronosServer cluster consists of workers launched at each node and a single gate at a dedicated machine. Gate receives client queries and coordinates workers responsible for data storage and processing. All workers have the same hierarchy of data directories on their local filesystems. A worker stores only a subset of all dataset files and only a portion of the whole namespace relevant to the data it possesses. A file may be replicated on several workers for fault tolerance and load balancing. It is not required to keep all workers up and running for the whole system to be operational.

The gate is unaware of file locations until a worker reports them to it. This is done for better scalability and fault tolerance. Upon startup workers connect to gate and receive the list of all available datasets and their file naming rules. Workers scan their local filesystems to discover datasets and their time intervals by parsing dataset file names. Workers transmit to gate the list of time intervals for each dataset they store. Gate keeps this information in worker pool – in-memory data structure used during query planning that maps time intervals to their respective owners (workers).

4   New Delegation Approach

4.1   ChronosServer Raster Data Processing Commands and Their Distributed Execution

ChronosServer syntax of a raster data processing command is the same as launching a tool from a command line. Command names coincide with names of existing command line tools. ChronosServer command options have the same meaning and names as for the tool but without options related to file names or paths. Commands and tools also support options with long names having the same meaning.

For example, NCO consists of several standalone command line tools: ncap2 (Arithmetic Processor v.2), ncks (Kitchen Sink), ncatted (Attribute Editor – metadata manager, Fig. 1), etc. [28, 29]. Metadata are crucial component of any raster data.

```
ncatted [-a ...] [--bfr sz] [-D nco_dbg_lvl] [--glb ...] [-h]
        [--hdr_pad nbr] [-l path] [-O] [-o out.nc] [-p path] [-R]
        [-r] [-t] in.nc [[out.nc]]
-a,    variable_name,mode,attribute_type,attribute_value
       mode = a,c,d,m,o (append, create, delete, modify, overwrite)
       att_typ = f,d,l/i,s,c,b (float, double, long, short, char, byte)
    --bfr_sz, --buffer_size sz       Buffer size to open files with
-D, --dbg_lvl, --debug-level lvl     Debug-level is lvl
    --glb nm=val                     Global attribute to add
-h, --hst,        Do not append to "history" global attribute
    --hdr_pad     Pad output header with nbr bytes
-l, --lcl         Local storage path for remotely-retrieved files
-o, out.nc        Output file name (or use last argument)
-O, --ovr         Overwrite existing output file, if any
-p, --path path   Path prefix for all input filenames
-R, --rtn         Retain remotely-retrieved files after use
-r, --revision    Compile-time configuration and program version
-t, --typ_mch,    Type-match attribute edits
in.nc [[out.nc]]  Input file name [[Output file name]]
```

Fig. 1. Parameters of the NCO ncatted tool (all listed) and Chronos ncatted command (in bold)

NetCDF and many other formats store metadata as attributes (key-value pairs). For example, attribute named "_FillValue" holds a constant used to mark raster cells with missing values (e.g., –9999).

By default, command is applied to the whole available dataset time interval and spatial coverage. They may be restricted by "select" query with alias dataset name specification. New virtual dataset will contain subset of the original dataset. Its name (alias) may be used in the subsequent commands. It is helpful to test a series of commands on a dataset sample to check hypotheses about the anticipated results before submitting large-scale query involving large data volumes to save time.

For example, ChronosServer command for "_FillValue" attribute deletion from dataset "r2.pressure.surface" is "ncatted -a _FillValue,r2.pressure.surface,d,,". Instead of "variable_name" – the term specific for NetCDF format, ChronosServer ncatted accepts a dataset name to be independent of a concrete format. Usually the same attributes are duplicated in all files of a dataset.

The gate receives and parses command line options, verifies their correctness and absence of malicious instructions since they are passed to operating system shell. The dataset or its subset is locked for reading/writing depending on the command. Several commands may work concurrently if they do not block each other. Gate selects workers on which dataset files with the required time/space intervals are located and sends them the modified command (see below). Workers complement command line with full paths to dataset files according to time and space limitations and launch the tool on each file.

In the simple case above, ChronosServer invokes several instances of the NCO ncatted tool on the cluster nodes where at least one dataset file is located (pres.sfc.1979. nc, …, pres.sfc.2015.nc). The execution command line for file pres.sfc.1979.nc is "<path to ncatted.exe>-a missing_value,pres,d,, <data path>\pres.sfc.1979.nc". The file path and "pres" were automatically put by worker and gate correspondingly. The latter is the NetCDF variable name that stores R2 surface pressure (NetCDF3 format does not have hierarchical namespace and stores data in structures called "variables").

Workers also collect standard output of the tool which is sent to gate after its completion. Running tool on a different cluster node in case of a hardware failure is under development. Gate reports to the user once it receives success messages from all workers involved in the command execution. Report contains the merged standard outputs from each run of the tool and total elapsed time.

4.2   Distributed Apply-Combine-Finally Execution Scheme (Under Development)

Raster operations can be broadly classified as global (involve all data), local (pixel-wise), focal (cell values from a rectangular window are required to compute new cell value), zonal (same as focal but spatial region is defined by a function) [30]. Thus, some operations cannot be completed autonomously using data on a single cluster node. For example, R2 data interpolation for 1980 year from 6 to 3 h time step requires grids for December 1979 and January 1981. Also, computation of maximum mean winter pressure involves all files potentially located on different cluster nodes.

In this case, user specifies commands: APPLY command1 INTERVALS intervals COMBINE command2 FINALLY command3. ChronosServer ensures that files on each involved node contain data in given temporal and spatial intervals (e.g., in case of winter means, there should be nodes with data for all winter months for at least two consecutive years: 1980–1981, 1981–1982, 1982–1983, etc. to be able to compute the mean). This may require data movement between cluster nodes. After the intervals requirement is met, command1 is executed autonomously on the data intervals on corresponding cluster nodes. Since some nodes may have several disjoint intervals, their intermediate results may be combined on the same node to reduce network traffic with command2 if it is possible (e.g. compute maximum of the means of intervals 1980–1981 and 1982–1983 that happened to reside on the same node). All results are gathered on a single node and command3 is applied to obtain final result (e.g., find maximum of means or maximums of maximums if combine phase was applied).

Unlike existing schemes [31, 32], the proposed distributed execution scheme takes into account peculiarities inherent to raster operations, geospatial data and ChronosServer file-based storage model. For example, respective intervals must be specified to guarantee the raster operation (command1) is possible to accomplish within a single node. It is widely recognized that numerous data processing tasks are much easier to parallelize once actions are expressed in functional style, not "for" loops.

4.3   Benefits of the Proposed Delegation Approach

While in-situ approach leverages benefits of already existing sophisticated file formats, delegation approach leverages benefits of already existing standalone command line tools.

- *Avoid learning new language.* ChronosServer provides command line syntax that is well-known to every console user instead of a new SQL dialect.
- *Steep learning curve.* Users work with ChronosServer as if with console tools they have accustomed to with only minor changes to already familiar tools' options.
- *Documentation reuse.* Most of the tool's documentation is applicable to the corresponding ChronosServer command due to exactly the same meaning and behavior.
- *Output conformance.* Output files are formatted as if a tool was launched manually.
- *Language independence.* ChronosServer may use tools written in any programming language.
- *Community support.* Bugs in tools are fixed by their developers as well as new functionality added, usage suggestions via mail lists are obtained regardless of ChronosServer context.
- *Zero-knowledge development (0-know dev.).* Developers of existing and emerging tools do not have to know anything about ChronosServer in order the tool could be used in ChronosServer.

The main difficulty of the proposed approach lies in the correct specification of the "intervals", "combine" and "finally" clauses. Meta-commands are a possible simplification of the problem. They consist of a single command line which translates to predefined apply-combine-finally clauses.

5   Performance Evaluation

To date, the only freely available distributed raster DBMS is SciDB, not operating in-situ. It lacks many core raster operations. Thus, only the performance of some basic ChronosServer and SciDB raster operations are compared. It is of special interest to evaluate ChronosServer against SciDB since the latter is currently being most actively popularized among similar DBMS at top journals and conferences [33–35].

Test of source data volume comprised only 100.55 MB in NetCDF3 format since it is impossible to import large data volumes into SciDB in a reasonable time frame (Sect. 5.2). In addition, experiments were carried out on a single machine for two reasons. First, raster operations being evaluated have linear scalability. Increasing machine number by a factor of N should roughly increase the performance also by N. Second, unlike ChronosServer, SciDB cluster deployment is very labor-intensive. Comparison of both systems running on computer cluster is left for future work.

However, small test data volume and single machine turned out to be sufficient for representative results (Table 1). Table 1 summarizes experimental results while details are given in following subsections.

Table 1. ChronosServer and SciDB performance comparison

Operation	Execution time, seconds			Ratio, SciDB/ ChronosServer	
	SciDB	ChronosServer			
		Cold	Hot	Cold	Hot
Data import	720.13	19.82	7.96	36.33	90.47
Max	13.46	4.43	3.10	3.04	4.34
Min	12.87	4.71	3.33	2.73	3.86
Average	21.42	4.71	3.23	4.55	6.63
Wind speed calc.	25.75	3.50	2.10	7.36	12.26
Chunk $100 \times 20 \times 16$	56.19	1.68	0.374	33.45	150.24
Chunk $10 \times 10 \times 8$	222.11	1.98	1.15	112.18	193.14

5.1   Test Raster Data and Experimental Setup

Eastward (U-wind) and northward (V-wind) wind speed (Fig. 2) at 10 meters above surface from NCEP/DOE AMIP-II Reanalysis (R2) were used for experiments [27]. These are 6-hourly forecast data (4-times daily values at 00.00, 06.00, 12.00 and

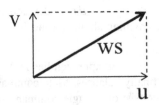

Fig. 2. Wind speed vector (ws) and its eastward (u) and northward (v) speed vectors.

18.00). Data are 3-dimensional on 94 latitudes × 192 longitudes Gaussian grid in NetCDF3 format. Dataset does not contain missing values. Single file is approximately 50.2 GB; total data volume is 3.63 GB for the whole available time interval 1979–2015. Due to SciDB limitations (Sect. 5.2), only data for 1979 year were used (about 100.55 MB as mentioned earlier).

Both ChronosServer and SciDB were run on Ubuntu 14.04 inside VirtualBox on Windows 10. Note that ChronosServer is also capable to run natively on Windows, unlike SciDB. The machine is equipped with SSD (OCZ Vertex 4). VirtualBox was assigned 4 GB RAM and 2 CPU cores (Intel Core i5-3210 M, 2.50 GHz per core). SSD speed inside VirtualBox: 4573.28 MB/sec and 222.04 MB/sec (cached and buffered disk reads respectively as reported by hdparm utility); 350 MB/sec disk write as reported by dd utility.

ChronosServer has 100% Java code, ran one gate and one worker, Java 1.7.0_75, OpenJDK IcedTea 2.6.4 64 bit, max heap size 978 MB (-Xmx), NCO v4.6.0 (May 2016). SciDB is mostly written on C++, v15.12 was used (latest, Apr. 2015) with recommended parameters: 0 redundancy, 4 instances per machine, 4 execution and prefetch threads, 1 prefetch queue size, 1 operator threads, 128 MB array cache, etc.).

Two types of query runs were evaluated: cold (query executed first time on given data) and hot (repeated query execution on the same data). Time reported in Table 1 is the average of three runtimes of the same query. Respective OS commands were issued to free pagecache, dentries and inodes each time before executing cold query to prevent data caching at various OS levels. Table 1 does not report cold and hot runs for SciDB since it did not reveal any significant difference in runtime between them. In contrast, ChronosServer does not cache data but benefits from native OS caching and demonstrates significant speedup for hot runs. This is particularly useful for continues experiments with the same data. The need for this type of experiments occurs quite often (e.g., tuning certain parameters, refer to Sect. 5.5 for an example).

5.2   SciDB Data Import and ChronosServer Data Discovery

Importing data into SciDB involves considerable efforts on software development for each dataset being considered. SciDB does not yet provide out-of-the-box import tool from formats other than CSV. The overall import procedure is very time-consuming and error-prone (both due to complicated raster formats and related possible coding bugs as well as inherent floating point calculations). Lack of documentation and complex query syntax (Appendix B) may elongate data import for several weeks.

For SciDB "data import" row, Table 1 reports only time taken to automatically import U-wind speed data for 1979 year (50.2 MB) from NetCDF3 format into SciDB. It does not report the time spent for Java program development to actually perform the import. Importing V-wind speed for 1979 also takes approximately the same amount of time. Estimate time is 14.76 h to automatically import U- and V-wind speed for 1979–2015. Thus, only U- and V-wind speed vectors for 1979 are considered for performance evaluation in the next subsections. This small data sample turned out to be representative for convincing results.

SciDB data import from NetCDF3 included reading original data file, preparing string representation of 94 × 192 grid for each time step in a format ingestible by SciDB, saving string to CSV file, and invoking SciDB tool to import grid from CSV.

On the contrary, to add new data under ChronosServer management, it is sufficient to copy data files on a cluster node and add a short entry in ChronosServer XML file specifying rules for file naming and a handful of some other information. ChronosServer will discover files as described in Sect. 3.1. Worker discovers files at startup. Table 1 "Data import" row for ChronosServer reports time of its "cold" and "hot" startup. The former startup mode rediscovers completely from scratch all existing as well as any newly added data. The latter mode assumes no new data were added since previous startup. Both measured times include complete startup time of one gate and one worker, metadata transfer from gate to worker (registered datasets), data discovery by worker, logging and any other startup overhead.

ChronosServer is able to discover 803 datasets with total volume of 6.78 GB in file formats NetCDF-3, -4, HDF-4, -5, Grib-2 (a collection of diverse satellite and climate reanalysis products) in 20 and 8 seconds for cold and hot startups respectively. This is 36× and 90× faster than SciDB imports just 50.2 MB of data.

5.3   Simple Statistics

Table 1 rows for max, min and average report time taken by the systems to calculate maximum, minimum and average U-wind speed for 1979 year for each 94 × 192 grid cell. Computation involves traversing 1460 time steps. ChronosServer is about 3 to 6 times faster than SciDB.

5.4   User-Defined Arithmetic Expressions

Both ChronosServer and SciDB support user-defined arithmetic expressions that could be applied to raster data. As an example, wind speed (ws) at each grid cell and time point is calculated from its eastward (u) and northward (v) components as $ws = sqrt(u^2 + v^2)$ (Fig. 2).

In this case, ChronosServer is 7 to 12 times faster than SciDB (Table 1).

It is worth noting, that SciDB query for wind speed calculation is very complex (Appendix B), unlike that for ChronosServer (Appendix A).

5.5   Multidimensional Chunking

Chunking is the process of partitioning original array (raster) onto a set of smaller subarrays called chunks (Fig. 3). Chunks are autonomous, possibly compressed arrays with contiguous storage layout. A chunk is usually read/written completely from/to disk in one request to storage subsystem. Chunking is one of the classical approaches to significantly accelerate disk I/O when only a portion of raster is read. Consider reading a 6 × 2 slice from a 2D array (Fig. 3). For a row-major storage layout, two vertically adjacent cells are located far apart each other. A possible solution is to read 6 portions

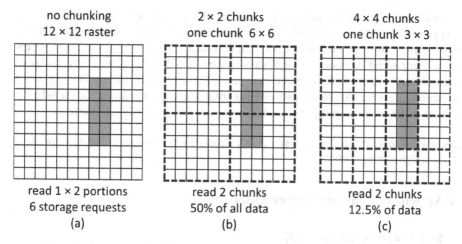

no chunking
12 × 12 raster

2 × 2 chunks
one chunk 6 × 6

4 × 4 chunks
one chunk 3 × 3

read 1 × 2 portions
6 storage requests
(a)

read 2 chunks
50% of all data
(b)

read 2 chunks
12.5% of data
(c)

Fig. 3. Chunking: row-major storage layout, read 6 × 2 slice

sized 1×2 which requires 6 I/O requests and disk seeks (Fig. 3a). For a compressed array, much larger part of it might be required to be read and uncompressed before getting the requested portion. In contrast, only chunks containing required data are read from disk from a chunked raster. However, inappropriate chunk shape may result in large I/O overhead (Fig. 3b). Good chunk shape allows to reduce communication with storage layer, disk seeks and I/O volume (Fig. 3c).

Since many raster operations are mostly I/O bound [28], chunk shape is one of the crucial performance parameters for a dataset [34]. Chunk shape depends on data characteristics and workload. Optimal chunk shape usually does not exist for all access patterns. It is also difficult to guess good chunk shape a priori: chunk shape is often tuned experimentally. Thus, raster DBMS must be capable to quickly alter chunk shape in order to support experimentation as well as to adapt to dynamic workloads.

To estimate chunking speed of both systems, U-wind speed data for 1979 were chunked with two different chunk shapes $ts \times lats \times lons$: $100 \times 20 \times 16$ and $10 \times 10 \times 8$, ts, $lats$ and $lons$ are chunk sizes along time, latitude and longitude axes respectively.

ChronosServer is up to 193 times faster than SciDB (Table 1). Presented timings are the average for 3 consecutive runs as mentioned earlier (each cold run precedes OS cache clear). In practice, ChronosServer could be even faster: 973 ms execution time (less than a second) could be obtained for a hot run leading to $228.3\times$ speedup.

6 Conclusion

The paper presented new approach of delegating in-situ raster data processing to existing command line tools. The approach has numerous benefits and is under development as an extension to ChronosServer – inherently distributed, file-based system for high performance raster data dissemination [7, 8]. This paper also presented first results on performance evaluation of ChronosServer against SciDB – one of the most popular, distributed state-of-the-art raster DBMS [33–35]. Raster operations were executed on 100.55 MB wind speed data from NCEP/DOE AMIP-II Reanalysis. This was governed by SciDB which is unable to import large data volumes in a reasonable time frame. However, this small data sample turned out to be sufficient for representative comparison. ChronosServer always outperforms SciDB. Also, query syntax of ChronosServer is much easier and cleaner compared to SciDB. Max, min and average operations are $3\times$ to $6\times$ faster, user-defined arithmetic expression was shown to be $7\times$ to $12\times$ faster while altering 3D chunk shape is about $33\times$ to $228.3\times$ faster.

Acknowledgements. This work was partially supported by Russian Foundation for Basic Research (grant #16-37-00416).

A Appendix. ChronosServer Queries

Max U-wind speed (Sect. 5.3):

```
ncap2 -alias u,r2.wind.10m.u
      -alias umax,r2.wind.10m.uv.umax
      -s "$(umax)=$(u).max($time)"
```

Calculate wind speed (Sect. 5.4):

```
ncap2 -alias u,r2.wind.10m.u
      -alias v,r2.wind.10m.v
      -alias ws,r2.wind.10m.uv.ws
      -s "$(ws)=sqrt($(u)*$(u) + $(v)*$(v));"
```

Alter chunk shape to $10\times10\times8$ (Sect. 5.5):

```
ncks -4 --cnk_map dmn --cnk_plc g2d --cnk_dmn time,10
     --cnk_dmn lat,10 --cnk_dmn lon,8
     r2.wind.u10m.u r2.wind.u10m_ch10x10x8
```

B Appendix. SciDB Queries

Initial SciDB array for U-wind speed:

```
r2_u10m<value:float>
[time=0:*,1,0,lat=0:93,94,0,lon=0:191,192,0]
```

Max U-wind speed (Sect. 5.3):

```
store(aggregate(r2_u10m, max(value), lat, lon),
r2_u10m_max);
```

Calculate wind speed (Sect. 5.4):

```
store(project(apply(join(r2_u10m, r2_v10m), ws,
float(sqrt(r2_u10m.value * r2_u10m.value + r2_v10m.value
* r2_v10m.value))), ws), r2_ws10m);
```

Alter chunk shape to $10 \times 10 \times 8$ (Sect. 5.5):

```
store(redimension(r2_u10m, <value:float>
[time=0:*,10,0,lat=0:93,10,0,lon=0:191,8,0]),
r2_u10m_10x10x8);
```

According to the answer of SciDB developers on their forum (question posted by the author of this paper in August 2016), above query is currently the fastest way to alter chunk size in SciDB: http://forum.paradigm4.com/t/fastest-way-to-alter-chunk-size/.

References

1. Launching DigitalGlobe's Maps API | Mapbox. https://www.mapbox.com/blog/digitalglobe-maps-api/
2. Grawinkel, M., et al.: Analysis of the ECMWF storage landscape. In: 13th USENIX Conference on File and Storage Technologies, 16–19 February 2015. Santa Clara, CA (2015). https://usenix.org/system/files/login/articles/login_june_18_reports.pdf
3. GeoTIFF. http://trac.osgeo.org/geotiff/
4. ImageMagic: History. http://imagemagick.org/script/history.php
5. NCO Homepage. http://nco.sourceforge.net/
6. The Orfeo ToolBox on Open Hub. https://www.openhub.net/p/otb
7. Rodriges Zalipynis, R.A.: ChronosServer: real-time access to "native" multi-terabyte retrospective data warehouse by thousands of concurrent clients. Inform. Cybern. Comput. Eng. **14**(188), 151–161 (2011)
8. ChronosServer. http://www.wikience.org/chronosserver/
9. Raster Data Management, Queries, and Applications (Chapter 5). http://postgis.net/docs/manual-2.2/using_raster_dataman.html
10. Baumann, P., Dumitru, A.M., Merticariu, V.: The array database that is not a database: file based array query answering in rasdaman. In: Nascimento, M.A., Sellis, T., Cheng, R., Sander, J., Zheng, Y., Kriegel, H.-P., Renz, M., Sengstock, C. (eds.) SSTD 2013. LNCS, vol. 8098, pp. 478–483. Springer, Heidelberg (2013). doi:10.1007/978-3-642-40235-7_32
11. RasDaMan features. http://www.rasdaman.org/wiki/Features
12. NetCDF. http://www.unidata.ucar.edu/software/netcdf/docs/
13. Wang, Y., Nandi, A., Agrawal, G.: SAGA: array storage as a DB with support for structural aggregations. In: SSDBM 2014, June 30–July 02 (2014)
14. Wang, L., et al.: Clustered workflow execution of retargeted data analysis scripts. In: CCGRID (2008)
15. Buck, J.B., Watkins, N., LeFevre, J., Ioannidou, K., Maltzahn, C., Polyzotis, N., Brandt, S.: SciHadoop: array-based query processing in Hadoop. In: Proceedings of SC (2011)

16. Wang, Y., Jiang, W., Agrawal, G.: SciMATE: a novel mapreduce-like framework for multiple scientific data formats. In: Proceedings of CCGRID, pp. 443–450, May 2012

17. Malensek, M., Pallickara, S.: Galileo: a framework for distributed storage of high-throughput data streams. In: Proceedings of the 4th IEEE/ACM International Conference on Utility and Cloud Computing (2011)

18. ArcGIS for Server | Image Extension. http://www.esri.com/software/arcgis/arcgisserver/extensions/image-extension

19. Oracle Spatial and Graph. http://www.oracle.com/technetwork/database/options/spatialandgraph/overview/index.html

20. Georaster: Import very large images with sdo_ge... | Oracle Community. https://community.oracle.com/thread/3820691?start=0&tstart=0

21. Paradigm4: Creators of SciDB. http://scidb.org/

22. Interpolation - SciDB usage - SciDB Forum. http://forum.paradigm4.com/t/interpolation/1283

23. TileDB - Scientific data management made fast and easy. http://istc-bigdata.org/tiledb/index.html

24. Hadoop Streaming. wiki.apache.org/hadoop/HadoopStreaming

25. GitHub - Paradigm4/streaming: Prototype Hadoop streaming-like SciDB API. https://github.com/Paradigm4/streaming

26. Zhang, Y., et al.: SciQL: bridging the gap between science and relational DBMS. In: IDEAS 2011, September 21–23. Lisbon, Portugal (2011)

27. NCEP-DOE AMIP-II Reanalysis. http://www.esrl.noaa.gov/psd/data/gridded/data.ncep.reanalysis2.html

28. Zender, C.S.: Analysis of self-describing gridded geoscience data with netCDF Operators (NCO). Environ. Model Softw. 23, 1338–1342 (2008)

29. Zender, C.S., Mangalam, H.: Scaling properties of common statistical operators for gridded datasets. Int. J. High Perform. Comput. Appl. 21(4), 458–498 (2007)

30. Geospatial raster data processing. http://rgeo.wikience.org/pdf/slides/rgeo-course-04-raster_processing.pdf

31. Wickham, H.: The split-apply-combine strategy for data analysis. J. Stat. Softw. 40, 1–29 (2011)

32. Yang, H.C., Dasdan, A., Hsiao, R.L., Parker, D.S.: Map-reduce-merge: simplified relational data processing on large clusters. In: ACM SIGMOD, June 12–14, Beijing (2007)

33. Stonebraker, M., Brown, P., Zhang, D., Becla, J.: SciDB: a database management system for applications with complex analytics. Comput. Sci. Eng. 15, 54–62 (2013)

34. Cudre-Mauroux, P., et al.: A demonstration of SciDB: a science-oriented DBMS. Proc. VLDB Endow. 2(2), 1534–1537 (2009)

35. Planthaber, G., Stonebraker, M., Frew, J.: EarthDB: scalable analysis of MODIS data using SciDB. In: BigSpatial, pp. 11–19 (2012)

Dynamics of Formation and Fine Structure of Flow Pattern Around Obstacles in Laboratory and Computational Experiment

Yu. D. Chashechkin[1]([✉]), Ya. V. Zagumennyi[2], and N.F. Dimitrieva[2]

[1] A.Yu. Ishlinskiy Institute for Problems in Mechanics of the RAS,
Moscow, Russia
yulidch@gmail.com
[2] Institute of Hydromechanics of NAS of Ukraine, Kiev, Ukraine
zagumennyi@gmail.com, dimitrieva@list.ru

Abstract. Non-stationary dynamics and structure of stratified and homogeneous fluid flows around a plate and a wedge were studied on basis of the fundamental equations set using methods of laboratory and numerical modeling. Fields of various physical variables and their gradients were visualized in a wide range of the problem parameters. Eigen temporal and spatial scales of large (vortices, internal waves, wake) and fine flow components were defined. The same system of equations and numerical algorithm were used for the whole range of the parameters under consideration. The computation results are in a good agreement with the data of laboratory experiments.

Keywords: Fundamental system · Laboratory experiment · High-resolution computations · Flow around obstacles

1 Introduction

Since the pioneering papers by d'Alembert [1, 2] and Euler [3, 4] calculations of flow patterns around obstacles with evaluation of forces, acting on their surfaces, occupy a leading position in the theoretical and experimental fluid mechanics. Stability of the interest to the problem is supported by its fundamental content and complexity, as well as by diversity and importance of its practical applications. A particular attention is paid to calculation of flow around obstacles with a rather simple shape, e.g. plate, cylinder, sphere, etc., which symmetry is used for simplification of the governing equations [5].

Due to the complexity of analysis of the problem the traditional system of continuity and Navier-Stokes equations in the homogeneous fluid approximation is replaced by various model systems, among which the boundary layer and turbulence theories are the most widely used. However, new systems are characterized by their own symmetries not coinciding with these of the initial system of equations derived on basis of the general physical principles [6]. Accordingly, the physical meaning of the quantities denoted by the same symbols and the nature of their relations are changed in new sets. Such transformations make it difficult to verify experimentally the results and compare different mathematical models to each other.

V. Voevodin and S. Sobolev (Eds.): RuSCDays 2016, CCIS 687, pp. 41–56, 2016.
DOI: 10.1007/978-3-319-55669-7_4

The techniques of experimental and theoretical studies of fluid flows conducted at the Laboratory of Fluid Mechanics IPMech RAS are based on the fundamental system including equations of state and transport of substance, momentum and energy for inhomogeneous fluids [5].

In the environment, i.e. the Earth's hydrosphere and atmosphere, and industrial devices, fluid density, as a rule, is not constant due to inhomogeneity of either soluble substances or suspended particles concentration or temperature and pressure distributions. Under the action of buoyancy forces fluid particles with different density move vertically and form a stable continuous stratification which is characterized by buoyancy scale, $\Lambda = |d \ln \rho/dz|^{-1}$, frequency, $N = \sqrt{g/\Lambda}$, and period, $T_b = 2\pi/N$, which are supposed to be constant in space and can vary from a several seconds in laboratory conditions and up to ten minutes in the Earth's atmosphere and hydrosphere [7].

In the present paper we study numerically and experimentally flow patterns of a continuously stratified water solution of the common salt $NaCl$ around 2D obstacles with aspect ratio of about 20. Two kinds of obstacles were examined including rectangular plates with different sizes and a wedge, which have been objects of thorough studies for the last century [8–10]. Main attention was paid to study the flow formation and visualization of a spatial structure of different physical variables field in a wide range of the flow condition.

2   Governing Equations, Basic Scales and Simulation Conditions

Mathematical modeling of the problem is based on the fundamental system of equation for multicomponent inhomogeneous incompressible fluid in the Boussinesq approximation [5] taking into account the buoyancy and diffusion effects of stratified components. In the study of slow, as compared to the speed of sound, flows of fluids characterized by high thermal conductivity, one can account in calculations only for variations in density associated with concentration of the stratified component neglecting temperature variations. Thus, the governing equations take the following form [8, 10]

$$\rho = \rho_{00}(\exp(-z/\Lambda) + s), \operatorname{div} \mathbf{v} = 0,$$
$$\frac{\partial \mathbf{v}}{\partial t} + (\mathbf{v}\nabla)\mathbf{v} = -\frac{1}{\rho_{00}}\nabla P + v\Delta\mathbf{v} - s \cdot \mathbf{g}, \quad \frac{\partial s}{\partial t} + \mathbf{v} \cdot \nabla s = \kappa_S \Delta s + \frac{v_z}{\Lambda}. \quad (1)$$

Here, s is the salinity perturbation including the salt compression ratio, $\mathbf{v} = (v_x, v_y, v_z)$ is the vector of the induced velocity, P is the pressure except for the hydrostatic one, $v = 10^{-2} \text{ cm}^2/\text{s}$ and $\kappa_S = 1.41 \cdot 10^{-5} \text{ cm}^2/\text{s}$ are the kinematic viscosity and salt diffusion coefficients, t is time, ∇ and Δ are the Hamilton and Laplace operators respectively.

The proven solvability of the two-dimensional fluid mechanics equations enables calculating flows around obstacles for both strongly ($\Lambda = 9.8 \text{ m}$, $N = 1 \text{ s}^{-1}$, $T_b = 6.28 \text{ s}$) and weakly ($\Lambda = 24 \text{ km}$, $N = 0.02 \text{ s}^{-1}$, $T_b = 5.2 \text{ min}$) *stratified* fluids, and, as well, *potentially* ($\Lambda = 10^8$ km, $N = 10^{-5} \text{s}^{-1}$, $T_b = 7.3$ days) and *actually homogeneous* media ($\Lambda = \infty$, $N = 0 \text{ s}^{-1}$, $T_b = \infty$). In the case of *potentially* homogeneous fluid, density variations are so small that cannot be registered by existing technical instruments but the original mathematical formulation (1) is retained. In the case of *actually* homogeneous fluid, the fundamental system of equations is degenerated on the singular components [7].

The experiments and calculations were carried out in two stages. Initially, an obstacle with impermeable boundaries is submerged with minimum disturbances into a quiescent stratified environment. Physically reasonable initial and boundary conditions in the associated coordinate system are no-slip and no-flux on the surface of the obstacle for velocity components and total salinity respectively, and vanishing of all perturbations at infinity.

Diffusion-induced flow is formed due to interruption of the molecular transport of the stratifying agent on the obstacle [8]. The calculated flow is then taken as initial condition of the problem

$$\mathbf{v}|_{t \leq 0} = \mathbf{v}_1(x, z), \quad s|_{t \leq 0} = s_1(x, z), \quad P|_{t \leq 0} = P_1(x, z), \quad v_x|_\Sigma = v_z|_\Sigma = 0,$$
$$\left[\frac{\partial s}{\partial \mathbf{n}}\right]\Big|_\Sigma = \frac{1}{\Lambda}\frac{\partial z}{\partial \mathbf{n}}, \quad v_x|_{x,z \to \infty} = U, \quad v_z|_{x,z \to \infty} = 0, \tag{2}$$

where, U is the uniform free stream velocity at infinity, $\mathbf{n}$ is external normal unit vector to the surface, Σ, of an obstacle which can be either a plate or a wedge with length, L, and height, h, or, $2h$.

The system of equations and the boundary conditions (1)–(2) are characterized by a number of parameters, which contain length (Λ, L, h) or time $\left(T_b, \ T_U^L = L/U\right)$ scales and parameters of the body motion or dissipative coefficients.

Large dynamic scales, which are internal wave length, $\lambda = U T_b$, and viscous wave scale, $\Lambda_\nu = \sqrt[3]{g\nu}/N = \sqrt[3]{\Lambda(\delta_N^\nu)^2}$, characterize the attached internal wave fields structure [8, 10].

The flow fine structure is characterized by universal microscales, $\delta_N^\nu = \sqrt{\nu/N}$, $\delta_N^{\kappa_S} = \sqrt{\kappa_S/N}$, defined by the dissipative coefficients and buoyancy frequency, which are analogues of the Stokes scale on an oscillating surface, $\delta_\omega^\nu = \sqrt{\nu/\omega}$ [5]. Another couple of parameters such as Prandtl's and Peclet's scales are determined by the dissipative coefficients and velocity of the body motion, $\delta_U^\nu = \nu/U$ and $\delta_U^{\kappa_S} = \kappa_S/U$ [7, 8].

Relations of the basic length scales produce dimensionless parameters such as Reynolds, $\text{Re} = L/\delta_U^\nu = UL/\nu$, internal Froude, $\text{Fr} = \lambda/2\pi L = U/NL$, Péclet, $\text{Pe} = L/\delta_U^{\kappa_S} = UL/\kappa_S$, sharpness factor, $\xi_p = L/h$ or fullness of form, $\xi_S = S/Lh$, where S is the cross-sectional area of an obstacle, and, as well, specific relations for a stratified medium. The additional dimensionless parameters includes length scales ratio, $C = \Lambda/L$, which is an analogue of the reverse Atwood number $\text{At}^{-1} = (\rho_1 + \rho_2)/(\rho_1 - \rho_2)$ for a continuously stratified fluid.

Such a variety of length scales with their significant differences in values indicates complexity of internal structure even of such a slow flow generated by small buoyancy forces, which arise due to the spatial non-uniformity of molecular flux of the stratifying agent.

The large length scales prescribe size selection for observation or calculation domains, which should contain all the structural components studied, such as upstream perturbations, downstream wake, internal waves, vortices, while the microscales determine grid resolution and time step. At low velocities of the body motion, the Stokes scale is a critical one, while at high velocities the Prandtl's scale is dominant.

3 Laboratory Modeling of Flows Around a Plate

Experiments were carried out on the stands "Laboratory mobile pool" (LPB) and "Experimental stand for modeling of surface manifestations of underwater processes" (ESP) which belong to the Unique research facility "Hydrophysical complex for modeling of hydrodynamic processes in the environment and their impact on the underwater technical facilities, as well as the contaminant transfer in the ocean and atmosphere (URF "HPhC IPMech RAS")" [11]. The stands include transparent tanks with windows made of optical glass which allows using high-resolution optical observation devices such as the schlieren instruments IAB-451 and IAB-458. Models were fixed on transparent knives to the towing carriage, which moved along the guide rails mounted above the tank. Optical control of buoyancy profile and conductivity probe measurement of buoyancy frequency were carried out before starting the experiment. The following experiment was conducted after decay of all perturbations, which were registered by contact and optical instruments.

Scale of the phenomena under study was limited to the size of the viewing area of the schlieren instrument, which had diameter of 23 cm in these experiments. The spatial resolution, limited by the optical characteristics of the instrument itself and the recording equipment, which is being improved constantly with the development of computer technology, did not exceed 0.05 cm in these experiments. The classical "Vertical narrow slit-Foucault knife" and "Vertical slit-thread" techniques were used for flow visualization. In these methods, colour and brightness variations of a flow image are defined by value of the horizontal component of refraction index gradient [10, 11]. Refraction index of the sodium chloride water solutions is proportional to density, so schlieren images present patterns of the horizontal component of fluid density gradient.

Technical capabilities of the LPB and ESP stands enable visualizing both the main large-scale components of flows around obstacles, including upstream perturbations, downstream wakes, vortices, internal waves, and the fine structure elements such as high-gradient interfaces and filament in the both strongly and weakly stratified fluids.

Contours of phase surfaces of the upstream perturbations and attached internal waves in Fig. 1 are quite adequately described by the existing analytical and numerical models, which take into account geometry and velocity of body movement [8, 10].

Geometry of the high gradient interfaces, forming the environment fine structure, is very diverse and depends on the shape and velocity of the body movement and the stratification parameters, as well.

At small velocities of the body movement, sharp interfaces outline the density wake. These envelopes did not contact with the body at the poles but touch it inside the rear part (Fig. 1a). With increase in the plate velocity, the envelopes break down into

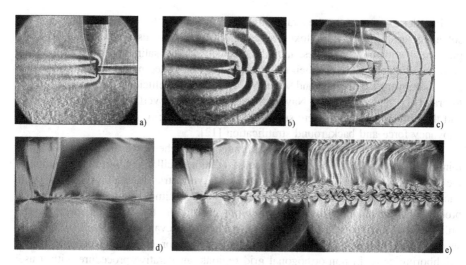

Fig. 1. Schlieren images of the flow pattern around a plate: (a–c) – placed vertically, $U =$ 0.03; 0.18; 0.29 cm/s (h = 2.5 cm, T_b = 12.5 s); (d, e) – placed horizontally $U = 2.3$; 4.9 cm/s (L = 2.5 cm, T_b = 7.5 s).

filaments coexisting with internal waves in the central part of the downstream wake (Fig. 1b). With a further increase in velocity of the body movement shapes of the filaments are changed becoming similar to the forms of the phase surfaces of internal waves over a wide barrier, which generates waves with large amplitudes (Fig. 1c).

Another type of the fine structure was observed in the wake past a thin horizontal plate. Short filaments, forming a transverse structure (Fig. 1d), are well expressed here. With increase in velocity of the plate movement the filaments are lengthened and fill each compact vortex of the downstream wake, which is covered with its own high gradient envelope. With a further increase in the velocity the whole flow pattern takes a vortex structure (Fig. 1e) requiring for its resolution more precise instruments and high speed sensors.

4   Method for Numerical Simulation of Flow Around Obstacles

Numerical solution of the system (1) with the boundary conditions (2) was constructed using our own solver *stratifiedFoam* developed within the frame of the open source computational package OpenFOAM based on the finite volume method. The package, which was originally developed for numerical calculation of 3D problems in fluid mechanics, can effectively simulate 2D problems, as well, that is technically done by selection of only a single computational cell in the third dimension and specification of 'empty' boundary conditions on the front and back boundaries of the calculation domain.

In order to account for the stratification and diffusion effects, the standard *icoFoam* solver for unsteady Navier–Stokes equations in homogeneous viscous fluid was supplemented with new variables, including density ρ and salinity perturbation s, and corresponding equations for their calculation. We also added new auxiliary parameters such as buoyancy frequency and scale, N, Λ, diffusion coefficient, κ_S, acceleration due to gravity, g, and others. The Navier–Stokes equation for vertical velocity component and the diffusion equation were supplemented with the terms characterizing effects of buoyancy force and background stratification [12].

To interpolate the convective terms a limited TVD-scheme was used, which ensures minimal numerical diffusion and absence of solution oscillations. For discretization of the time derivative a second-order implicit asymmetric three-point scheme with backward differencing was used, which ensures a good time resolution of the physical process. For calculating the diffusion terms, based on the Gauss theorem within orthogonal grid sections, a surface normal gradient was evaluated at a cell face using a second order normal-to-face interpolation of the vector connecting centers of two neighboring cells. In non-orthogonal grid regions, an iterative procedure with a user specified number of cycles was used for non-orthogonal error correction due to a grid skewness.

Meshing of the computational domain was performed using the open integrable platform SALOME, which allows creating, editing, importing, and exporting CAD (Computer Aided Design) models, as well as building a grid for them, using different algorithms and connecting physical parameters with geometry of the problem under study. For computational grid construction, the standard OpenFOAM utilities, such as 'blockMesh', 'topoSet', and 'refineMesh', were used, as well. The main OpenFOAM C++ class 'polyMesh', which handles a grid, is constructed using the minimum amount of information necessary to determine the partition elements and parameters such as vertices, edges, faces, cells, blocks, external boundaries, etc. By choosing an appropriate type of computational grid, i.e. structured or unstructured, orthogonal or non-orthogonal, consistent with boundaries of a domain or inconsistent ones, each of which normally has its own advantages and disadvantages, one can provide successful searching for solution of a problem under study. Therefore, methods for grid construction were chosen individually for a particular problem based on values of typical length scales and geometric complexity of a problem under consideration [13].

An unstructured grid, which consists usually of 2D triangles or 3D tetrahedrons, can be applied to domains with arbitrary geometry without any restrictions on form and number of boundaries of a computational domain. Due to a high level of automation, it is possible to reduce significantly duration of grid reconstruction. However, the main disadvantage of such a grid is an irregular structure of data, which requires sophisticated methods for numerical solution of problems. Numerical algorithms are complicated, as well, by usage of unstructured grids, requiring additional memory to store data on connections between grid cells. Furthermore, increase in number of tetrahedral cells, compared to that of hexahedral type, imposes more requirements for operational memory resources.

Structured grids matching the external boundaries of a computational domain are believed to be the most effective ones, which enable implementing computational algorithms with a higher order accuracy and reducing both computation time and amount of RAM required. By creating a curvilinear mesh, one can align grid lines with boundaries of a domain and, thereby, simplify specification of boundary conditions. However, a number of additional terms usually appear in governing equations due to a corresponding coordinate transformation. At the same time, procedures of a regular grid construction require a certain level of skills, efforts and computing resources, and can be applied only to a rather simple geometry of computational domain.

If it is impossible to construct a single mesh for a whole computational domain, grid is divided into a certain number of blocks. Complexity of such an approach consists in need for implementation of merging the solutions obtained in different subdomains. However, the technique for construction of a block-structured computational grid provides wide opportunities for using efficient numerical methods inside individual blocks with a regular grid structure.

Computation domain for the problems under consideration is a rectangle, which is divided into seven blocks. An obstacle is located in the central part of the computation domain, which is a plate or a horizontal wedge with length, $L = 10$ cm, and thickness, $h = 0.5$ cm, or height of the base, $h = 2$ cm, respectively (Fig. 2).

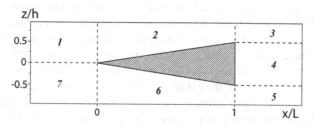

Fig. 2. Scheme of the computational domain partitioning into blocks.

The procedure for spatial discretization of the problem is parameterized that enables reducing significantly duration of grid reconstruction when changing parameters of the problem. The geometry of the computational domain allows constructing a block-structured hexahedral computational grid with nodes aligned at the block interfaces. Test computations with different grid resolutions confirm the need for resolving the smallest micro-scales of the problems, since a relatively coarse grid with total number of cells, $N_c = 5 \cdot 10^5$, gives unstable solution. Thus, numerical simulation of even 2D problems on continuously stratified fluids flows around impermeable obstacles require high-performance computing.

Algorithm for discretization of the computational domain involves mesh grading towards the obstacle (Fig. 3a). Near the body, the aspect ratio of a grid cell is approximately equal to unity, which has a positive effect on convergence of the solution. The main disadvantage of this method consists in necessity of cells rearrangement in all the subdomains at once if a grid is reconstructed, which leads to significant increase in computation time. In order to improve the quality of the computational domain discretization the OpenFOAM utilities, such as 'topoSet' and 'refineMesh', are additionally used, which enable selecting computational subdomains of interest and locally refining

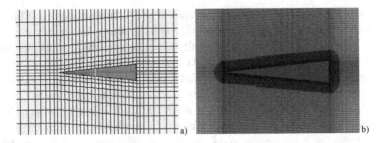

Fig. 3. The scheme of the computational domain partitioning: (a) – with a linear grid refinement, (b) – with an additional local partition.

them in accordance with prescribed scales and selected directions (Fig. 3b). The minimum mesh size of $2.5 \cdot 10^{-3}$ cm near impermeable boundaries satisfactorily resolves the diffusion microscale $\delta_N^{\kappa_S}$ with a relatively small total number of grid cells of 10^6 order.

However, partitioning of the grid cells even in a small part of the calculation domain requires a corresponding decrease in time step that, in turn, leads to an increase in duration of the computations. An essential disadvantage of additional local partitioning is a significant change in size of the grid cells on the boundary of subdomains that can affect the calculation results. By checking the constructed computational grid with the utility 'checkMesh' we make sure of its compliance with a set of constraints associated with the topology of the external boundaries and geometric characteristics of the grid cells, i.e. aspect ratio, skewness, twisting, non-orthogonality, etc.

Discretization of the boundary conditions (2) was carried out using the standard and extended utilities of the OpenFOAM package. The boundary condition on the surface of an obstacle for salinity perturbation gradient, which depends on orientation of the normal unit vector to the surface, was implemented by forming a non-uniform list of the field values using the standard, extended and self-elaborated utilities of the package. 'Empty' boundary conditions were set on the front and back faces of the computational domain, which exclude computations of the 2D problem in the third dimension.

To solve the resulting system of linear equations different iterative methods were used such as conjugate gradient method with PCG preconditioning applied to symmetric matrices and biconjugate gradient method with PBiCG preconditioning used for asymmetric matrices. As preconditioners for symmetric and asymmetric matrices DIC and DILU procedures were chosen, which are based on simplified procedures of incomplete Cholesky and LU factorization respectively. For coupling equations for momentum and mass conservation a steady well-convergent algorithm PISO (Pressure-Implicit Split-Operator) was used, which works in the most effective way for transient problems.

The need for a high spatial resolution of the problem results in a quite large number of computational cells that makes it irrational performing computations on a single-processor computer. Decomposition of the computational domain for a parallel run is carried out by a simple geometric decomposition in which the domain is split into pieces in certain directions with an equal number of computational cells in each block.

Such an approach allows setting a high spatial resolution of the computational domain and studying the problem in a wide range of the basic parameters for a quite reasonable time. The computations were performed in parallel using computing resources of the supercomputer "Lomonosov" of the Scientific Research Supercomputer Complex of MSU (SRCC MSU) and the technological platform UniHUB, which provides direct access to the Joint Supercomputer Center Cluster of the RAS (JSCC RAS).

The calculations were terminated when the integral characteristics or their statistical evaluations took values of steady-state regime. The spatial dimensions of the computational cells were chosen from the condition of adequate resolution of the fine flow components associated with the stratification and diffusion effects, which impose significant restrictions on the minimum spatial step. In high-gradient regions of the flow, at least several computational cells must fit the minimal linear scale of the problem. Calculation time step, Δt, is defined by the Courant's condition, $Co = |\mathbf{v}|\Delta t/\Delta r \leq 1$, where Δr is the minimal size of grid cells and $\mathbf{v}$ is the local flow velocity. Additional control was ensured by comparison of independent calculations for fluids with different stratification.

5   Calculation Results

5.1   The Structure of Diffusion-Induced Flow on a Motionless Plate

In contrast to the stationary solutions by Prandtl [14], which lose regularity on horizontal surface, the complete solution of the Eq. (1) for transient flow remains finite for any sloping angle of the surface. A cellular flow structure is kept in patterns of all the physical variables, but thicknesses of interfaces are specific for each parameter.

The calculated pattern of diffusion-induced flow on the horizontal plate, which simulates the central section of impermeable obstacle with an arbitrary shape (Fig. 4a), consists of a layered sequence of symmetrically arranged horizontal vortex cells [13].

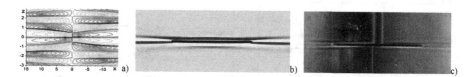

Fig. 4. The diffusion-induced flows on an obstacle with horizontal boundaries: (a) – calculated pattern of streamlines around the plate; (b, c) – patterns of density gradient field around the horizontal disc, calculated numerically and schlieren-visualized in a laboratory tank.

Uniformity of the streamline pattern indicates similarity of the velocity profile along the most length of the plate except for the narrow transition regions around its edges. Values of fluid velocity and vorticity decrease sharply with distance from the plate surface.

Even a small deviation from the horizontal position of the plate leads to violation of the flow symmetry and formation of new circulation systems including ascending and descending jets along the upper and lower sides of the plate respectively, and a system of compensating circulating cells, as well [13]. The calculated and observed patterns of the density gradient fields on the plate and disc are in a good agreement (Fig. 4b and c).

5.2 Diffusion-Induced Flow on a Wedge

An impermeable obstacle immersed in a stably stratified fluid at rest forms a complex system of flows including the main thin jets along the sloping sides of the obstacle with the adjacent compensating counterflows [15]. With a distance from the obstacles, the layered structures are enlarged and the maximum fluid velocity decreases. Pattern of the horizontal component of salinity gradient shows a multiscale flow structure, including extended side flows and wavy structures (Fig. 5).

Positive values of a visualized field are coloured green, while negative ones are in blue. Differences in values of a field between neighboring isolines are the same. The side boundaries of a wedge-shaped obstacle with length, $L = 10$ cm, and maximal thickness, $h = 2$ cm, can be either straight or curved in form of a concave or convex arc [16].

The additional fine-structure flow components, such as dissipative-gravitational waves, are formed in form of rosettes around the corner points of the wedge, where maximum value of the longitudinal component of salinity gradient, $|\partial s/\partial x|_{max} = 4 \cdot 10^{-2}$, is observed, while the corresponding value of salinity perturbation s is an order of 10^{-5}.

Structure of the horizontal component of salinity gradient field depends essentially on sign of the curvature of the wedge surface, the sharper the edges of the wedge the more pronounced the visualized beams of strips with alternating signs (Fig. 5a). At the same time, the maximum values are weakly dependent on curvature of the wedge sides. For a convex wedge (Fig. 5b) when angle between the base and the side edge is close to 90° the beam of fine structure elements is widened.

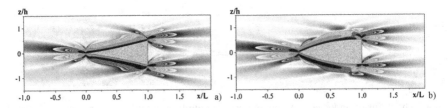

Fig. 5. Field of the longitudinal component of salinity gradient, $\partial s/\partial x$, around a wedge with concave (a) and convex (b) side boundaries ($L = 10$ cm, $h = 2$ cm, $T_b = 6.28$ s, $\tau = t/T_b = 16$)

In the pattern of pressure perturbation field there is a deficit pronounced in front of the tip, manifestation and height of which depends on curvature of the wedge side (Fig. 6). This deficit zone causes pressure difference which results in self-motion of a free wedge with neutral buoyancy observe in laboratory experiments.

The initial structure of the medium formed by diffusion-induced flows is changed dramatically with start of forced motion of a body even with the lowest velocity

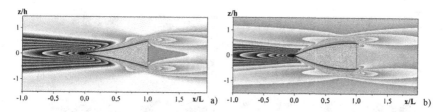

Fig. 6. Fields of pressure perturbation, P, around a wedge with concave (a) and convex (b) side boundaries ($L = 10$ cm, $h = 2$ cm, $T_b = 6.28$ s, $\tau = t/T_b = /16$ $\tau = 16$).

comparable with the typical velocities of the diffusion-induced flow (Fig. 7). Upstream perturbations, rosettes of transient and extended fields of attached internal waves and downstream wake past extreme points of the body are formed in a continuously stratified fluid. Number of the attached internal waves observed, which do not penetrate into the wake behind the body, increases with time.

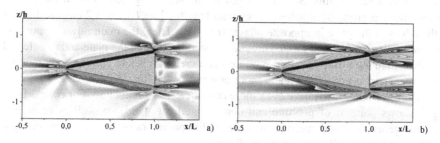

Fig. 7. Evolution of the horizontal component of salinity gradient perturbation field after start of motion of a wedge with straight boundaries ($L = 10$ cm, $h = 2$ cm, $T_b = 6.28$ s, $U = 0.001$ cm/s): (a, b) – $\tau = t/T_b = 2$; 50.

Structure of the calculated patterns of flow around a wedge is in an agreement with the schlieren visualization of refraction coefficient gradient field in a laboratory tank by "color shadow method" with a horizontal slit and grating for the bodies with other geometrical forms [11].

5.3 Flow Around a Rectangular Strip

The fundamental system of equations allows calculating a complete flow pattern and determining forces acting on a moving obstacle with a high degree of accuracy in a wide range of parameters without additional hypotheses and constants. This allows tracing a consistent restructuring of the flow due to "inclusion" of new components such as internal waves at start of motion, non-stationary vortices in the wake past an obstacle or rolling vortices along its surface (Fig. 8).

For all the velocities of the body motion, the flow field is characterized by a complicated internal structure. In the flow pattern around motionless body dissipative

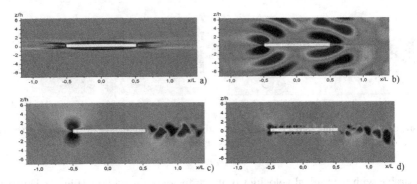

Fig. 8. Field of the vertical component of velocity around the plate ($N = 1.2\,\mathrm{s}^{-1}$, $L = 10$ cm, $h = 0.5$ cm): (a) – diffusion-induced flow for $U = 0$; (b–d) – $U = 1.0$; 5.0; 80 cm/s, $\mathrm{Re} = 10^3$; $5 \cdot 10^3$; $8 \cdot 10^4$, $\mathrm{Fr} = U/NL_x = 0.1$; 0.5; 6.7, $\lambda = UT_b = 5.2$; 26; 418 cm.

gravity waves are manifested at the edges of the strip (Fig. 8a). Around the slowly moving body a group of attached waves with lengths, $\lambda = UT_b = 5.2$ cm, are formed at the edges of the plate in opposite phases (Fig. 8b). Then, the main flow components become vortices, which are formed around the leading edge of the plate and manifested downstream in the wake (Fig. 8c). With further increase in velocity of the body motion, the flow pattern becomes more and more non-stationary (Fig. 8d).

Patterns of other fields have even more complicated fine structure. For comparison, Fig. 9 shows patterns of pressure and density gradients fields, as well as baroclinic vorticity generation and mechanical energy dissipation rates for both the strongly stratified and potentially homogeneous fluids.

Complexity of the patterns of the pressure and density gradients fields shown in Fig. 9 is attributable to the properties of differential operators generating two groups of spots for each vortex core, which correspond to the regions of perturbation growth and attenuation respectively. At the same time, advantage of studying these patterns consists in a more complete visualization of the structural details, which makes it possible to identify small-scale elements against the background of larger ones.

The transverse dimensions of the flow components, which are determined by the values of kinetic coefficients in the given mathematical formulation, are minimal in the patterns of the density gradient fields. The pressure gradient fields are generally smooth, but close to the leading edge there are large variations due to the simultaneity of generation of the both large (internal waves, vortices) and fine flow components (Fig. 9a, b).

In the pattern of the horizontal component of pressure gradient, $\partial P/\partial x$, (Fig. 9a) a sequence of spots with different signs is manifested like in the vertical component of velocity field in Fig. 8d. At the time distribution of the vertical component of pressure gradient, $\partial P/\partial z$, shown in Fig. 9b, demonstrates a more rarefied set of vertically oriented spots with different colours. Here, the vortices are well resolved and, in particular, there are ten cores above the plate, as it is seen from the patterns in Fig. 9, which is greater by one compared that to the case of the horizontal component of velocity field.

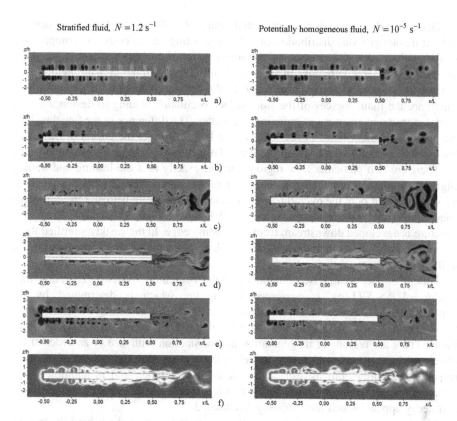

Fig. 9. Calculated patterns of instantaneous fields near the plate ($L = 10$ cm, $h = 0.5$ cm, $U = 80$ cm/s) in the stratified (left column $N = 1.2\,\text{s}^{-1}$, Fr = 6.7) and potentially homogeneous fluids (right column $N = 10^{-5}\,\text{s}^{-1}$, Fr = $8 \cdot 10^{5}$): (a, b) – horizontal and vertical components of pressure gradient, (c, d) – horizontal and vertical components of density gradient, (e) – baroclinic vorticity generation rate, (f) – mechanical energy dissipation rate.

Even a more complicate flow pattern is resolved near the leading edge. In the weakly stratified fluid, the perturbations degenerate more slowly as compared to that in the strongly stratified medium and scales of the vortex structures are noticeably smaller.

The patterns of the components of density gradient are presented in Fig. 9c, d. In the horizontal component of the density gradient field, thin layers with the both signs are localized on the vortex shells, which are combined into compact spots behind the body. In the weakly stratified fluid, the structures of the vertical component of density gradient are oriented mostly horizontally and form its own system of spiral curls typical for vortex elements. The local patterns of the physical variables in Fig. 9c, d are substantially different, locations of the centers of the regions, which they outline, not coinciding.

It should be specially noted, the differences in the geometry and fine structure of the pressure and density gradients fields, which determine the spatial and temporal variations of such a dynamic parameter as vorticity generation rate of the flow and, hence, change in value of the vorticity itself.

Due to the differences in the spatial and temporal scales of the flow components, a general inhomogeneous distribution of the forces acting on the body, i.e. compression on the leading edge and stretch at the initial section of the side surface, is here supplemented with large variations in space and time. Further in the pressure field, deficit zones (side stretching) are expressed due to the passage of centers of large vortices, which are the main sources of the side surface oscillation leading to development of such dangerous phenomena as buffeting and flutter. The difference in the fine details of the pressure gradient fields in the wake of the body is due to the effects of buoyancy forces, suppressing fluid transfer in the vertical direction, and the fine flow components.

There are two particular flow regions located exactly near the leading and trailing edges of the plate, where the main generation of vorticity vector, $\mathbf{\Omega} = \text{rot } \mathbf{v}$, occurs as a consequence of both the overall reorganization of the velocity field and the baroclinic effects. Geometry of the density gradients field is complicated as the layered flow structure is developed downstream from the leading edge. In the weakly stratified fluid, range of oscillations of the vortices' size in the wake increases with distance from the trailing edge when stabilizing buoyancy forces are absent.

Pattern of the streamwise component of the baroclinic vorticity generation rate, $d\mathbf{\Omega}/dt = \nabla P \times \nabla(1/\rho)$, is presented in Fig. 9e, which is determined by the non-collinearity of pressure ∇P and density gradients $\nabla \rho$ according to the Bjerknes theorem. This field is the most complex and structured one in flows of inhomogeneous fluids. There are regions of its generation and dissipation with size of an order of the plate's thickness manifested above and below the leading edge and in front of the body respectively. Further downstream, the structures get thinner, and in addition to the remaining vortex elements a number of multiple thin zones of vorticity amplification and decay appear, which are gradually lengthened.

Fine linear structures are predominantly manifested in the wake behind the trailing edge of the plate, which are oriented mainly in the direction of the flow and deformed by the large irregularities. In the strongly stratified fluid, the perturbations of the both signs are expressed along the whole length of the plate, but in the potentially homogeneous one they are only in the first quarter. There are regions of vorticity generation and dissipation manifested in the areas of interaction of the vortices with the free stream in the wake of the body. The geometry of the baroclinic vorticity generation rate field explains the formation dynamics of vortex flow fine structure and the mechanism for splitting of the fields into a set of layered structures observed in the schlieren images of the wake flows [10, 11].

Field of the mechanical energy dissipation rate, $\varepsilon = 0.5\rho\nu(\partial v_i/\partial x_k + \partial v_k/\partial x_i)^2$, presented in Fig. 9f, is different from zero in a relatively narrow zone in front of the body, where the horizontal flow turns flowing the plate, and reaches maximal values in the vortex structures at the first third of the plate's length. Regions of dissipation, observed at the second half of the plate, are larger in the potentially homogeneous fluid, compared to that for the strongly stratified one. It should be noted a qualitative difference in geometry of the vorticity generation rate field with its pronounced fine structures (Fig. 9c) and a relatively smooth distribution of the energy dissipation rate field (Fig. 9f).

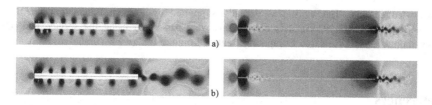

Fig. 10. Patterns of the pressure field around the horizontal plate ($L = 10$ cm, $h = 0.5$ cm, $U = 80$ cm/s): thick ($h = 0.5$ cm, left) and thin ($h = 0.05$ cm, right) ones: (a, b) – $N = 1.2\,\mathrm{s}^{-1}$ and $N = 0$.

All the instantaneous flow patterns including the vorticity generation fields in Fig. 9 are in a continuous evolution. The variation of the velocity pattern in the kinematic description is caused by the generation and break-up of new elements, such as vortices due to the non-synchronized variation of the physical parameters with thermodynamic nature, in particular, density and pressure gradients.

In the dynamic description, the generation of new elements with their own kinematics and spatial and temporal scales is associated with the high order and nonlinearity of the fundamental system of equations. Even in the linear approximation, the complete solutions of this system contain several functions, which can be treated in the nonlinear models as analogues of the components interacting with each other and generating new types of disturbances [7].

The pressure perturbation field shows a strong dependence on the vertical dimension of the obstacle. In the wake past the thick plate, a number of large vortices develop (Fig. 10, left column), while past the thin plate the transverse streaky structures are manifested similar to ones observed in the experiments (Fig. 1). A more complete collection of flow patterns past the plates with different shapes are presented in [8, 17].

6 Conclusion

Based on the open source software 2D numerical simulations of incompressible stratified (strongly and weakly) and homogeneous (potentially and actually) fluids flows were performed. The method allows analyzing in a single formulation the dynamics and fine structure of flow patterns past obstacles in a wide range of stratification and flow parameters.

Transient flow patterns past obstacles were analyzed, and physical mechanisms were determined, which are responsible for formation of vortices in regions with high pressure and density gradients near the edges of an obstacle. The calculation results are in a qualitative agreement with the data from laboratory experiments.

Flow around obstacles is a complex, multiscale, and transient physical process, which requires additional detailed experimental and theoretical study accounting for the effects of diffusion, thermal conductivity, and compressibility of the medium with control of the observability and solvability criteria for all the physical parameters and structural components of the flows under study.

Acknowledgements. The work was partially supported by Russian Foundation for Basic Research (grant 15-01-09235). The calculations were performed using the service UniHUB (www.unihub.ru) and Research Computing Centre "Lomonosov" (www.parallel.ru).

References

1. D'Alembert, J.-R.: Traitè de l'èquilibre et de movement des fluids. Paris, David. 458 p. (1744)
2. D'Alembert J.-R.: Réflexions sur la cause générale des vents. Paris (1747)
3. Euler, L., Robins, B.: Neue Grundsätze der Artillerie enthaltend die Bestimmung der Gewalt des Pulvers nebst einer Untersuchung über den Unterscheid des Wiederstands der Luft in schnellen und langsamen Bewegungen. Aus d. Engl. übers. u. mit Anm. v. L. Euler. Berlin, Haude, 720 S. (1745)
4. Euler, L.: Principes généraux du mouvement des fluids. Mémoires de l'Academié royalle des sciences et belles letters. Berlin. vol. 11 (papers of 1755 year). (1757)
5. Landau, L.D., Lifshitz, E.M.: Fluid Mechanics: Course of Theoretical Physics, vol. 6, p. 731. Pergamon Press, Oxford (1987)
6. Baidulov, V.G., Chashechkin, Y.: Invariant properties of systems of equations of the mechanics of inhomogeneous fluids. J. Appl. Math. Mech. **75**(4), 390–397 (2011). http://www.sciencedirect.com/science/journal/00218928
7. Chashechkin, Y.D.: Differential fluid mechanics – harmonization of analytical, numerical and laboratory models of flows. In: Neittaanmäki, P., Repin, S., Tuovinen, T. (eds.) Mathematical Modeling and Optimization of Complex Structures. CMAS, vol. 40, pp. 61–91. Springer, Heidelberg (2016). doi:10.1007/978-3-319-23564-6_5
8. Zagumennyy, Y.V., Chashechkin, Y.D.: Pattern of unsteady vortex flow around plate under a zero angle of attack. Fluid Dyn. **51**(3), 53–70 (2016). doi:10.7868/S056852811603018X
9. Schlichting, H.: Boundary-Layer Theory, p. 535. McGraw-Hill, New York (1955)
10. Bardakov, R.N., Mitkin, V.V., Chashechkin, Y.: Fine structure of a stratified flow near a flat-plate surface. J. Appl. Mech. Technol. Phys. **48**(6), 840–851 (2007). doi:10.1007/s10808-007-0108-6
11. Chashechkin, Y.D.: Structure and dynamics of flows in the environment: theoretical and laboratory modeling. In: Actual Problems of Mechanics. 50 years of the A.Y. Ishlinskiy Institute for Problems in Mechanics of the RAS, pp. 63–78. (2015) (in Russian)
12. Dimitrieva, N.F., Zagumennyi, Y.V.: Numerical simulation of stratified flows using OpenFOAM package. In: Proceedings of the Institute for System Programming RAS, vol. 26, no. 5, pp. 187–200 (2014). doi:10.15514/ISPRAS-2014-26(5)-10
13. Zagumennyi, I.V., Chashechkin, Y.D.: Diffusion-induced flow on a strip: theoretical, numerical and laboratory modeling. Procedia IUTAM. **8**, 256–266 (2013). doi:10.1016/j.piutam.2013.04.032
14. Prandtl, L.: Führer durch die Strömungslehre. Braunschweig: Vieweg. 638 p. (1942)
15. Dimitrieva, N.F., Chashechkin, Y.: Numerical simulation of the dynamics and structure of a diffusion-driven flow on a wedge. Comput. Contin. Mech. **8**(1), 102–110 (2015). doi:10.7242/1999-6691/2015.8.1.9
16. Zagumennyi, Ia.V., Dimitrieva, N.F.: Diffusion-induced flow on a wedge-shaped obstacle. Phys. Scr. **91**, Article No.084002 (2016). doi:10.1088/0031-8949/91/8/084002
17. Chashechkin, Y.D., Zagumennyi, I.V.: Hydrodynamics of horizontal stripe. Probl. Evol. Open Syst. **2**(18), 25–50 (2015). (The Republic of Kazakhstan)

EnOI-Based Data Assimilation Technology for Satellite Observations and ARGO Float Measurements in a High Resolution Global Ocean Model Using the CMF Platform

Maxim Kaurkin[1,2,3(✉)], Rashit Ibrayev[1,2,3],
and Alexandr Koromyslov[3,4]

[1] Institute of Numerical Mathematics RAS, Moscow, Russia
kaurkinmn@gmail.com
[2] P.P. Shirshov Institute of Oceanology RAS, Moscow, Russia
[3] Hydrometeorological Centre of Russia, Moscow, Russia
[4] M.V. Lomonosov Moscow State University, Moscow, Russia

Abstract. A parallel implementation of the ensemble optimal interpolation (EnOI) data assimilation method for the high resolution general circulation ocean model is presented. The data assimilation algorithm is formulated as a service block of the Compact Modeling Framework (CMF 3.0) developed for providing the software environment for stand-alone and coupled models of the Global geophysical fluids. In CMF 3.0 the direct MPI approach is replaced by the PGAS communication paradigm implemented in the third-party Global Arrays (GA) toolkit, and multiple coupler functions are encapsulated in the set of simultaneously working parallel services. Performance tests for data assimilation system have been carried out on the Lomonosov supercomputer.

Keywords: Earth system modeling · Data assimilation · Ensemble optimal interpolation · Coupler

1 Introduction

The up-to-date ocean models with ultra-high spatial resolution and the assimilation of continuously incoming observations allow to implement an operational forecast of three-dimensional state of the marine environment, similar to the meteorological weather forecasts. As well ocean modeling is an important component in the climate change researches and monitoring of the environmental state.

The solution of this problem is impossible without using of supercomputing technology due to the large amount of receiving and processing information. The amount of resources required to perform the quick calculations in high spatial resolution models today estimated 10^2–10^3 processing cores for short-term forecasts, and 10^4–10^5 - for the medium and long term forecasts. Satellite observations of surface temperature and surface topography with a resolution of 1 km (e.g. NASA AQUA, AVISO project, etc.) are already available. This conforms to a flow of gigabytes of observation data per day. It is important to assimilate the information correctly and

© Springer International Publishing AG 2016
V. Voevodin and S. Sobolev (Eds.): RuSCDays 2016, CCIS 687, pp. 57–66, 2016.
DOI: 10.1007/978-3-319-55669-7_5

quickly, especially when the spatial resolution of global ocean models will be about 1 km. The issue of computational efficiency becomes especially critical when the model and the assimilation system functions in the operational mode for the medium-term and short-term forecasts construction. Output delay of modern satellite observations is about a couple of hours, a high spatial resolution of ocean models can simulate the actions of eddy structures and the assimilation of satellite observations serves theirs timely detection, that, consequently allows to predict dangerous natural phenomenon such as storms and typhoons. Therefore during the development and implementation of data assimilation methods you need to pay close attention to the scalability of using methods and their ability to process large amounts of information.

Currently, there are few data assimilation algorithms, which are used in problems of weather forecast and in operational oceanology. Used approaches can be divided into variational (3d-Var, 4d-Var) [1] and dynamic-stochastic (mainly ensemble Kalman filter - EnKF [2] and ensemble optimum interpolation - EnOI [3]). The main advantage of assimilation methods based on EnKF and EnOI over variational approach (3d-Var, 4d-Var) is the fact that EnKF and EnOI don't require the construction of model's adjoint operator, that it is often very difficult. The data assimilation based on the ensemble approach in general can be implemented using a ocean model like a "black box". Such methods can have efficient parallel realization and quite applicable for global models, while the 4d-Var method due to the computational complexity today is not used in any Global ocean model of the high spatial resolution [4].

The purpose of this work is implementation EnOI method as a Data Assimilation Service (DAS) of framework CMF3.0 to use in models of high spatial resolution on massive parallel computers with shared memory, replacing the integrated in the ocean model previously used simple method multivariate optimum interpolation (MVOI) [5].

2   The Compact Computing Platform for Modeling CMF 3.0 and Service DAS

Along with the development of models of individual components of the Earth system, the role of the instruments organizing their coordinated work (couplers, and the coupling frameworks) becomes more and more important. The coupler architecture depends on the complexity of the models, on the characteristics of interconnections between models and on computer environment. Historically the development of couplers follows the development of coupled atmosphere-ocean models. The first systems combined physical components directly and don't require additional code. As models became more complex and segregate in individual programs there is appear a need to separate one service component - coupler, which was engaged in data interpolation between different model grids of components. At the first stage it was a simple set of procedures to sending fields through the file system, and then was segregated in a separate sequential program, the analogue of the central hub for communication between all models. As the increasing resolution of model's grids the sequential coupler's algorithms became ineffective, it was completely replaced by entirely parallel architecture.

The compact computing platform for modeling CMF is a framework for the coupled modeling of the Earth system and its high-resolution components on parallel computing systems. By using abstract interfaces the main program and the coupler become completely independent from the number of connectable models - to work in coupling system it is enough for user to create a derived class of his component [6].

Version CMF 2.0 of compact framework for modeling proved its aptness for creating high-resolution models, it had several directions for improvement. Firstly, although pure MPI-based messaging is quite fast, it needs explicit work with sending and receiving buffers. Additionally, development of nested regional sea submodels becomes quite difficult using only MPI-routines. CMF 2.0 test results showed that we can easily sacrifice some performance and choose better (but perhaps less efficient) abstraction to simplify messaging routines.

2.1   PGAS-Communicator

In version CMF 3.0 [7] it is used a Global Arrays library (GA) [8], which realizes the paradigm PGAS (Partitional Global Adress Space). The library gives you access to the global index of the array as if it is all available in the local memory.

CMF 3.0 contains class Communicator that encapsulates the logic of work with the library and provides API for put/get operations in different component's parts of global data. It turned out that this approach allows not only simplifies the interaction of nested components, but also provides a convenient replacement of the data exchange system based on a direct MPI-approach between components and coupler.

As a result, all exchanges between parts of the system are implemented using class Communicator. It contains a hash-table to store all the information about arrays, including their status and meta-data.

2.2   The New Architecture of Coupling Model

As the complexity of the coupled system grows, it is need more convenient way of the components combining. Originally appeared to web applications SOA (Service Oriented Architecture) provides a good foundation for the solution of this problem.

In the CMF 3.0 all models send their requests to common queue (Fig. 1). Service components receive from this queue only messages that can be processed, get the data from a virtual global arrays and perform required actions. Architecture allows minimizing the dependencies between physical and service components and make development easier. Moreover, since all services inherit a general base class Service, the addition of a new service is easy. Now CMF 3.0 includes the following independent parallel services: CPL (Coupler service for operations of data interpolation in transit between the model components on different grids), IOD (Input/Output Data service to work with files), DAS (data assimilation service).

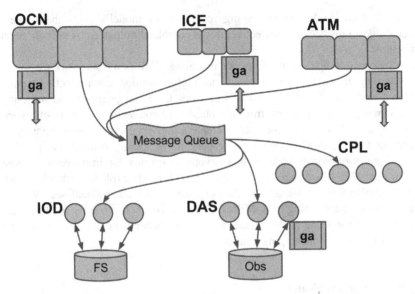

Fig. 1. DAS (Data Assimilation Service) in the architecture of the compact framework CMF 3.0. There are three components in this example: ocean model (OCN), ice model (ICE), atmosphere model (ATM). They send requests to the common message queue, where they are retrieved by coupler (CPL), data assimilation (DAS), input and output data (IOD) services. The data itself is transferred through the mechanism of global arrays, which are also used for interprocessor communication in the components and in the DAS service.

In the CMF 3.0 was implemented service of data assimilation DAS for providing the system of data assimilation work based on EnOI method (Sect. 3).

2.3   Coupler: Interpolation

Despite the fact that the logic of coupler interpolation subroutines remained the same, PGAS abstraction allowed greatly simplify the code. Now all the data, necessary to coupler process, are obtained from neighbors using the Communicator class. Disadvantage of this approach is the decreasing of performance associated with the inability to use deferred MPI-operation and the availability of the library's GA own costs.

To test the system used time estimate, including sending a request in queue, sending the data, interpolation and pushing data into final destination arrays, i.e. simulates the performance of all system. The tests were conducted on the Lomonosov supercomputer.

Graph shows practically linear scalability (Fig. 2) [7]. Finally, the absolute values ~2–3 s on coupling (on 20–50 CPL cores) per modeling day satisfy the practical purposes of the experiments.

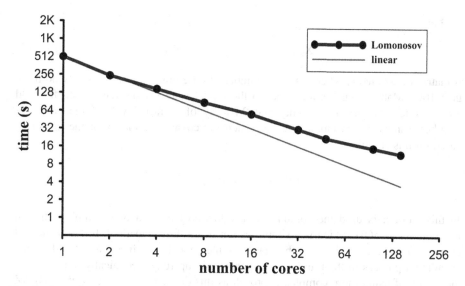

Fig. 2. Operating time of the test for CMF 3.0 in seconds depending on the size (cores number) of the coupler's communicator (CPL service) on the Lomonosov supercomputer [7]. The X axis - number of cores used by the CPL service, and on the Y - time, spending on the interpolation operation performed by this service.

3   Description of DAS Service

3.1   EnOI Method: Basic Equations

Basic equations for methods EnOI and EnKF [2, 3] are follows:

$$x_a = x_b + K(y_{obs} - Hx_b) \tag{1}$$

$$K = BH^T(HBH^T + R)^{-1} \tag{2}$$

Where x_b, x_a are the model solution vectors of size n before and after data assimilation (background and analysis, correspondent), where n - the number of grid points, which has the order of 10^8 (for the ocean model with a resolution of 0.1 degrees); y_{obs} - the vector of size m of observations, where m - number of observation points ($\sim 10^4$); K (n * m) is Kalman gain matrix; R (m * m) is the variance matrix of the observational error (it is diagonal if the data at observation points do not depend on each other), Hx_b is the projection of the background vector onto the observational space by the observational operator H; Matrix B is the co-variance matrix of the model errors. Its rigorous definition usually is not given, but instead describes the method of its calculation.

   In the method of multivariate optimum interpolation (MVOI) the elements of the matrix B defined by function depending on the distance between grid points. The main idea of the EnKF (and EnOI) method is that the covariance matrix B is obtained from the ensemble of the state vectors of the model (sample).

Let

$$X_b^{en} = \left[x_b^1 \ldots x_b^{en}\right] - \left[\overline{x_b} \ldots \overline{x_b}\right] \tag{3}$$

is matrix of size n*en, where en - the number of ensemble elements (usually no more than 100). Matrix columns are equal to the model state vectors (composed from 3d arrays of temperature and salinity and 2d array of surface level) of the ensemble members minus the ensemble average. Then the covariance matrix of the model B, based on this sample:

$$B^{en} = \frac{1}{en - 1} X_b^{en} (X_b^{en})^T \tag{4}$$

In this work it is used the computationally low-cost method of ensemble optimum interpolation (EnOI), which is a simplification of the EnKF method. In EnOI method elements of the ensemble are the states of the model, obtained and stored in the calculation process in the previous few years (spin-up run). Technically implemented the ability of using more complex approach as this done, for example, in the TOPAZ project [9], when the model starts from a hundred (number of elements of the ensemble) different initial conditions, then on the basis of a hundred forecasts the covariance matrix B is build. It is easy to see that computational cost of this approach is proportional to the number of elements in the ensemble, but these calculations can be performed parallel as one hundred model predictions are made independently.

3.2   Features of the Parallel Implementation of EnOI Method

As well as any existing service of the software system CMF 3.0, data assimilation is performed on separate computing cores. This allows to structure the system co-simulation better, in order to make, each software component solve its own problem. At the same time the model of the ocean does not take part in the assimilation. Only, results of the model calculation in the form of vector elements of the ensemble are used. On their basis the covariance matrix of model B is calculated, to be more accurate, a matrix (HBH^T + R), which has a smaller dimension m * m. Data from the ocean model is submitted to the service (usually once a modeling day) without the request to the file system, which is important because the size of three-dimensional state arrays for the ocean model with 0.1° resolution requires several gigabytes.

Problems and Causes of Allocation Data Assimilation into a Separate Service DAS.

1. Observational data (satellite or drifters data) almost always is distributed on the estimated area of the ocean very irregularly, so if you use the same two-dimensional processor decomposition of the ocean for them, the load on the computing cores will also be distributed unevenly.
2. Covariance matrices occupy a significant amount of memory, and it is better to keep them on the computing cores separately from model components.
3. DAS service can be shared between different components and this corresponds to the concept where each software component solves its own problem.

General Algorithm of EnOI Assimilation.

- Constructing a new 1d domain decomposition where each sub-domain is assigned to some processor of DAS (Fig. 3). The sub-domains are uniform according to observation data and are not uniform according number of grid nodes. Sub-domains have the form of horizontal stripes.
- Each "stripe" gets only its data of observation (x_b) and only its part of the global model state vector.
- Projection of model state vector in the observation points using bilinear interpolation is calculated for each "stripe" (Hx_b).
- Calculating the vector of observations innovation (y_{obs}-Hx_b) locally for each "strip" also.
- Asynchronous reading the ensemble of states (if necessary) in matrix X_b (distributing across DAS cores) building the projection of state vectors in the observation point Hx_b.
- In order to invert the matrix ($HBH^T + R$) its singular decomposition is used, it is calculated using a parallel procedure `pdegsvd` from ScaLAPACK Library (Intel MKL package).
- The array x_a is sent back to the ocean component and used as an initial condition for the further integration of the ocean model.

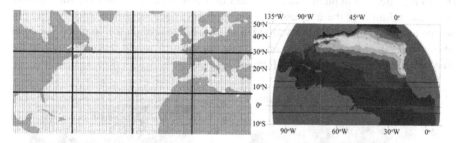

Fig. 3. A two-dimensional field decomposition used in the INMIO ocean model and a one-dimensional decomposition used in the DAS service. The dots show example of observation data.

Remarks

Matrix H for bilinear interpolation is constructed for each DAS service call, as long as the observational data is available each time in new points. As a result, every time a new domain decomposition is built.

In order to implement the algorithm described above, the function calls from the package BLAS and LAPACK (using Intel MKL) through the API Global Arrays library are used. They have simplified the programming of interprocessor communications greatly.

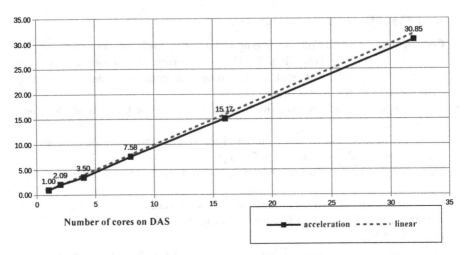

Fig. 4. Scalability of EnOI method in the context of the DAS service at the assimilation of 10^4 points on the Lomonosov supercomputer.

3.3 Service DAS Testing

Due to the effective implementation of the EnOI method as a parallel software service DAS, the solution of the data assimilation problem is scaled almost linearly (Fig. 4). So, the assimilation of 10^4 observation points of satellite data on the 16 processor cores takes about 20 s instead of 5 min on a single core, which would be comparable to the time spent on daily ocean models forecast for 200 cores, which is unacceptable.

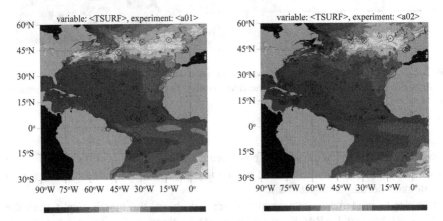

Fig. 5. The Sea Surface Temperature (SST) on 2008-06-29 in the North Atlantic region is calculated using INMIO ocean model in the basic experiment, and the experiment with the data assimilation by EnOI method. Circles show the location of ARGO drifters received data about profiles of temperature and salinity. The size of circles is proportional to the difference between the temperature of drifters and a modeling temperature. A cross mark in the circle means that the modeling temperature is lower than a drifter temperature, a point mark in a circle - the model temperature is higher.

As an example of the data assimilation sea surface temperature (SST) in the INMIO model [10] for the North Atlantic is shown (Fig. 5) without and with data assimilation using DAS service [11]. By analyzing the size of circles, which are proportional to the difference between the modeling SST and measurement data, it is possible to make a conclusion about the effectiveness of assimilation. A big difference of these circles is very well noticeable near the equator and the North Atlantic current, i.e. in those areas where the dynamics of the ocean is significant expressed. The assimilation corrects the model temperature in the right direction, in accordance with observation data.

4 Conclusion

This work presents the implementation of data assimilation algorithm based on EnOI method as a service block of the original Compact Modeling Framework developed for providing the software environment for the high-resolution stand-alone and coupled models of the Global geophysical fluids.

CMF 3.0 utilizes Service Oriented Architecture design which allows one to divide coupler responsibilities into separate services and greatly simplify the entire communication system through the use of PGAS abstraction. To test the performance of data assimilation system and coupler were conducted tests on the Lomonosov supercomputer, which confirmed the numerical efficiency of the proposed numerical software.

The study was performed by a Grant #14-37-00053 from the Russian Science Foundation in Hydrometcentre of Russia.

References

1. Agoshkov, V.I., Zalesnyi, V.B., Parmuzin, E.I., Shutyaev, V.P., Ipatova, V.M.: Problems of variational assimilation of observational data for ocean general circulation models and methods for their solution. Izv. Atmos. Oceanic Phys. **46**(6), 677–712 (2010)
2. Evensen, G.: Data Assimilation, The Ensemble Kalman Filter, 2nd edn. Springer, Berlin (2009)
3. Xie, J., Zhu, J.: Ensemble optimal interpolation schemes for assimilating Argo profiles into a hybrid coordinate ocean model. Ocean Model. **33**, 283–298 (2010)
4. GODAE OceanView Science Team. Work Plan 2009–2013. http://www.godae-oceanview.org
5. Kaurkin, M.N., Ibrayev, R.A., Belyaev, K.P.: Data assimilation into the ocean dynamics model with high spatial resolution using the parallel programming methods. Russ. Meteorol. Hydrol. **7**, 47–57 (2016). (in Russian)
6. Kalmykov, V., Ibrayev, R.: CMF - framework for high-resolution Earth system modeling. In: 2015 CEUR Workshop Proceedings vol, 1482, pp. 34–40 (2015) (in Russian)
7. Kalmykov, V.V., Ibrayev, R.A.: A framework for the ocean-ice-atmosphere-land coupled modeling on massively-parallel architectures. Numer. Methods Program. **14**(2), 88–95 (2013)
8. Nieplocha, J., Palmer, B., Tipparaju, V., Krishnan, M., Trease, H., Apra, E.: Advances, applications and performance of the global arrays shared memory programming toolkit. IJHPCA **20**(2), 203–231 (2006)

9. Sakov, P., et al.: TOPAZ4: an ocean-sea ice data assimilation system for the north Atlantic and Arctic. Ocean Sci. **8**(4), 633–656 (2012)
10. Ibrayev, R.A., Khabeev, R.N., Ushakov, K.V.: Eddy-resolving 1/10° model of the world ocean. Izv. Atmos. Oceanic Phys. **48**(1), 37–46 (2012)
11. Kaurkin, M.N., Ibrayev, R.A., Belyaev, K.P.: Data assimilation ARGO data into the ocean dynamics model with high spatial resolution using Ensemble Optimal Interpolation (EnOI). Oceanology **56**(6), 1–9 (2016)

Experience of Direct Numerical Simulation of Turbulence on Supercomputers

Kirill V. Belyaev[1], Andrey V. Garbaruk[1], Mikhail L. Shur[1],
Mikhail Kh. Strelets[1(✉)], and Philippe R. Spalart[2]

[1] Saint-Petersburg Polytechnic University, Saint-Petersburg, Russia
strelets@mail.rcom.ru
[2] Boeing Commercial Airplanes, Seattle, USA

Abstract. Direct Numerical Simulation, i.e., numerical integration of the unsteady 3D Navier-Stokes equations is the most rigorous approach to turbulence simulation, which ensures an accurate prediction of characteristics of turbulent flows of any level of complexity. However its application to complex real-life flows, e.g. the flows past airplanes, cars, etc., demands huge computational resources and, even according to optimistic assessments of the rate of growth of computer power, will become possible only in the end of the current century. Nonetheless, already today the most powerful supercomputers allow DNS of some flows of high practical importance. The present paper demonstrates this by an example of DNS of high Reynolds number transonic flow over a circular cylinder with axisymmetric bump. The flow features shock wave formation and its interaction with a turbulent boundary layer on the cylinder surface, the phenomenon being of great importance for the aerodynamic design of the commercial airplanes.

Keywords: Direct Numerical Simulation · Turbulence · Massively parallel computations · Shock-boundary layer interaction

1 Introduction

Turbulent flows present the most common form of motion of gases and liquids, and a reliable prediction of their characteristics is one of the highest priorities for aeronautics and many other industries (ground transportation, ship-building, turbo-machinery, chemical technology, ecology, etc.). The most rigorous (based on the "first principles") approach to solving this problem is the so called Direct Numerical Simulation (DNS), i.e., numerical integration of the unsteady 3D Navier-Stokes equations which ensures an adequate representation of any flows within the continuum medium assumption. Hence, in principle, DNS provides a high fidelity tool for prediction of turbulent flows of any level of complexity. However its application to computation of real-life flows demands huge computational resources. This is caused by an extremely wide range of the time and spatial scales of turbulence, which should be accurately resolved in the framework of the DNS approach at high Reynolds numbers typical of aerodynamic flows. As a result, even according to rather optimistic assessments of the rate of growth of the computer power, DNS of the flow past a commercial airplane or a car will

© Springer International Publishing AG 2016
V. Voevodin and S. Sobolev (Eds.): RuSCDays 2016, CCIS 687, pp. 67–77, 2016.
DOI: 10.1007/978-3-319-55669-7_6

become possible only in the second half of the current century [1]. Nonetheless, already today the most powerful of existing supercomputers allow performing DNS of some flows of high practical importance.

In the present paper such a possibility is demonstrated by an example of DNS of the transonic flow past a circular cylinder with an axisymmetric bump studied in the experiments of Bachalo and Johnson [2] (B-J flow hereafter). This flow is characterized by formation of a shock wave which interacts with the turbulent boundary layer on the bump surface resulting in its separation and subsequent reattachment of the separated shear layer to the cylinder surface. This phenomenon is typical of the cruise flight of the commercial airliners (Mach number of the flow M = 0.7 – 0.9) and is directly related to a so called transonic buffet, i.e., loss of stability of the flow past an airliner's wing and onset of intensive self-sustaining oscillations of the flow, which, in turn, may cause resonating loads on the wing leading to its destruction. Hence elucidation of the mechanism of the Shock-Boundary Layer Interaction and an accurate prediction of characteristics of this phenomenon, which is only possible using by DNS, are not only of great physical interest but also of significant practical value.

The DNS performed in the present study reproduces all the conditions of the experiments [2], which were conducted in the wind-tunnel of the NASA Ames Research Center. It has been carried out on two supercomputers. The first one is the Mira cluster (IBM Blue Gene/Q system) of the Argonne National Laboratory, USA (6th place in the Top500 List, June 2016). It has 49,152 Power PC A2 nodes (1.6 GHz) with 16 four-thread cores each and the total shared RAM of 16 GB per node. The second supercomputer is the cluster Tornado of the St.-Petersburg Polytechnic University (the 159th place in the List). It has 612 nodes (Intel Xeon E5-2697v3 14C 2.6 GHz), each having 28 cores and the shared RAM of 64 GB per node. On both clusters, simulations were performed with the use of a massively parallel in-house CFD code Numerical Turbulence Simulation (NTS code).

The paper is organized as follows. Section 2 provides a brief description of the NTS code, Sect. 3 presents results of the computations illustrating efficiency of the code performance on both computers, and Sect. 4 discusses some results of the DNS of the considered flow and presents their comparison with the experimental data [2].

2   NTS Code

This is a finite volume CFD code accepting structured multi-block overlapping grids of Chimera type. It is aimed at computing turbulent flows and ensures a possibility of simulating both incompressible and compressible flows in a wide range of Mach number. Approaches to turbulence representation implemented in the code include the Reynolds Averaged Navier-Stokes equation (RANS), Large Eddy Simulation (LES), hybrid RANS-LES approaches and DNS. The code has passed through an extensive verification by comparisons with other well established CFD codes (CFL3D code of NASA, GGNS and BCFD codes of The Boeing Company, TAU code of DLR, ANSYS-CFX, and ANSYS-FLUENT), and as of today is considered as one of the most reliable and efficient research CFD codes for aerodynamic applications. Examples of previous DNS studies carried out with the use of the NTS code can be found in [3, 4]. Below we briefly dwell upon the numerics implemented in the code for solution of the compressible Navier-Stokes equations and on the approaches for massively parallel computations used in the present study (a more detailed code description can be found in [5]).

2.1 NTS Code Numerics

For DNS of compressible flows considered in the present study, the code performs numerical integration of the unsteady 3D Navier-Stokes equations with the use of the implicit flux-difference scheme of Roe's type [6] based on the MUSCL approach. The inviscid and viscous components of the fluxes in the governing Navier-Stokes equations are approximated with the use of centered 4th and 2nd order schemes respectively. The time-integration is performed with the use of the second order three-layer backward scheme with sub-iterations in pseudo-time. For solution of the resulting set of discrete equations, the code employs Gauss-Seidel algorithm with relaxation by planes.

2.2 Algorithm Parallelization

For parallel computations of the flow variables in the separate grid-blocks or in groups of blocks, the NTS code employs a so called hybrid, Message Passing Interface (MPI)/ Open Multi Processing (OpenMP), concept of parallelization. This implies the use of both MPI library (distributed memory parallelization technology) and OpenMP instructions (shared memory technology). Other than that, OpenMP-instructions are used for additional parallelization of computations inside individual blocks. An important advantage of the hybrid parallelization concept over the use of only MPI or only OpenMP strategies is its flexibility, i.e., simplicity of adapting to computers with different architectures.

For enhancement of efficiency of the computations on supercomputers with a large number of nodes/cores, an additional grid partitioning is employed, which is not dictated by peculiarities of geometry of the flow in question. In order to keep the order of spatial approximation at the boundaries of the additional ("artificial") grid-blocks same as that for the internal grid cells, the neighboring artificial blocks overlap by three cells (see Fig. 1). The optimal size of the grid-blocks is 50,000–100,000 cells (further decrease of the block-size results in a degradation of the parallelization efficiency).

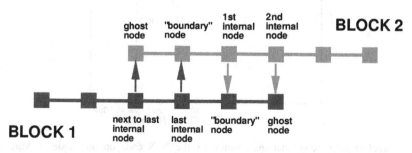

Fig. 1. Schematic illustrating overlapping of artificial grid-blocks in NTS code

It should be noted that parallelization is used not only in the code solver but also in its input and output routines and that, in order to avoid memory restrictions typical of many computers, each MPI process carries out I/O to its own separate file.

Finally, all the operations performed at the pre- and post-processing stages are parallelized as well. This, again, is done to avoid the memory restrictions, since the memory per node of the existing computers is insufficient not only for storing the solutions obtained on the grids used in the present study (several billion cells) but even such grids themselves.

The code is written in FORTRAN-90 and does not use any other libraries but the standard MPI и OpenMP ones.

3 Parallelization Efficiency

In this section we present results of some preliminary tests carried out for assessing the efficiency of parallelization of the NTS code on the Mira and Tornado computers.

An objective of the first test was evaluating efficiency of the OpenMP parallelization. For that, measuring was performed of the time required for the single-block grid computations on one node of the computers with the use of different numbers of Open MP processes. The size of the block was about 100,000 cells (as mentioned above, this size is close to the optimal one). Results of the test are shown in Fig. 2. The figure suggests that the acceleration of the computations with increase of the number of Open MP processes on the both computers is nearly the same and amounts to about 20 times for 30 OpenMP processes, i.e., about 67% of the ideal.

The second test (Weak-Scaling Parallel Performance Efficiency Test) was aimed at an assessment of the efficiency of the MPI parallelization in the NTS code. The test consists in performing computations, in which an increase of the problem size (the number of grid-blocks) is accompanied by a proportional increase of the number of the

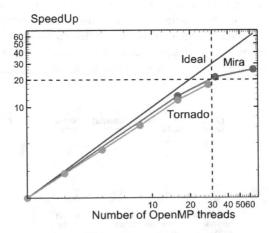

Fig. 2. Acceleration of computations ensured by the NTS code on one node of Mira and Tornado computers with increase of the number of OpenMP processes

computational nodes used, i.e., the computations are carried out with a constant computational load per node. Considering the large difference of the number of nodes of the Mira and Tornado clusters (49,152 and 612 respectively), the test was carried out with the use of 512 to 16,384 nodes on Mira and 16 to 512 nodes on Tornado. Results of this test are presented in Table 1 and in Fig. 3. One can see that even for the maximum number of nodes (16,384 on Mira and 512 on Tornado), the efficiency of the MPI parallelization remains not less than about 70%.

Table 1. Results of Weak-Scaling Parallel Performance Efficiency Test (dependence of relative time per one iteration on the number of used nodes at a constant computational load per node)

Number of nodes		Efficiency (inverse relative time per iteration)	
Mira	Tornado	Mira	Tornado
512	16	1.0	1.0
1024	32	0.95	0.99
2048	64	0.875	0.98
4096	128	0.82	0.94
8192	256	0.78	0.89
16384	512	0.71	0.82

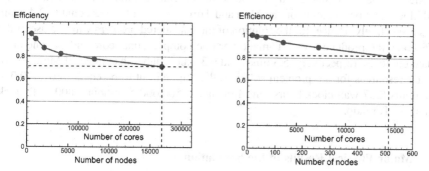

Fig. 3. Dependence of efficiency of MPI parallelization in the NTS code on the number of computational nodes and cores (weak scaling test) on Mira (left) and Tornado (right) computers

To summarize, the results of the preliminary tests suggest that the efficiency of parallelization of the NTS code on the both computers is rather high. This has made it possible to carry out DNS of the B-J flow in a reasonable time.

4 Description of the Production Simulation and Its Results

4.1 Problem Size, Computational Productivity, and Consumed Computational Resources

The grid built for DNS of the B-J flow [2] ensures sufficient resolution of the entire spectrum of the spatial scales of the flow down to the Kolmogorov length-scale at the experimental flow parameters: the Mach $M_\infty = U_\infty/a_\infty = 0.875$ (U_∞ is the free-stream velocity, a_∞ is the speed of sound) and the Reynolds number Re $= \rho_\infty U_\infty c/\mu_\infty = 6.7 \cdot 10^6$ (ρ_∞ is the density, μ_∞ is the dynamic viscosity, c is the length of the bump on the cylinder). The size of the grid is $9602 \times 850 \times 1024$ cells in the streamwise, radial, and azimuthal directions, respectively. Accounting for the need of overlapping of the neighboring artificial grid-blocks (see Sect. 2.2), this amounts to 8.7 billion cells total (note that to the authors' knowledge, these are the first aerodynamic simulations in Russia carried out on such large grids).

The time-step in the simulation was also defined based on Kolmogorov's time scale and was set equal to $1.25 \cdot 10^{-4} (c/U_\infty)$. The full physical time of the simulation including the time necessary for reaching the statistically developed regime and the time required for accumulating the data sufficient for reliable statistics amounted to about $5 (c/U_\infty)$. Thus, the total number of time steps in the simulation was equal to $4 \cdot 10^4$. With the number of sub-iterations in pseudo-time at every time step equal to 10, this amounts to $4 \cdot 10^5$ total iterations. The first half of the simulation ($2 \cdot 10^5$ iterations) was carried out on the Mira cluster and then the simulation was continued on the Tornado cluster. The computation on Mira was performed with the use of 16,384 nodes or 262,144 cores, and on Tornado 586 nodes, i.e., 16,408 cores were used.

Measurements conducted in the course of the computations have shown that the wall-clock time per iteration on the Mira and Tornado machines was equal to 2.4 s and 7.2 s respectively. Hence, for the solution of the considered problem with the use of the NTS code, the productivity of Tornado per one computational core and per one node turned out to be, respectively, 5.3 times and 9.3 times higher than those of Mira. As for the full resources for the problem solution, they turned out to be equal to 61,440,000 core-hours (267 wall-clock hours) on Mira and 11,500,000 core-hours (800 wall-clock hours) on Tornado.

4.2 Major Physical Results of the Simulation

The value of the performed simulation consists, first of all, in generation of a huge numerical database which significantly supplements the experimental database [2]. Namely, in the experiment, only some mean and major statistical flow characteristics (Reynolds stresses) were measured in several flow cross-sections, whereas the performed DNS provides detailed information on the unsteady turbulent structures allowing computation of the high order turbulence statistics which is needed, e.g., for development of Reynolds Stress Transport models for the RANS equations.

As an example, Fig. 4 shows a picture of the experimental visualization of the surface flow obtained with the use of the oil-film technique (so called "oil flow

Fig. 4. A picture of experimental "oil flow pattern" (left) and a zoomed in fragment of the instantaneous isosurface of swirl in the vicinity of the bump crest colored by streamwise velocity component (blue arrows show flow direction) (Color figure online)

pattern") and a zoomed in fragment of the flow visualization from the DNS in the form of an instantaneous isosurface of the "swirl" quantity λ_2 (the second eigenvalue of the velocity gradient tensor) colored by the value of the streamwise velocity. The figure clearly demonstrates that the experimental picture, in fact, provides only the location of the line of the flow separation caused by the shock-boundary layer interaction, whereas the visualization of the DNS results gives comprehensive information on the extremely complex structure of turbulence in the boundary layer. This structure is characterized by the presence of turbulent eddies with a wide range of resolved spatial scales, thus considerably supplementing the experimental information.

Some other forms of visualization of results of the computations, illustrating wide possibilities ensured by the DNS in terms of investigation of the internal structure of turbulence and the nature of the shock-boundary layer interaction in the considered flow are presented in Figs. 5, 6 and 7.

In particular, Figs. 5 and 6 present instantaneous fields of the vorticity magnitude on the surface of the B-J body and in the cross-section of the flow located somewhat upstream of the bump ($x/c = -0.1$). Figure 5 visibly illustrates a radical transformation of the near-wall turbulence along the cylinder and, particularly, reveals a crucial difference of its structure in the regions of zero, favorable, and adverse pressure gradient, in the separation and reattachment zones, and in the area of the relaxation of the reattached boundary layer. Figure 6 demonstrates a radical alteration of the turbulent structures in the boundary layer as the distance to the solid wall is increased.

Finally, Fig. 7 compares a picture of the experimental holographic interferogram in the vicinity of the bump and an instantaneous field of the magnitude of the pressure gradient in a meridian plane in this region computed with the use of the DNS. One can see that the experimental interferogram displays only the wave-pattern of the flow, whereas the results of the DNS, along with this, give a clear idea of the rich structure of the flow turbulence. It includes both fine-grained turbulence and relatively large nearly coherent turbulent eddies in the area of the shock-boundary layer interaction and in the separation bubble downstream of the shock caused by this interaction.

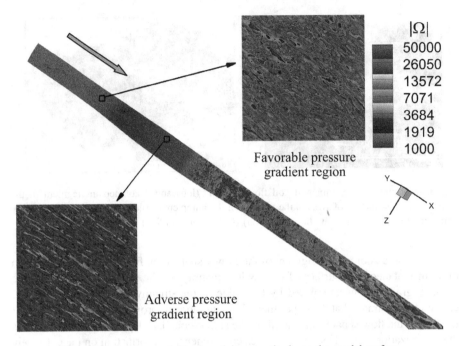

Fig. 5. Instantaneous field of vorticity magnitude on the model surface

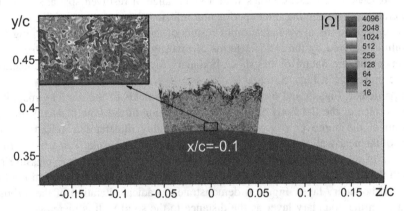

Fig. 6. Instantaneous field of vorticity magnitude in a flow cross-section located upstream of the bump (at $x/c = -0.1$)

The last figure (Fig. 8) is aimed at demonstrating accuracy of the performed DNS. It shows the longitudinal distributions of the mean (time and azimuthal direction averaged) skin-friction and pressure coefficients, $C_f = 2\tau_w/(\rho_\infty U_\infty^2)$ and $C_P = 2(p - p_\infty)/(\rho_\infty U_\infty^2)$, predicted by DNS and a comparison of the latter with the corresponding experimental distribution (in the experiment, the friction coefficient was not measured). The fairly good agreement of the DNS with the data provides evidence

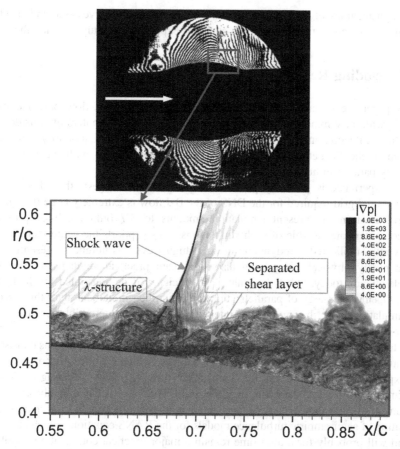

Fig. 7. Picture of holographic interferogram of the flow-field from the experiment [2] (up; white arrow shows the flow direction) and zoomed in fragment of instantaneous field of magnitude of pressure gradient from DNS (bottom)

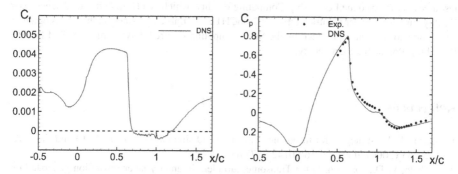

Fig. 8. Streamwise distributions of friction (left) and pressure (right) coefficients

of the high accuracy of the DNS predictions, on one hand, and serves as an independent confirmation of correctness of the non-standard experimental setup, on the other.

5   Concluding Remarks

In this paper we describe an experience of DNS of transonic flow past a circular cylinder with axisymmetric bump carried out on two supercomputers of considerably different architecture and power (Mira of the Argonne National Laboratory, USA, and Tornado of the St.-Petersburg Polytechnic University, Russia) with the use of the massively parallel in-house CFD code NTS.

The experience is generally positive. In particular, it suggests that although the computational grid required for the DNS of the B-J flow is extremely large (the size of the grid used in the present simulation amounts to 8.7 billion cells), up-to-date supercomputers are capable of such simulations. This supports the claim that DNS of some very complicated aerodynamic flows at high Reynolds numbers, presenting not only a purely physical interest but also significant practical value is, in principle, possible already today. Other than that, high efficiency is demonstrated for the hybrid MPI – OpenMP strategy of parallelization of the implicit numerical algorithm used in the simulations for the numerical integration of the compressible unsteady 3D Navier-Stokes equations.

In terms of flow physics, the primary value of the performed simulations consists in accumulation of a comprehensive numerical database which significantly complements the experimental database on the considered flow [2]. The resulting combined experimental-numerical database may be used both for detailed analysis of the mechanism of the shock-boundary layer interaction phenomenon and for creation and calibration of semi-empiric turbulence models for the RANS equations which currently are and will probably for a long time remain a major practical computational tool for aerodynamic design.

Acknowledgments. An award of computer time was provided by the Innovative and Novel Computational Impact on Theory and Experiment (INCITE) program. This research used resources of the Argonne Leadership Computing Facility, which is a DOE Office of Science User Facility supported under Contract DE-AC02-06CH11357. Large resources on the supercomputer RSC Tornado were also provided by the Supercomputing Centre "Polytechnichesky" of the St.-Petersburg Polytechnic University.

References

1. Spalart, P.R.: Strategies for turbulence modeling and simulations. Int. J. Heat Fluid Flow **21**, 252–263 (2000). doi:10.1016/S0142-727X(00)00007-2
2. Bachalo, W.D., Johnson, D.A.: Transonic, turbulent boundary-layer separation generated on an axisymmetric flow model. AIAA J. **24**, 437–443 (1986). doi:10.2514/3.9286

3. Spalart, P.R., Strelets, M.K., Travin, A.K.: Direct numerical simulation of large-eddy-break-up devices in a boundary layer. Int. J. Heat Fluid Flow **27**, 902–910 (2006). doi:10.1016/j.ijheatfluidflow.2006.03.014

4. Spalart, P.R., Shur, M.L., Strelets, M.K., Travin, A.K.: Direct simulation and RANS modelling of a vortex generator flow. Flow Turbul. Combust. **95**, 335–350 (2015). doi:10.1007/s10494-015-9610-8

5. Shur, M.L., Strelets, M.K., Travin, A.K.: High-order implicit multi-block Navier-Stokes code: ten-year experience of application to RANS/DES/LES/DNS of turbulent flows. In: 7th Symposium on Overset Composite Grids & Solution Technology, Huntington Beach, California (2004). http://cfd.spbstu.ru/agarbaruk/c/document_library/DLFE-42505.pdf

6. Roe, P.L.: Approximate Riemann solvers, parameter vectors, and difference schemes. J. Comput. Phys. **43**, 357–372 (1981). doi:10.1016/0021-9991(81)90128-5

GPU-Accelerated Molecular Dynamics: Energy Consumption and Performance

Vyacheslav Vecher[1,2(✉)], Vsevolod Nikolskii[1,3], and Vladimir Stegailov[1]

[1] Joint Institute for High Temperatures of RAS, Moscow, Russia
vecher@phystech.edu,thevsevak@gmail.com,v.stegailov@hse.ru
[2] Moscow Institute of Physics and Technology (State University),
Dolgoprudny, Russia
[3] National Research University Higher School of Economics, Moscow, Russia

Abstract. Energy consumption of hybrid systems is an actual problem of modern high-performance computing. The trade-off between power consumption and performance becomes more and more prominent. In this paper, we discuss the energy and power efficiency of two modern hybrid minicomputers Jetson TK1 and TX1. We use the Empirical Roofline Tool to obtain peak performance data and the molecular dynamics package LAMMPS as an example of a real-world benchmark. Using the precise wattmeter, we measure Jetsons power consumption profiles. The effectiveness of DVFS is examined as well. We determine the optimal GPU and DRAM frequencies that give the minimum energy-to-solution value.

Keywords: Nvidia Jetson · LAMMPS · Energy efficiency

1 Introduction

The method of molecular dynamics (MD) is a modern and powerful method of computer simulations allowing to calculate the behavior of millions of atoms. Based on the integration of classical Newton's equation of motion, this method allows one to estimate the evolution of systems consisted of particles that obey the laws of classical mechanics. With the increase in available computation power the size and complexity of models increase as well. Usually, it leads to significant improvements in the accuracy of results. However, the growth of the computational demands has led to a situation when MD simulations are considered among the main tasks for parallel computing.

The technological progress in the development of graphical accelerators makes them quite powerful computation devices with relatively low price. GPUs outperform the conventional processors in the constantly increasing number of computational tasks in the price-to-performance ratio. This trend leads to the situation when the use of graphical accelerators for MD computation has become a common practice [1–9].

However, the increase of power consumption and heat generation of computing platforms is also a very significant problem, especially in connection

V. Voevodin and S. Sobolev (Eds.): RuSCDays 2016, CCIS 687, pp. 78–90, 2016.
DOI: 10.1007/978-3-319-55669-7_7

with the development of exascale systems. Measurement and presentation of the results of performance tests of parallel computer systems become more and more often evidence-based [10], including the measurement of energy consumption [11], which is crucial for the development of exascale supercomputers [12].

The purpose of this work is to evaluate the efficiency of MD algorithms in terms of power consumption on Nvidia Tegra K1 and X1 systems-on-chip (SoCs).

2   Related Work

The power and energy consumption have been under consideration for a long time. For example, we could mention the work [13] that showed the way of lowering the energy consumption of processors by reducing the number of switching operations. Joseph and Martonosi [14] investigated the problem of energy consumption in its relationship with code optimization for 32-bit embedded RISC processors. Russel and Jacome [15] discussed a more complex model of evaluation of power consumption in real-time. The work [16] showed the evaluation of energy consumption at the OS level.

The development of portable devices gave additional impulse to this filed (e.g. see Zhang et al. [17]).

An important aspect of GPGPU technologies that makes them beneficial is the energy efficiency. A lot of efforts have been invested into the low-level models for modeling the energy consumption of GPUs. In recent review [18], the key aspects of accelerator-based systems performance modeling have been discussed. The McPAT model (Multicore Power, Area and Timing) [19] is considered as one of the cornerstone ideas in this area. Another approach called GPUWattch [20] is aimed at the prediction of energy consumption and its optimization through careful tuning on the basis of series of microtests. These approaches make possible to accurately predict the power consumption of CPU and/or GPU with accuracy of the order of 5–10%. However, the use of a specific model of energy consumption (like McPAT or GPUWattch) for new types of hardware and software is a very significant effort to be undertaken. Therefore, direct experimental measurements of power consumption and energy usage are instructive.

The work of Calore et al. [21] discloses some aspects of relations between power consumption and performance for Tegra K1 device running the Lattice Boltzmann method algorithms. Our preliminary results on energy consumption for minicomputers running MD benchmarks have been published previously for Odroid C1 [22] and Nvidia Jetson TK1 and TX1 [23]. In the work of Gallardo et al. [24] one can find the performance analysis and comparison of Nvidia Kepler and Maxwell and Intel Xeon Phi accelerators for the hydrodynamic benchmark LULESH. An energy-aware task management mechanism for the MPDATA algorithms on multicore CPUs was proposed by Rojek et al. [25].

The method of determining the power consumption of large computer systems are constantly improving [12]. The work of Rajovic et al. [26] shows the possible way of designing HPC systems from modern commodity SoCs and presents the energy consumption analysis of the prototype cluster created.

3   Software and Algorithms

3.1   A Peak Load Benchmark: Empirical Roofline Toolkit

The performance of heterogeneous systems can be evaluated in different ways. The consideration of only the theoretical peak performance can be instructive (e.g. see [27,28]) but is not sensitive to details of algorithms. This approach is justified for compute-bound algorithms only. For memory-bound algorithms, the memory bandwidth is to be addressed.

This idea has led to the creation of the Roofline model [29]. The model introduces a special characteristic called "arithmetic intensity". It quantifies the ratio of the number of arithmetic operations to the amount of data transferred. Obviously, the limiting factor for algorithms with large arithmetic intensity is the peak performance of a processor, while the memory bandwidth limits the performance of algorithms with intensive data transfers.

The main outcome of the Roofline model is a graph of performance restrictions for algorithms with different arithmetic intensities. One can estimate the performance of a system under consideration for a particular algorithm from such a roofline plot. For example, the typical arithmetic intensity for Lattice Boltzmann methods is less than one Flops/byte, whether for particle methods this parameter is usually around ten Flops/byte.

One can use the Empirical Roofline Toolkit (ERT) [30] for evaluation of memory bandwidth and peak computing power taking into account memory hierarchy of today's complex heterogeneous computing systems. The core idea of the ERT algorithm consists in performing cycles of simple arithmetic operations on the elements of an array of specified length. The algorithm varies the size of the array and the number of operations on the same element of the array (ERT_FLOPS) in nested loops (Fig. 1). The change of the data array size helps

```
for(int i=0; i<n; ++i){
        if (ERT_FLOPS==1){
                b1 = a[i] + alpha;
        }
        if (ERT_FLOPS==2){
                b1 = a[i]*b1 + alpha;
        }
        if (ERT_FLOPS==4){
                b1 = a[i]*b1 + alpha;
                b2 = a[i]*b2 + alpha;
        }
    ...
};
```

Fig. 1. An illustration for the ERT_FLOPS parameter

to detect the presence of caches. The change of operations number on one element of array helps to identify automatic vectorization effects.

3.2 Classical Molecular Dynamics: LAMMPS

The LAMMPS package [31] is used in this work as an example of a real-life application. It is a flexible tool for building models of classical MD in materials science, chemistry and biology. LAMMPS is not the only MD package that is ported to the hybrid architecture (for example HOOMD [32] was originally designed with the perspective to run it on GPU accelerators). Two GPU MD algorithms implemented in LAMMPS are considered in this work: USER-CUDA [33] and GPU [34, 35].

To evaluate the performance of the hardware available, we use the Lennard-Jones fluid model (108,000 atoms, the density $0.8442\sigma^{-3}$, the cut-off radius 2.5σ, NVE-ensemble, 250 timesteps).

4   Hardware

4.1   Tested Platforms

We consider two different generation of Nvidia Tegra SoCs installed in Jetson TK1 and TX1 platforms. These SoCs consist of several ARM-cores and GPU on a single chip. These platforms are designed for embedded applications (robots, drones) and optimized for minimum power consumption with relatively high performance.

These devices are aimed to be energy effective and usually operate in the dynamic voltage and frequency scaling (DVFS) mode. In this mode, the GPU memory and core frequencies change during the runtime that allows to reduce the power consumption of hardware significantly. In the measurements, we disable the DVFS mode by setting manually the GPU and DRAM frequencies. However, we make several measurements with the DVFS mode enabled.

Nvidia Jetson TK1. Nvidia Jetson TK1 is a developer board based on the 32-bit Tegra K1 SoC with LPDDR3 (930 MHz). Tegra K1 CPU complex includes 4 Cortex-A15 cores running at 2.3 GHz, the 5-th low power companion Cortex core designed to replace the basic cores in the low load mode to reduce power consumption and heat generation. The chip includes one GPU Kepler streaming multiprocessor (SM) running at 852 MHz (128 CUDA cores). Each Cortex-A15 core has 32 KB L1 instruction and 32 KB L1 data caches. 4-core cluster has 2 MB of shared L2 cache.

The program environment of the device consists of Linux Ubuntu 14.04.1 LTS (GNU/Linux 3.10.40-gdacac96 armv7l). The toolchain includes GCC ver. 4.8.4 and CUDA Toolkit 6.5.

Nvidia Jetson TX1. Jetson TX1 is based on the 64-bit Tegra X1 SoC with LPDDR4 memory (1600 MHz). Tegra X1 includes 4 Cortex-A57 cores running at 1.9 GHz, 4 slower Cortex-A53 in big.LITTLE configuration and two GPU Maxwell SMs running at 998 MHz (256 CUDA cores). Each Cortex-A57 core has 48 KB L1 instruction cache, 32 KB L1 data cache and 2 MB of shared L2 cache.

The operation system is Linux Ubuntu 14.01.1 LTS with 64-bit core built for aarch64. Nevertheless we use the 32-bit toolchain and software environment (same as for Nvidia Jetson TK1), except for the newer CUDA Toolkit 7.0.

In summer 2016, the 64-bit userspace and toolchain have been released. Preliminary tests show that the new 64-bit software can be noticeably faster in some rare cases only.

4.2 Energy Consumption Measurement Technique

We use the SmartPower digital wattmeters with the integrated DC source to measure the energy consumption of the Jetson boards. The wattmeter provides voltage in the range from 3 to 5.25 V and measures the current and power consumption every 0.2 s with a nominal error of less than 0.01 V. The wattmeter shows the data on the display in real time and allows to transfer the data via USB to the PC for further analysis.

Because both Jetson platforms have nominal voltage values higher than 5.25 V, we connect several SmartPower wattmeters in a sequential way to achieve higher voltage (see Fig. 2). To confirm the accuracy of the achieved output voltage, we use the precise Tektronix TDS2014C oscilloscope. In this way we show that the average error in the power consumption measurements are about 1%.

The nominal voltages are 12 V for Jetson TK1 and 19 V for Jetson TX1. However, we discover that both devices can operate at much lower voltages: down to 6 V for TK1 and 8 V for TX1.

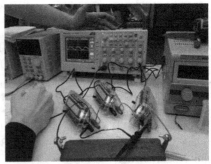

Fig. 2. The Jetson TX1 and TK1 boards with the ODROID-C1 in the center (left photo). The sequentially connected SmartPower wattmeters with the oscilloscope (right photo)

The measurement of energy consumption in a particular test consists in the simultaneous execution of the logging program on the PC as well as the necessary tests on the Jetson. The Jetson boards do not have any connected peripherals except for the LAN cable.

We should note that other methods of measuring power consumption exist as well. For example, built-in hardware counters can be used that allow a user to accurately determine the chip power consumption. However, we are not aware of the similar counters in Tegra SoCs and therefore measure directly the power consumption of the entire development board.

5 Measurements Results

5.1 Energy Consumption for the ERT Benchmark

We use the results of the ERT launches to determine the ratio of the peak performance (in GFlops) to the average power consumption during the benchmark.

On Fig. 3 one can see an example of the power consumption profiles for both Jetson boards. Using the measured value of energy consumed for the benchmark launch and the peak performance obtained, one can determine the energy efficiency of systems considered.

Since the first 10 s of the ERT benchmark are spent on rebuilding the binary, this part of the log is not included in consideration.

The total energy consumption for the GPU single precision ERT benchmark for TK1 is 5.9 W, with the maximum achieved performance in single precision of 209.9 GFlops. This gives us the ratio of 35.5 GFlops/W. For TX1, the energy consumption is 6.28 W, which is slightly above TK1. However, the newer device is significantly superior in terms of achieved maximum performance (485.1 GFlops) that gives a higher performance of 77.2 GFlops/W in single precision.

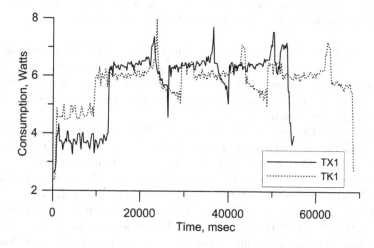

Fig. 3. An example of Jetsons power consumption profiles during ERT launches

Both Jetson minicomputers demonstrate not very impressive results in double precision: 2.1 GFlops/W for TK1 and 2.7 GFlops/W for TX1. The reason for this is the significantly lower double precision performance with the energy consumption level comparable to the single precision case.

On the other hand, the ERT launches on the ARM cores of TX1 show that the Cortex-A57 core has much lower efficiency: 0.8 GFlops/W for double precision and 4 GFlops/W for single precision.

5.2 Energy Consumption for the LAMMPS Benchmark

On Fig. 4, one can see typical power consumption profiles during LAMMPS launches. The area under the graph represents the amount of energy spent for the calculation.

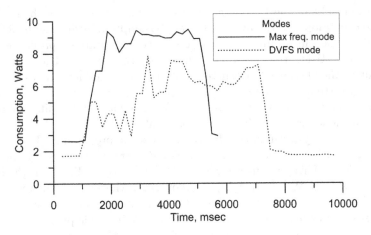

Fig. 4. Power consumption profiles for the launches of LAMMPS with USER-CUDA for TX1 in the maximum frequency mode and in the DVFS mode

As noted above, both Jetson systems support the DVFS energy optimization mode. Therefore, we consider how the results change in the case of the activated DVFS mode. Thus, for each launch of the standard Lennard-Jones benchmark in the fixed maximum frequency mode, we have performed the same launch but with the DVFS mode enabled.

Figure 4 shows the comparison of the LAMMPS energy consumption in DVFS and fixed maximum frequencies modes. One can see that the DVFS-enabled launch takes more time with clearly lower power consumption level.

To answer the question whether or not DVFS is beneficial, we calculate the consumed energy per one LAMMPS launch with and without DFVS.

The results presented on Fig. 5 show that the total energy consumption values for the LAMMPS calculations with the DVFS mode enabled are roughly equal or higher than the corresponding values in the case of the DVFS mode disabled. However, the times-to-solution with DVFS are much higher than in the case

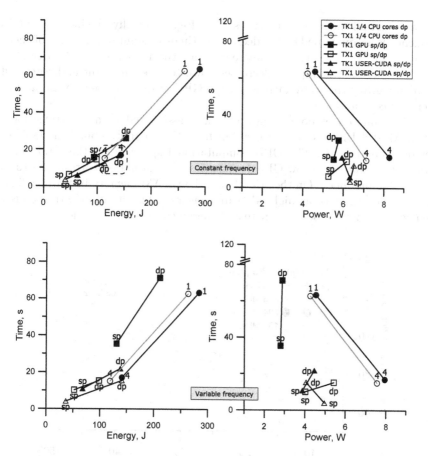

Fig. 5. Time-to-solution and the corresponding energy consumption values for the CPU MD algorithm, GPU and USER-CUDA variants in the maximum frequency mode and in the DVFS mode (single and double precision)

of the fixed maximum frequency. Therefore, the usage of DVFS in the cases considered does not improve energy efficiency.

6   GPU and DRAM Frequencies Variation Effect

We use the USER-CUDA accelerated LAMMPS variant as a benchmark that shows the effect of GPU and DRAM frequencies variation on the execution time and the total energy consumption. The frequency of the GPU are changed from one launch to another and the DRAM frequency is fixed for the whole group of launches. For each group of experiments, we set the Jetson TK1 GPU frequency to the following values (in MHz): 72, 108, 180, 252, 324, 396, 468, 540, 612, 648, 684, 708, 756, 804 and 852. The DRAM frequencies are fixed for each group at the values of 924, 396, 204 and 102 MHz. For each launch, the power consumption is measured.

One can find the measurement results in Fig. 6. Initially, the increase of GPU frequency is accompanied by the decrease in times-to-solution. In terms of energy consumption, the situation is different. With the increase of GPU frequency, the TK1 energy consumption decreases down to a certain limit and reaches its minimum. After that, any increase of the GPU frequency leads to the increase of energy consumption.

This minimum of power consumption is associated with the transition of the LAMMPS USER-CUDA algorithm from the compute-bound mode to the memory-bound mode. The GPU computational speed limits the total performance of the system at low GPU frequencies. This situation corresponds to the compute-bound mode. On the other hand, the DRAM memory bandwidth limits the total performance at high GPU frequencies. This situation corresponds to the memory-bound mode. An increased energy consumption in the latter case is

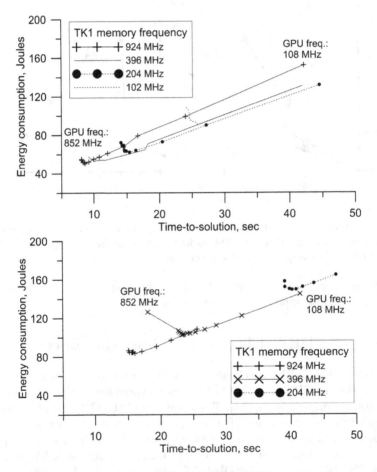

Fig. 6. LAMMPS power consumption on different TK1 memory frequencies with USER-CUDA (upper plot) and GPU (lower plot) packages.

not a desirable effect because this growth of consumption is not associated with any significant speedup of the calculation.

Also, we notice that lowering the DRAM memory frequency shifts the point of minimum consumed energy to lower GPU frequencies, as it might be expected.

7   Summary

We have described the energy consumption of the minicomputers Nvidia Jetson TK1 and TX1 based on the hybrid systems-on-chip Nvidia Tegra K1 and X1.

The peak load benchmarks have been performed with the Empirical Roofline Toolkit (with the CPU and GPU versions). The CPU version has shown 4 GFlops/W for single precision and 0.8 GFlops/W for double precision for one Cortex-A57 core of Jetson TX1. The GPU version for Kepler TK1 and Maxwell TX1 has shown 35.5 GFlops/W and 77.2 GFlops/W respectively for single precision and 2.1 GFlops/W and 2.7 GFlops/W for double precision.

Two CUDA-accelerated MD algorithms implemented in LAMMPS have been used for energy consumption benchmarks (in single and double precision). DVFS has been found inefficient for energy efficiency improvement in the cases considered.

By changing GPU and DRAM frequencies on TK1, we have shown the transition of the both CUDA-based MD algorithms from the compute-bound to memory-bound mode. We have located the minima of the energy-to solution with respect to the set of GPU and DRAM frequencies considered.

In the future, we plan to conduct a similar analysis for systems with desktop or server GPU accelerators. In addition, we are going to consider the case of more complex molecular dynamics models, e.g. with the Coulomb interaction.

Acknowledgments. HSE and MIPT provided funds for purchasing the hardware used in this study. The work was supported by the grant No. 14-50-00124 of the Russian Science Foundation.

References

1. Morozov, I., Kazennov, A., Bystryi, R., Norman, G., Pisarev, V., Stegailov, V.: Molecular dynamics simulations of the relaxation processes in the condensed matter on GPUs. Comput. Phys. Commun. **182**(9), 1974–1978 (2011). doi:10.1016/j.cpc.2010.12.026
2. Budea, A., Derzsi, A., Hartmann, P., Donko, Z.: Shear viscosity of liquid-phase yukawa plasmas from molecular dynamics simulations on graphics processing units. Contrib. Plasma Phys. **52**(3), 194–198 (2012). doi:10.1002/ctpp.201100083
3. French, W.R., Pervaje, A.K., Santos, A.P., Iacovella, C.R., Cummings, P.T.: Probing the statistical validity of the ductile-to-brittle transition in metallic nanowires using GPU computing. J. Chem. Theory Comput. **9**(12), 5558–5566 (2013). doi:10.1021/ct400885z

4. Fu, H., Zheng, L., Yang, M.: Accelerating modified shepard interpolated potential energy calculations using graphics processing units. Comput. Phys. Commun. **184**(4), 1150–1154 (2013). doi:10.1016/j.cpc.2012.12.005

5. Wu, Q., Yang, C., Tang, T., Xiao, L.: MIC acceleration of short-range molecular dynamics simulations. In: Proceedings of the First International Workshop on Code OptimiSation for MultI and Many Cores, COSMIC 2013, pp. 2:1–2:8. ACM, New York (2013). doi:10.1145/2446920.2446922

6. Wu, Q., Yang, C., Tang, T., Xiao, L.: Exploiting hierarchy parallelism for molecular dynamics on a petascale heterogeneous system. J. Parallel Distrib. Comput. **73**(12), 1592–1604 (2013). doi:10.1016/j.jpdc.2013.07.015

7. Filho, T.M.R.: Molecular dynamics for long-range interacting systems on graphic processing units. Comput. Phys. Commun. **185**(5), 1364–1369 (2014). doi:10.1016/j.cpc.2014.01.008

8. Minkin, A.S., Knizhnik, A.A., Potapkin, B.V.: GPU implementations of some many-body potentials for molecular dynamics simulations. Adv. Eng. Softw. (2016). doi:10.1016/j.advengsoft.2016.05.013

9. Nguyen, T.D.: GPU-accelerated Tersoff potentials for massively parallel molecular dynamics simulations. Comput. Phys. Commun. **212**, 113–122 (2017). doi:10.1016/j.cpc.2016.10.020

10. Hoefler, T., Belli, R.: Scientific benchmarking of parallel computing systems: twelve ways to tell the masses when reporting performance results. In: Proceedings of the International Conference for High Performance Computing, Networking, Storage and Analysis, SC 2015, pp. 73:1–73:12. ACM, New York (2015). doi:10.1145/2807591.2807644

11. Pruitt, D.D., Freudenthal, E.A.: Preliminary investigation of mobile system features potentially relevant to HPC. In: Proceedings of the 4th International Workshop on Energy Efficient Supercomputing, E2SC 2016, pp. 54–60. IEEE Press, Piscataway (2016). doi:10.1109/E2SC.2016.13

12. Scogland, T., Azose, J., Rohr, D., Rivoire, S., Bates, N., Hackenberg, D.: Node variability in large-scale power measurements: perspectives from the Green500, Top500 and EEHPCWG. In: Proceedings of the International Conference for High Performance Computing, Networking, Storage and Analysis, SC 2015, pp. 74:1–74:11. ACM, New York (2015). doi:10.1145/2807591.2807653

13. Su, C.L., Tsui, C.Y., Despain, A.M.: Low power architecture design and compilation techniques for high-performance processors. In: Compcon Spring 1994, Digest of Papers, pp. 489–498 (1994). doi:10.1109/CMPCON.1994.282878

14. Joseph, R., Martonosi, M.: Run-time power estimation in high performance microprocessors. In: Proceedings of the 2001 International Symposium on Low Power Electronics and Design, ISLPED 2001, pp. 135–140. ACM, New York (2001). doi:10.1145/383082.383119

15. Russell, J.T., Jacome, M.F.: Software power estimation and optimization for high performance, 32-bit embedded processors. In: Proceedings International Conference on Computer Design. VLSI in Computers and Processors (Cat. No. 98CB36273), pp. 328–333 (1998). doi:10.1109/ICCD.1998.727070

16. Li, T., John, L.K.: Run-time modeling and estimation of operating system power consumption. SIGMETRICS Perform. Eval. Rev. **31**(1), 160–171 (2003). doi:10.1145/885651.781048

17. Zhang, L., Tiwana, B., Qian, Z., Wang, Z., Dick, R.P., Mao, Z.M., Yang, L.: Accurate online power estimation and automatic battery behavior based power model generation for smartphones. In: Proceedings of the Eighth IEEE/ACM/IFIP International Conference on Hardware/Software Codesign and System Synthesis, CODES/ISSS 2010, pp. 105–114. ACM, New York (2010). doi:10.1145/1878961.1878982

18. Lopez-Novoa, U., Mendiburu, A., Miguel-Alonso, J.: A survey of performance modeling and simulation techniques for accelerator-based computing. IEEE Trans. Parallel Distrib. Syst. **26**(1), 272–281 (2015). doi:10.1109/TPDS.2014.2308216

19. Li, S., Ahn, J.H., Strong, R.D., Brockman, J.B., Tullsen, D.M., Jouppi, N.P.: McPat: an integrated power, area, and timing modeling framework for multicore and manycore architectures. In: Proceedings of the 42nd Annual IEEE/ACM International Symposium on Microarchitecture, MICRO 42, pp. 469–480. ACM, New York (2009). doi:10.1145/1669112.1669172

20. Leng, J., Hetherington, T., ElTantawy, A., Gilani, S., Kim, N.S., Aamodt, T.M., Reddi, V.J.: GPUWattch: enabling energy optimizations in GPGPUs. SIGARCH Comput. Archit. News **41**(3), 487–498 (2013). doi:10.1145/2508148.2485964

21. Calore, E., Schifano, S.F., Tripiccione, R.: Energy-performance tradeoffs for HPC applications on low power processors. In: Hunold, S., et al. (eds.) Euro-Par 2015. LNCS, vol. 9523, pp. 737–748. Springer, Heidelberg (2015). doi:10.1007/978-3-319-27308-2_59

22. Nikolskiy, V., Stegailov, V.: Floating-point performance of ARM cores and their efficiency in classical molecular dynamics. J. Phys.: Conf. Ser. **681**(1), 012049 (2016). http://stacks.iop.org/1742-6596/681/i=1/a=012049

23. Nikolskiy, V.P., Stegailov, V.V., Vecher, V.S.: Efficiency of the Tegra K1 and X1 systems-on-chip for classical molecular dynamics. In: 2016 International Conference on High Performance Computing Simulation (HPCS), pp. 682–689 (2016). doi:10.1109/HPCSim.2016.7568401

24. Gallardo, E., Teller, P.J., Argueta, A., Jaloma, J.: Cross-accelerator performance profiling. In: Proceedings of the XSEDE16 Conference on Diversity, Big Data, and Science at Scale, XSEDE16, pp. 19:1–19:8. ACM, New York (2016). doi:10.1145/2949550.2949567

25. Rojek, K., Ilic, A., Wyrzykowski, R., Sousa, L.: Energy-aware mechanism for stencil-based MPDATA algorithm with constraints. Concurr. Comput.: Pract. Exp. (2016). doi:10.1002/cpe.4016

26. Rajovic, N., Rico, A., Mantovani, F., Ruiz, D., Vilarrubi, J.O., Gomez, C., Backes, L., Nieto, D., Servat, H., Martorell, X., Labarta, J., Ayguade, E., Adeniyi-Jones, C., Derradji, S., Gloaguen, H., Lanucara, P., Sanna, N., Mehaut, J.F., Pouget, K., Videau, B., Boyer, E., Allalen, M., Auweter, A., Brayford, D., Tafani, D., Weinberg, V., Brömmel, D., Halver, R., Meinke, J.H., Beivide, R., Benito, M., Vallejo, E., Valero, M., Ramirez, A.: The Mont-Blanc prototype: an alternative approach for HPC systems. In: Proceedings of the International Conference for High Performance Computing, Networking, Storage and Analysis, SC 2016, pp. 38:1–38:12. IEEE Press, Piscataway (2016). http://dl.acm.org/citation.cfm?id=3014904.3014955

27. Stegailov, V.V., Orekhov, N.D., Smirnov, G.S.: HPC hardware efficiency for quantum and classical molecular dynamics. In: Malyshkin, V. (ed.) PaCT 2015. LNCS, vol. 9251, pp. 469–473. Springer, Heidelberg (2015). doi:10.1007/978-3-319-21909-7_45

28. Smirnov, G.S., Stegailov, V.V.: Efficiency of classical molecular dynamics algorithms on supercomputers. Math. Models Comput. Simul. **8**(6), 734–743 (2016). doi:10.1134/S2070048216060156

29. Williams, S., Waterman, A., Patterson, D.: Roofline: an insightful visual performance model for multicore architectures. Commun. ACM **52**(4), 65–76 (2009). doi:10.1145/1498765.1498785

30. Lo, Y.J., Williams, S., Straalen, B., Ligocki, T.J., Cordery, M.J., Wright, N.J., Hall, M.W., Oliker, L.: Roofline model toolkit: a practical tool for architectural and program analysis. In: Jarvis, S.A., Wright, S.A., Hammond, S.D. (eds.) PMBS 2014. LNCS, vol. 8966, pp. 129–148. Springer, Heidelberg (2015). doi:10.1007/978-3-319-17248-4_7

31. Plimpton, S.: Fast parallel algorithms for short-range molecular dynamics. J. Comput. Phys. **117**(1), 1–19 (1995). doi:10.1006/jcph.1995.1039

32. Glaser, J., Nguyen, T.D., Anderson, J.A., Lui, P., Spiga, F., Millan, J.A., Morse, D.C., Glotzer, S.C.: Strong scaling of general-purpose molecular dynamics simulations on GPUs. Comput. Phys. Commun. **192**, 97–107 (2015). doi:10.1016/j.cpc.2015.02.028

33. Trott, C.R., Winterfeld, L., Crozier, P.S.: General-purpose molecular dynamics simulations on GPU-based clusters. arXiv e-prints (2010). http://arxiv.org/abs/1009.4330

34. Brown, W.M., Wang, P., Plimpton, S.J., Tharrington, A.N.: Implementing molecular dynamics on hybrid high performance computers – short range forces. Comput. Phys. Commun. **182**(4), 898–911 (2011). doi:10.1016/j.cpc.2010.12.021

35. Brown, W.M., Kohlmeyer, A., Plimpton, S.J., Tharrington, A.N.: Implementing molecular dynamics on hybrid high performance computers – particle-particle particle-mesh. Comput. Phys. Commun. **183**(3), 449–459 (2012). doi:10.1016/j.cpc.2011.10.012

Implementation and Evaluation of the PO-HEFT Problem-Oriented Workflow Scheduling Algorithm for Cloud Environments

Gleb Radchenko[✉], Ivan Lyzhin, and Ekaterina Nepovinnyh

Department of System Programming,
South Ural State University, Chelyabinsk, Russia
{gleb.radchenko,lyzhinia}@susu.ru,
kwadraterry@gmail.com

Abstract. Modern computational experiments imply that the resources of the cloud computing environment are often used to solve a large number of tasks, which differ only in the values of a relatively small set of simulation parameters. Such sets of tasks may occur while implementing multivariate calculations aimed at finding the simulation parameter values, which optimize certain characteristics of the computational model. Applications of this type make a large percentage of modern HPC systems load, which implies a need for methods and algorithms for efficient allocation of resources in order to optimize systems for solving such problems. The aim of this work is to implement a PO-HEFT problem-oriented scientific workflow scheduling algorithm and to compare it with other workflow scheduling algorithms.

Keywords: Problem-oriented environment · Workflow · Cloud · Simulation · Scheduling algorithm · HEFT · PO-HEFT

1 Introduction

The last thirty years a revolutionary turn in the fundamentals of science and engineering takes place. Computational methods have become the "third branch" of the scientific approach, along with theory and experiment. Computational methods are used in applications that involve data analysis and visualization of experiments results [1]. The use of supercomputer simulation and data mining provides a qualitatively new level of results in all fields of knowledge, allowing for numerical studies of physical, biological, social and others processes, providing a real alternative for expensive (or impossible) experiments [2, 3].

The typical scenario of computational experiment is a repetitive cycle consisting of:

1. transferring data to a supercomputer for analysis or simulation;
2. performing calculations;
3. storing results of calculations.

Thus, a typical computational experiment scenario can be implemented by the so-called "workflows". In [4] the following workflow definition is proposed: the

© Springer International Publishing AG 2016
V. Voevodin and S. Sobolev (Eds.): RuSCDays 2016, CCIS 687, pp. 91–105, 2016.
DOI: 10.1007/978-3-319-55669-7_8

automation of the processes, which involves the orchestration of a set of services, agents and actors that must be combined to solve a problem or to define a new service. The most common way of workflow representation is a directed graph, where nodes correspond to the data processing actions, and edges represent data dependencies [5]. However, most of all, workflows are represented as Directed Acyclic Graph (DAG) [6] or even as a series of actions (pipeline). The Directed Cyclic Graphs (DCG) are much more difficult to implement because they require support of iterative computational processes during their planning and execution [7].

In our case, the specifics of implementation of computational experiments are often imply that resources of a distributed computing environment are often used to solve numerous tasks from a narrow domain, differing only by values of a relatively small set of simulation parameters.

To improve the utilization of computing resources, problem-oriented distributed computing environments are created, providing the solution of problems in specific subject areas (mathematics, computational fluid dynamics, computational chemistry, computational engineering, etc.) [8–10]. This restriction allows to use the information about the subject area to predict problem characteristics during the scheduling and resources allocation stages, increasing the efficiency of available computing resources usage [11].

The article [11] proposes an approach to the implementation of problem-oriented scheduling of computing resources based on the PO-HEFT algorithm. The aim of this work is to test the implementation of PO-HEFT algorithm and analyze its efficiency in comparison with existing algorithms of workflow scheduling.

The paper is organized as follows. Section 2 provides an overview of current approaches to scheduling and resource utilization prediction during workflow execution. In addition, we present analysis of existing platforms for the simulation of distributed computing environments, including cloud systems and workflow systems. Section 3 presents the implementation of a scheduling algorithm PO-HEFT. Next, in Sect. 4 the architecture of problem-oriented environment simulation system is presented. Section 5 is devoted to the implementation of the comparative evaluation of problem-oriented workflow scheduling algorithms. In conclusion, a summary of the work and the possible directions of further research are presented.

2   Review of Workflows Scheduling Methods

Workflow management algorithms can be divided into two categories: workflow scheduling and workflow prediction algorithms. Workflow scheduling algorithms are responsible for distribution of workflow tasks between the available computational resources. Workflow prediction algorithms allow us to estimate the time required to perform certain tasks and the entire workflow. Workflow scheduling algorithms can use the prediction algorithms to achieve the optimal (or quasi-optimal) distribution of tasks between the resources.

2.1   Independent Tasks and Workflows Scheduling Algorithms

Authors of [12] provide a comparison of independent tasks scheduling algorithms, including such algorithms like Opportunistic Load Balancing (OLB), Minimum Execution Time (MET), Minimum Completion Time (MCT), The Min-Min Heuristic, The Max-Min Heuristic, etc. However, it is necessary to consider dependencies between the tasks for workflow scheduling.

Let us consider the basic algorithms of workflow scheduling. *Heterogeneous Earliest Finish Time (HEFT)* algorithm [13] is one of the most common workflow scheduling algorithms. Let $|T_x|$ denote the size of task T_x and let R denote a set of computing resources with an average computing power $|R| = \sum_{i=1}^{n} |R_i|/n$. Then the average time to complete the task on all available computational resources is:

$$E(T_x) = \frac{|T_x|}{|R|},\tag{1}$$

Let $\overline{T}_{xy}$ denote the amount of data transferred between tasks T_x and T_y and R - a set of available resources with an average power of data transfer $R = \sum_{i=1}^{n} \overline{R}_i/n$. Then the average cost of data transfer between tasks T_x and T_y for all available resources is:

$$D(T_{xy}) = \frac{\overline{T}_{xy}}{\overline{R}},\tag{2}$$

In this case, the priority of task may be defined as:

$$rank(T_x) = E(T_x) + \max_{T_y \in succ(T_x)} (D(T_{xy}) + rank(T_y)),\tag{3}$$

where $succ(T_x)$ denotes the set of tasks that depend on the task T_x.

Thus, the task priority is directly determined by the priority of all its dependent tasks. The procedure of assignment of tasks to the resources is implemented as follows: a task with a higher priority, if all the tasks on which it depends are completed, is assigned to the computational resource, providing less time for completion of this task.

Some science groups suggested modifications of the HEFT algorithm aimed at solving certain problems in the workflow planning. So *PDHEFT* [14] makes it possible to avoid the costs of data transfer between nodes through the duplication of tasks; *PO–HEFT* [11] provides an estimation of computational task based on its type and input parameters, predicting the time of task execution, scaling limits and the amount of transmitted data; *Dynamic Data-Resource-Selection (DDRS)* [15] provides a dynamic recalculation of priorities of tasks at certain intervals.

There are also several workflow scheduling algorithms that are not related to HEFT:

- *Problem Oriented Scheduling (POS)* [16] – provides a resource scheduling in distributed problem-oriented computing environments. A distinctive feature of the POS algorithm is that it considers the knowledge of domain specificity, enabling the execution of a workflow task on multiple processor cores with the restrictions on the scalability of this problem.
- *Dominant Sequence Clustering (DSC)* [17] – it consists of two phases. At the first phase, the algorithm searches the critical path in the graph and makes the clustering of problems along this path. Tasks of the same cluster are performed at the same node. Then, the first phase is applied recursively to the remaining tasks. In the second phase, some clusters are merged to reduce the amount of resources used.
- *Critical Path First (CPF)* [18] – it searches the critical path in the workflow graph (the path with the highest span time). Tasks on the critical path are assigned to the one computing unit to avoid the additional cost for communication. Other tasks are assigned to the remaining nodes without an increase of overall execution time.

2.2 Algorithms of Resource Utilization Prediction

Analysis of the main areas of research in the field of workflow scheduling in distributed computing environments shows that prediction of resource utilization in a computational environment is one of the most urgent issues today [19–21]. This is because the main problem of all the algorithms considered above is the complexity of obtaining information about the computational performance of tasks in the workflow before the task has been completed. Therefore, different techniques and algorithms for computational performance prediction are used together with the scheduling algorithms. A prediction object is a set of characteristics of the computational process, such as the time to complete a task or a workflow, the amount of resources needed to complete a task, the amount of input and output data, the degree of scaling and other parameters.

Existing approaches and techniques of assessment of task execution time can be divided into three categories: static, dynamic and hybrid. Static approaches estimate total workflow and its individual tasks before any execution has happened. In the dynamic approach, estimation of the task execution time is carried out immediately before its execution, considering the current state of the system and the available resources. However, the best approach is a hybrid approach that combines the advantages of both considered approaches.

The authors of [5] introduced the concept of a MAPE-K cycle ("Monitoring, Analysis, Planning, Performance and Knowledge") associated with the dynamic scheduling. Initially, the entire workflow is estimated and the initial plan for its implementation is built. Upon completion of each task, the estimated values are replaced with real, and a workflow reassessment and possibly plan rebuilding are performed. This allows the algorithm to immediately correct estimation errors and not distribute them to the entire workflow.

Authors of the article [22] propose to use a regression model for estimation of task execution time. This model shows the dependence of the number of CPU operations on the size of the input data. This model in combination with the static (CPU speed) and dynamic (e.g. the amount of free RAM) execution node parameters allow evaluation of the execution time of a task at the node computational node.

2.3 Workflow Simulation

Scientific Workflow Management System (SWMS) support the development, deployment and analysis of the results of execution of scientific workflows. They provide the automation of computational experiment cycle, making it easier for researchers and engineers to use distributed computing resources, allowing them to focus on solving practical problems, rather than on the computational process control [1]. SWMS enables definition of an abstract workflow; construction of a Workflow Execution Plan (WEP), considering the possibilities of optimization and parallel execution of individual actions and sub-flows; data sources management; workflow execution; and execution results gathering. There are several SWMS that are mostly used by research groups, like Pegasus [23, 24], Taverna [25, 26], Askalon [27], Kepler [28], Triana [29], etc.

To simulate the workflow scheduling algorithms some simulation systems can be applied. One of the most used systems for workflow management platforms simulation is the WorkflowSim [30], which is an extension of the CloudSim [31] simulator.

CloudSim is a tool for the simulation of cloud computing environments. It supports the simulation of the components of cloud computing environments like data centers, virtual machines, and resource allocation policies. CloudSim platform is highly extensible, and therefore there exist a number of independent projects such as CloudSimEx [32], WorkflowSim [30], Cloud2Sim [33], DynamicCloudSim [34] et al., which are focused on performance improvement or expansion of the system functionality.

3 Implementation of the PO-HEFT Algorithm

We propose a PO-HEFT list-based algorithm for problem-oriented scheduling of applications in cloud computing environments based on their execution profiles [11]. The proposed algorithm is based on the HEFT algorithm, but contains modifications in calculating the task rank and considers the incoming communication value of its parent task.

Cloud workflow execution system consists of a computing cluster, Infrastructure as a Service (IaaS) cloud platform deployed on this computer cluster, and workflow management platform that can provide execution of tasks in a certain order on a computing cluster. We assume that the following elements are also determined in the problem-oriented cloud computing environment [11, 35]:

- $\mathcal{F}$ – the set of all functions f, which are implemented in the domain of problem-oriented cloud computing environment;
- $\mathfrak{M}$ – the set of virtual machine images $\mathfrak{m}$, available for deployment at the nodes;
- $\pi : \mathfrak{m} \to \mathbb{Z}_{>0}$ – virtual machines performance characteristics;

- $\widehat{\tau}(f, \mathcal{I}^{in}, \pi)$ – the operator of execution time estimation of function f for a given set of input data objects $\mathcal{I}^{in}$ at a machine with performance π;
- $\widehat{v}(f, \mathcal{I}^{in})$ – the operator of output size estimation. It provides a prediction of the total size in bytes of all output data objects while executing function f with input parameters values $\mathcal{I}^{in}$.

Then the task T_x is a separate instance of the function $f_x \in \mathcal{F}$ with certain set of input data objects $\mathcal{I}^{in}$:

$$T_x = f_x(\mathcal{I}_x^{in}). \tag{4}$$

Let the R be a set of virtual machines available for deployment with average computational performance:

$$|R| = \sum_{i=1}^{n} \frac{|R_i|}{n} = \sum_{i=1}^{n} \frac{\pi_i}{n}. \tag{5}$$

In this case, the operator of estimation of execution time can be applied for estimation of the $E(T_x)$ execution time:

$$E(T_x) = \widehat{\tau}(f_x, \mathcal{I}_x^{in}, |R|) \tag{6}$$

In the model of problem-oriented services, the amount of data returned by each task T_x should be considered. We can use the operator of output size estimation $\widehat{v}(f, \mathcal{I}^{in})$ for prediction of this value. Consequently, for estimation of data transfer time between two tasks, the following formula can be used within the problem-oriented model:

$$D(T_{xy}) = \widehat{v}(f_x, \mathcal{I}_x^{in}) * \bar{R}_{xy}, \tag{7}$$

where $\bar{R}_{xy}$ is the bandwidth of data transfer channel in the cloud computing system.

Thus, the priority of task can be determined as:

$$rank(T_x) = E(T_x) + \max_{T_y \in succ(T_x)} (D(T_{xy}) + rank(T_y)), \tag{8}$$

where $succ(Tx)$ - the set of all tasks, which depends on task T_x.

Pseudocode of PO-HEFT algorithm is shown in Fig. 1.

To implement operators $\widehat{\tau}$ and $\widehat{v}$ we apply heuristic prediction, which is based on the analysis of the results of previous executions of similar tasks based on k-nearest neighbors algorithm:

Step 1. Select information about all previous executions of the function f from a database.

Step 2. For each previous execution, calculate a distance relative to the current execution based on values of input parameters. We will use standard Euclidean distance for calculating distance between values of input parameters:

$$\rho\left(\mathcal{I}^{in}, \mathcal{I}^{in}_j\right) = \sqrt{\sum_{i=0}^{n}\left(I^{in}_i - I^{in}_{ji}\right)}, \tag{9}$$

where I^{in}_i и I^{in}_{ji} – values of the i-th parameter for the current and j-th execution.

Step 3. Select k previous executions with a minimal distance.

Step 4. Calculate average execution time and average sizes of output files for selected k previous executions.

```
PROCEDURE: PO-HEFT
INPUT DATA: Task graph G(T, E), values of input parame-
ters Jin, set of resources R
OUTPUT DATA: Distribution list of tasks
BEGIN
    for each t ∈ T from task graph G
Estimate execution. time in accordance with operation (6)
    for each e ∈ E from task graph G
Estimate data transfer time in accordance with operation
(7)
    Run breadth first search in reverse order of tasks
and calculate rank of each task in accordance with (8)
    while in T there are uncompleted tasks
        TaskList get completed tasks from task graph G
        Assign task(TaskList, R)
        Update TaskDistributionList
END
PROCEDURE: Assign task
INPUT DATA: TaskList, Set of resources R
BEGIN
    Sort TaskList in decreasing order by rank
    for each t ∈ TaskList
    r ← get resource from R, able to complete the task t
earlier
    distribute task t to resource r
    update status of r
END
```

Fig. 1. PO-HEFT workflow applications scheduling algorithm for problem-oriented computational environments.

4 Simulation of Workflow Execution System

4.1 Architecture of the Workflow Execution System

Let us consider the structure of workflow execution system. It consists of the following components: DAX parser, prediction subsystem, scheduler, workflow engine, a database that stores the statistics from previous executions. In addition, a profiler is running on each virtual machine in a computing cluster. Profiler collects all the necessary statistics and stores it in a database. A detailed diagram of the entire system is shown in Fig. 2. At this work, prediction subsystem is implemented, which is then used to implement the algorithm PO-HEFT.

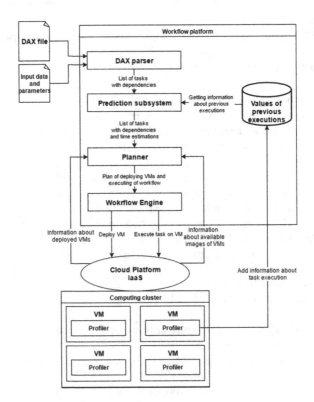

Fig. 2. Architecture of the workflow execution system

Let us examine in more detail the individual components that make up the workflow execution system.

1. *DAX parser* – a component that receives an XML file that contains information about the tasks, included in the workflow and their dependencies and returns a list of tasks with dependencies in some internal representation, with which other components of the system operate.

2. *Database of previous executions* – a database, containing statistics on the previous executions of individual tasks. Statistics on task includes input parameters, the size of the output data and the execution time.
3. *Prediction subsystem* – a system for estimation of the execution time and the size of the output data, based on input parameters and information about the previous executions from the database.
4. *Planner* – a component that accepts a task list with dependencies and execution time estimations; and returns workflow execution plan based on resources available in a cloud platform.
5. *Workflow Engine* – a component, which acts based on the workflow execution plan and sends the task to the node only if all parent tasks have been completed.
6. *Profiler* – component that runs on each virtual machine and collects statistics on the execution of tasks.

We have developed a prediction subsystem and database of previous executions. We used PostgreSQL as a database system. Prediction subsystem is written in Java programming language. The remaining components of the test stand for the comparative analysis of scheduling algorithms are implemented in the WorkflowSim simulator.

4.2   Model of the Executions History Database

To predict the execution time and the size of output data of these tasks, it is necessary to have statistics on the previous executions of tasks of this type. We developed a database, where the characteristics of the tasks' executions would be stored. This data will be used by the prediction system. Within the developed test stand, the following information about executions of tasks on a computing cluster is stored in the database (Fig. 3):

- tasks – names of tasks which can be executed;
- executions – statistics of tasks' executions, including execution time measured in milliseconds;
- input_sizes – sizes of input files;
- output_sizes – sizes of output files.

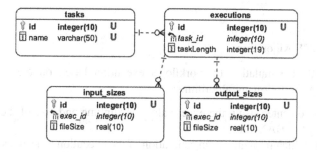

Fig. 3. Database scheme of executions history

5 Evaluation of Workflow Scheduling Algorithms

5.1 Individual Tasks Prediction Algorithm

The set of 45 workflows of the Pegasus Project [36] was used to test the prediction subsystem. Each workflow is represented by XML-file with «dax» («Directed Acyclic graph Xml»). The total number of tasks in all collected workflows was about 24 000. This set was divided into training and test parts. The training part has been loaded into the database of previous executions, then the prediction algorithm described in Sect. 3 was used on the test part. The value of k was set to 10. The relative error was calculated for each run, using the following formula:

$$\delta^i = \frac{|T^i_{predicted} - T^i_{real}|}{T^i_{real}}. \tag{10}$$

Thus, we calculated the average relative error of all predictions. All collected data was divided in different proportions in the training and test blocks, and for each partition testing for prediction of individual tasks was performed. The test results are shown in Table 1.

Table 1. Test results of prediction algorithm on individual tasks

Test No.	«training part/test part» partition ratio	Relative error of the execution task estimation (runtime)	Relative error of the output data estimation (outsize)
1	50/50	6.7%	6.3%
2	60/40	6.2%	5.8%
3	70/30	6.0%	5.6%
4	80/20	5.5%	5.3%
5	90/10	5.0%	4.8%

As seen from the test results, the accuracy of the prediction of the execution time of the tasks and the output size is 5 to 7%, which decrease with the increasing of training sample size.

5.2 PO-HEFT Algorithm

We have made a simulation of workflows execution based on the WorkflowSim platform using the following algorithms:

- *HEFT*, implemented using the exact values of runtime and size of the output data given from the DAX-files;
- *PO-HEFT*, which implements an estimation of the execution time and size of output data based on information received from the prediction subsystem;
- *RANDOM*, which provides a random distribution of tasks on the nodes.

HEFT and RANDOM algorithms implementations were used from the Work-flowSim package. We used the CyberShake (30, 50, 100 and 1000 of tasks), Epigenomics (24, 46, 100, and 997 tasks) and Inspiral (30, 50, 100 and 1,000 tasks) [37] workflows for the test. For each test, the largest workflow was taken and the smaller workflows gradually loaded into the database of previous executions to evaluate the quality of the prediction. In the first series of test runs, the testing was performed on a set of 5 identical virtual machines with the following parameters: MIPS (million instructions per second) = 1000 and Bandwidth (Megabytes per second) = 1000 running on the basis of a data center with the fully connected network topology.

To get consistent results, the RANDOM algorithm was executed for five times each time, after which the result is determined as the average value of the obtained results. Results of HEFT, PO-HEFT and RANDOM algorithms evaluation are shown in Table 2 and Fig. 4.

Table 2. Comparison of the algorithms in a homogeneous computing environment

Workflow	Algorithm					
	Workflow execution time (relative slowdown compared with HEFT)					
	HEFT	PO-HEFT		RANDOM		
CyberShake_1000	4 754	4 957	(+4.3%)	5 368	(+12.9%)	
Epigenomics_997	776 051	789 115	(+1.7%)	885 290	(+14.1%)	
Inspiral_1000	45 716	4 6791	(+2.3%)	50 860	(+11.3%)	

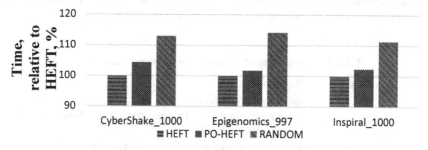

Fig. 4. Comparison of the algorithms in a homogeneous computing environment

Also, to evaluate the impact values of the parameter K, a series of experiments with the use of PO-HEFT algorithm was carried out using various parameter K values in the K-nearest algorithm. Test results of PO-HEFT algorithm depending on the amount of information in the database and a parameter K are shown in Table 3.

Table 3. Test results of PO-HEFT algorithm with different parameters on workflows CyberShake_1000, Epigenomics_997, Inspiral_1000

Workflow	K = 1	K = 3	K = 5	K = 10
CyberShake_1000	4 843	4 843	4 942	5 034
Epigenomics_997	789 076	789 053	789 394	789 087
Inspiral_1000	47 612	46 524	47 301	46 910

The results of the first series of experiments allow us to make the following conclusions:

- on the analyzed workflows, even the random distribution of tasks (RANDOM algorithm) on the compute nodes in a homogeneous computing environment gives relatively good planning results, only 10–15% worse than HEFT algorithm;
- the scheduling results of the PO-HEFT algorithm, on average, are 2–4% worse than the results of the HEFT algorithm;
- tasks in the analyzed workflows are characterized by a strong direct relationship between the amount of input data, the execution time and the size of output data. In this regard, PO-HEFT algorithm shows equally good results in average with the use of values of the parameter K in a range from 1 to 10 in the K-nearest neighbors algorithm.

In the second test series, five virtual machines were created with the parameters shown in Table 4.

Table 4. Parameters of virtual machines in the second series of experiment

	VM_0	VM_1	VM_2	VM_3	VM_4
MIPS	200	400	600	800	1 000
Bandwidth (MB/s)	200	400	600	800	1 000

For the second series of experiments, the values of the parameter K for K-nearest neighbors algorithm was equal to 10. The average test results of HEFT, PO-HEFT and RANDOM algorithms are shown in Table 5 and in Fig. 5.

Table 5. Comparison of algorithms in the second test series

Workflow	Algorithm				
	Workflow execution time (number in parentheses indicates the relative slowdown compared with HEFT)				
	HEFT	PO-HEFT		RANDOM	
CyberShake_1000	7 795	8 106	(+3.8%)	23 716	(+204%)
Epigenomics_997	1 294 702	1 331 352	(+2.8%)	3 995 491	(+209%)
Inspiral_1000	79 130	79 051	(−0.1%)	230 608	(+191%)

Based on these results, we can conclude that the PO-HEFT algorithm showed slightly lower (1–4%) performance of workflow planning in comparison with the HEFT algorithm. At the same time, HEFT algorithm in the planning process uses a priori information about the time of the task execution and the amount of output data. Besides, the efficiency of the algorithm PO-HEFT is several times better than the RANDOM algorithm when a cloud computing environment may use virtual machines with different computing capabilities.

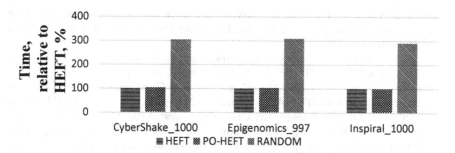

Fig. 5. Comparison of algorithms in the second test series

6 Conclusion

In this paper, the prototype of workflow prediction system in problem-oriented computing environments was developed and PO-HEFT algorithm was tested. In the absence of complete statistics of executions of real-world problems, a prototype has been tested only on abstract workflows, which has only the size of the input file, but does not have other parameters.

In the future, we plan to perform evaluation of the comparative effectiveness of PO-HEFT, POS, and DSC scheduling algorithms based on the information about the parameters of the executions of real workflows executed on resources of South Ural State University supercomputer. However, we can already say that the PO-HEFT algorithm can be implemented and is comparable in its effectiveness with the HEFT algorithm.

Acknowledgements. This work has been partially funded by the RFBR, research project No. 14-07-00420-a, by the Grant of the President of the Russian Federation No. MK-7524.2015.9 and by Act 211 Government of the Russian Federation, contract No. 02. A03.21.0011.

References

1. Deelman, E., Gannon, D., Shields, M., Taylor, I.: Workflows and e-science: an overview of workflow system features and capabilities. Futur. Gen. Comput. Syst. **25**, 528–540 (2009). doi:10.1016/j.future.2008.06.012
2. Glotzer, S.C.: International Assessment of Research and Development in Simulation-Based Engineering and Science. Imperial College Press, Covent Garden (2011)
3. Davis, P.K., Henninger, A.E.: Analysis, Analysis Practices, and Implications for Modeling and Simulation. Rand Corporation, Santa Monica (2007)
4. Fox, G.C., Gannon, D.: Special issue: workflow in grid systems. Concurr. Comput. Pract. Exp. **18**, 1009–1019 (2006)
5. Da Silva, R.F., Juve, G., Rynge, M., Deelman, E., Livny, M.: Online task resource consumption prediction for scientific workflows. Parallel Process. Lett. **25**, 25 (2015). doi:10.1142/S0129626415410030

6. Zhang, J., Chen, X., Li, J., Li, X.: Task mapper and application-aware virtual machine scheduler oriented for parallel computing. J. Zhejiang Univ. Sci. C. **13**, 155–177 (2012). doi:10.1631/jzus.C1100217

7. Yu, J., Buyya, R.: A taxonomy of workflow management systems for grid computing. J. Grid Comput. **3**, 171–200 (2005). doi:10.1007/s10723-005-9010-8

8. Afanasiev, A., Sukhoroslov, O., Voloshinov, V.: MathCloud: publication and reuse of scientific applications as RESTful web services. In: Malyshkin, V. (ed.) PaCT 2013. LNCS, vol. 7979, pp. 394–408. Springer, Heidelberg (2013). doi:10.1007/978-3-642-39958-9_36

9. Knyazkov, K.V., Kovalchuk, S.V., Tchurov, T.N., Maryin, S.V., Boukhanovsky, A.V.: CLAVIRE: e-Science infrastructure for data-driven computing. J. Comput. Sci. **3**, 504–510 (2012). doi:10.1016/j.jocs.2012.08.006

10. Radchenko, G., Hudyakova, E.: A service-oriented approach of integration of computer-aided engineering systems in distributed computing environments. In: Proceedings of UNICORE Summit 2012, Dresden, Germany, pp. 57–66 (2012)

11. Nepovinnykh, E.A., Radchenko, G.I.: Problem-oriented scheduling of cloud applications: PO-HEFT algorithm case study. In: Proceedings of the 39th International Convention on Information and Communication Technology, Electronics and Microelectronics, MIPRO 2016, pp. 196–201. IEEE Computer Society, Opatija (2016). doi:10.1109/MIPRO.2016. 7522134

12. Kannan, R., Rasool, R.U., Jin, H., Balasundaram, S.R.: Managing and Processing Big Data in Cloud Computing. IGI Global, Hershey (2016)

13. Topcuoglu, H., Hariri, S., Min-You, W.: Performance-effective and low-complexity task scheduling for heterogeneous computing. IEEE Trans. Parallel Distrib. Syst. **13**, 260–274 (2002). doi:10.1109/71.993206

14. Lee, Y.C., Zomaya, A.Y.: A productive duplication-based scheduling algorithm for heterogeneous computing systems. In: Yang, L.T., Rana, O.F., Martino, B., Dongarra, J. (eds.) HPCC 2005. LNCS, vol. 3726, pp. 203–212. Springer, Heidelberg (2005). doi:10. 1007/11557654_26

15. Pandey, S., Buyya, R.: Scheduling of scientific workflows on data grids. In: Proceedings CCGRID 2008 - 8th IEEE International Symposium on Cluster Computing and the Grid, pp. 548–553. Lyon, France (2008). doi:10.1109/CCGRID.2008.32

16. Sokolinsky, L.B., Shamakina, A.V.: Methods of resource management in problem-oriented computing environment. Program. Comput. Softw. **42**, 17–26 (2016). doi:10.1134/ S0361768816010084

17. Yang, T., Gerasoulis, A.: DSC: scheduling parallel tasks on an unbounded number of processors. IEEE Trans. Parallel Distrib. Syst. **5**, 951–967 (1994). doi:10.1109/71.308533

18. Lee, Y.C., Zomaya, A.Y.: Stretch out and compact: workflow scheduling with resource abundance. In: 13th IEEE/ACM International Symposium on ClusterCloud, and Grid Computing, pp. 219–226. Delft (2013). doi:10.1109/CCGrid.2013.55

19. Kliazovich, D., Pecero, J.E., Tchernykh, A., Bouvry, P., Khan, S.U., Zomaya, A.Y.: CA-DAG: communication-aware directed acyclic graphs for modeling cloud computing applications. In: IEEE 6th International Conference on Cloud Computing, pp. 277–284. IEEE, Santa Clara (2013). doi:10.1109/CLOUD.2013.40

20. Tchernykh, A., Schwiegelsohn, U., Alexandrov, V., Talbi, E.: Towards understanding uncertainty in cloud computing resource provisioning. Procedia Comput. Sci. **51**, 1772–1781 (2015). doi:10.1016/j.procs.2015.05.387

21. Tchernykh, A., Lozano, L., Schwiegelshohn, U., Bouvry, P., Pecero, J.E., Nesmachnow, S., Drozdov, A.Y.: Online bi-objective scheduling for IaaS clouds ensuring quality of service. J. Grid Comput. **14**, 5–22 (2016). doi:10.1007/s10723-015-9340-0

22. Wu, Q., Datla, V.V.: On performance modeling and prediction in support of scientific workflow optimization. In: 2011 IEEE World Congress on Services, pp. 161–168 (2011). doi:10.1109/SERVICES.2011.37

23. Deelman, E., Vahi, K., Juve, G., Rynge, M., Callaghan, S., Maechling, P.J., Mayani, R., Chen, W., Ferreira da Silva, R., Livny, M., Wenger, K.: Pegasus, a workflow management system for science automation. Futur. Gen. Comput. Syst. **46**, 17–35 (2015). doi:10.1016/j.future.2014.10.008

24. Gil, Y., Ratnakar, V., Deelman, E.: Wings for pegasus: creating large-scale scientific applications using semantic representations of computational workflows. In: Proceedings of the National Conference on Artificial Intelligence, Vancouver, British Columbia, Canada, vol. 22, no. 2, pp 1767–1774 (2007)

25. Wolstencroft, K., Haines, R., Fellows, D., Williams, A., Withers, D., Owen, S., Soiland-Reyes, S., Dunlop, I., Nenadic, A., Fisher, P., Bhagat, J., Belhajjame, K., Bacall, F., Hardisty, A., Nieva de la Hidalga, A., Balcazar Vargas, M.P., Sufi, S., Goble, C.: The taverna workflow suite: designing and executing workflows of web services on the desktop, web or in the cloud. Nucleic Acids Res. **41**, W557–W561 (2013). doi:10.1093/nar/gkt328

26. Missier, P., Soiland-Reyes, S., Owen, S., Tan, W., Nenadic, A., Dunlop, I., Williams, A., Oinn, T., Goble, C.: Taverna, Reloaded. In: Gertz, M., Ludäscher, B. (eds.) SSDBM 2010. LNCS, vol. 6187, pp. 471–481. Springer, Heidelberg (2010). doi:10.1007/978-3-642-13818-8_33

27. Wieczorek, M., Prodan, R., Fahringer, T.: Scheduling of scientific workflows in the ASKALON grid environment. ACM SIGMOD Rec. **34**, 56 (2005). doi:10.1145/1084805.1084816

28. Ludäscher, B., Altintas, I., Berkley, C., Higgins, D., Jaeger, E., Jones, M., Lee, E.A., Tao, J., Zhao, Y.: Scientific workflow management and the Kepler system. Concurr. Comput. Pract. Exp. **18**, 1039–1065 (2006). doi:10.1002/cpe.994

29. Shiroor, A., Springer, J., Hacker, T., Marshall, B., Brewer, J.: Scientific workflow management systems and workflow patterns. Int. J. Bus. Process Integr. Manag. **5**, 63 (2010). doi:10.1504/IJBPIM.2010.033175

30. Chen, W., Deelman, E.: WorkflowSim: a toolkit for simulating scientific workflows in distributed environments. In: 8th International Conference on E-Science, pp. 1–8. IEEE Computer Society, Chicago (2012). doi:10.1109/eScience.2012.6404430

31. Calheiros, R.N., Ranjan, R., Beloglazov, A., De Rose, C.A.F., Buyya, R.: CloudSim: a toolkit for modeling and simulation of cloud computing environments and evaluation of resource provisioning algorithms. Softw. Pract. Exp. **41**, 23–50 (2011). doi:10.1002/spe.995

32. Yang, M., Rutherfoord, B., Jung, E.: Learning cloud computing and security through cloudsim simulation. Inf. Secur. Educ. J. **1**, 62–69 (2014). doi:10.1145/2670739.2670747

33. Kathiravelu, P., Veiga, L.: Concurrent and distributed cloudsim simulations. In: IEEE 22nd International Symposium on Modelling, Analysis & Simulation of Computer and Telecommunication Systems, pp. 490–493. IEEE Computer Society, Paris (2014). doi:10.1109/MASCOTS.2014.70

34. Bux, M., Leser, U.: DynamicCloudSim: simulating heterogeneity in computational clouds. Futur. Gen. Comput. Syst. **46**, 85–99 (2015). doi:10.1016/j.future.2014.09.007

35. Radchenko, G.: Model of problem-oriented cloud computing environment. Proc. Inst. Syst. Program. RAS. **27**, 275–284 (2015). doi:10.15514/ISPRAS-2015-27(6)-17

36. Bharathi, S., Chervenak, A., Deelman, E., Mehta, G., Su, M.-H., Vahi, K.: Characterization of scientific workflows. In: Third Workshop on Workflows in Support of Large-Scale Science, pp. 1–10. IEEE Computer Society, Austin (2008). doi:10.1109/WORKS.2008.4723958

37. Mehta, G., Juve, G., Chen, W.: Workflow Generator. https://confluence.pegasus.isi.edu/display/pegasus/WorkflowGenerator

Layer-by-Layer Partitioning of Finite Element Meshes for Multicore Architectures

Alexander Novikov[✉], Natalya Piminova, Sergey Kopysov, and Yulia Sagdeeva

Institute of Mechanics, Ural Branch of the Russian Academy of Sciences,
34 ul. T. Baramzinoy, Izhevsk 426067, Russia
alexander.k.novikov@gmail.com,n.k.piminova@gmail.com,
s.kopysov@gmail.com,sagdeeva@yandex.ru

Abstract. In this paper, we present new partitioning algorithms for unstructured meshes that prevent conflicts during parallel assembly of FEM matrices and vectors in shared memory. The algorithms use a criterion that determines if any two mesh cells are neighboring. This neighborhood criterion is used to partition the mesh into layers, which are then combined into blocks and assigned to different parallel processes/threads. The proposed partitioning algorithms are compared with the existing algorithms on quasi-structured and unstructured meshes by the number of potential conflicts and by the load imbalance.

Keywords: Unstructured meshes · Mesh layers · Shared memory · Parallel FEM · Multicore processors

1 Introduction

The main objectives of mesh partitioning for parallel FEM are the following: (i) uniform data distribution, (ii) minimization of communications, (iii) improvement of data locality [1,2]. Geometric mesh partitioning algorithms, such as recursive coordinate bisection [3], recursive inertial bisection [4], and algorithms based on the space-filling curves [5], focus on load balancing. Graph and hypergraph partitioning algorithms [6] are applied to meshes to minimize communications by reducing the boundary between the resulting subdomains: multilevel division k-way [7], spectral bisection [8]. There are combined graph and geometric approaches [9], which target both (i)–(ii).

In parallel FEM algorithms for multi-core processors, where memory is shared between multiple processes/threads, good data locality and minimum resource contention are expected from the partition. When shared memory is used, global assembly operations become a bottleneck in parallel FEM algorithms [10]. In assembly operations, concurrent adding of the matrix or vector elements that correspond to the vertices common to multiple elements may result in conflicts between processes/threads and computational errors. Shared memory conflicts are resolved by different methods, such as atomic operations [11] and critical sections; replacing element-by-element assembly scheme by nodal assembly scheme; mesh coloring based on graph coloring; etc.

© Springer International Publishing AG 2016
V. Voevodin and S. Sobolev (Eds.): RuSCDays 2016, CCIS 687, pp. 106–117, 2016.
DOI: 10.1007/978-3-319-55669-7_9

Graph coloring is widely used in parallel linear solvers to reorder unknowns for more efficient parallel processing [12,13]. When applied to meshes, coloring algorithms generate sets of disconnected mesh cells [14]. The time complexity of the coloring algorithms is defined not only by the number of vertices, but also by the vertex degrees (valence) and the choice of the initial vertex, so that the graph coloring problem is NP-complete. To provide a uniform distribution of mesh cells over parallel processes/threads and a good data locality, after mesh coloring the cells should be additionally reordered.

In this work, we propose new layer-by-layer mesh partitioning algorithms, which eliminate concurrent computations in assembly operations over the elements that have common vertices. These algorithms divide mesh into one-cell-thick layers, which are then combined into subdomains and assigned to different threads or processor cores. These algorithms are designed to improve the performance of parallel element-by-element assembly. The proposed partitioning algorithms are compared with the existing algorithms on quasi-structured and unstructured meshes by the number of potential conflicts and by the load imbalance.

The paper is structured as follows. In Sect. 2, we outline the new layer-by-layer partitioning algorithm and introduce a neighborhood criterion. In Sect. 3, we describe how this criterion is used for partitioning the mesh into one-cell-thick layers. In Sect. 4, we combine the layers into subdomains, targeting a well-balanced load distribution. In Sect. 5, we compare the results of layer-by-layer and multilevel k-way graph partitioning in terms of memory conflicts. Section 6 concludes the paper.

2   Layer-by-Layer Partitioning of Finite Element Meshes

In this section, we introduce a neighborhood criterion, which is used to partition mesh into layers. A new algorithm of layer-by-layer partitioning of unstructured meshes is based on this criterion.

Usually, finite element meshes are unstructured so that the neighborhood relationships of nodes or cells cannot be defined by incremental numbering of nodes or cells. For element-by-element FEM, we propose the following neighborhood criterion:

$$\omega_k \in Adj(\omega_j), \text{ if } V(\omega_j) \bigcap V(\omega_k) \neq \emptyset, \tag{1}$$

where ω_j, ω_k are the cells of the mesh Ω; $Adj(\cdot)$ is the neighborhood operator; $V(\omega_j)$ and $V(\omega_k)$ are the subsets of mesh nodes that correspond to the cells ω_j and ω_k.

This neighborhood ratio reflects the data dependencies in element-by-element assembly more accurately than the dual or nodal graphs of the mesh. Indeed, with such a neighborhood criterion, each cell will have more neighbors, m_e, even in quasi-structured meshes like the one shown in Fig. 1.

We propose a new algorithm that partitions the mesh into thin layers using the neighborhood criterion. *Algorithm 1* can be summarized as follows:

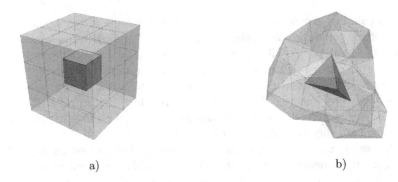

a)   b)

Fig. 1. Neighboring cells: (a) quasi-structured mesh, $m_e = 26$; (b) unstructured mesh, $m_e = 119$.

1. First, one-cell-thick layers are formed using the neighborhood criterion.
2. Then, the layers are combined into subdomains.

In the following sections, we describe these steps in more detail.

3   Forming Layers Using the Neighborhood Criterion

In order to partition the mesh into one-cell-thick layers using the criterion (1), we need to find the elements each node belongs to. Traditionally, unstructured meshes are stored in memory as a set of finite elements, which are defined by nodal connectivity. Therefore, the search of all elements the node belongs to requires two loops over the mesh cells, which totals to $O(2 \cdot m \cdot n_e)$ operations, where m is the number of finite elements; n_e is the number of nodes in an element.

The algorithm of forming mesh layers (*Algorithm 2*) can be summarized as follows:

1. Choose an initial layer $s_1 = \{\omega_j^{(i)}\}, j = 1, 2, \ldots, m_1$, where m_1 is the number of cells in s_1.
2. Consequent layers $s_i, i = 2, 3, \ldots, n_s$ are defined as follows:

$$s_i = \begin{cases} Adj(s_{i-1}), & i = 2; \\ Adj(s_{i-1}) \setminus s_{i-2}, & i > 2, \end{cases} \tag{2}$$

where $s_i\{\omega_j^{(i)}\}$ is the set of cells in layer i; $\Omega_j^{(i)}$ is the j-th cell in layer i; m_i is the number of cells in s_i. In the (2), an $Adj(s_i) = \bigcup_{j=1}^{m_i} Adj(\omega_j^{(i)})$ is the set of cells neighboring to s_i.

This algorithm requires selecting the initial layer s_1 and then defines sequent layers. The choice of the initial layer s_1 affects not only the number of layers n_s, but also the uniformity of distribution of cells in layers.

Figures 2 shows the layers formed for quasi-structured mesh (a fourth part of a membrane on Fig. 2a) and unstructured mesh (a frame construction on the Fig. 2b). Here s_1 is selected in the variant (x_{min}, y, z). Other cases of s_1 selection are shown on Fig. 3a and b.

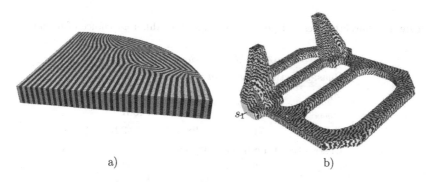

a)   b)

Fig. 2. Layers for unstructured meshes: (a) hexahedtral mesh, $m = 31744$, $n_s = 64$; (b) tetrahedral mesh, $m = 485843$, $n_s = 137$

Distribution of cells between layers depends on the choice of the initial layer in *Algorithm 2*. Table 1 shows the range of variation of the number of cells in the layers for different numbers of layers n_s for the quasi-structured and unstructured meshes shown in Figs. 2 and 3. The columns are marked by a choice of the initial layer s_1 (see Figs. 2a, b, and 3). For example, (x_{min}, y, z) means that at least one vertex in a cell of the layer s_1 has the abscissa equal to x_{min}.

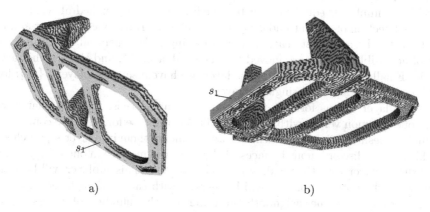

a)   b)

Fig. 3. Layers of unstructured mesh are formed with selected layer s_1 in coordinate directions: a) variant (x, y_{min}, z); b) variant (x, y, z_{min}).

For the quasi-structured mesh, 32 ideally balanced layers were obtained (column (x, y, z_{min})): $m_{max} = m_{min} = 15872$ cells. For the unstructured mesh (column (x, y_{min}, z)), the maximum number of cells in a layer is 46635 for $m = 485843$, and the minimum number of cells is 17 cells, the number of layers $n_s = 47$.

Table 1. Characteristics of layers constructed with different choice of initial layer

Parameters	(x_{min}, y, z)	(x, y_{min}, z)	(x, y, z_{min})
Quasi-structured mesh, $m = 507904$			
m_{min}	1056	1568	15872
m_{max}	6112	5120	15872
n_s	71	28	32
Unstructured mesh, $m = 485843$			
m_{min}	9	17	14
m_{max}	9202	71548	12457
n_s	137	47	140

The *Algorithm 2* assumes that there may be discontinuous layers. In general, the discontinuity of layers does not matter for element-by-element FEM schemes. However, if the Schur complement method [15] is used, the layers must be continuous, and any two neighbor cells must share a face with each other. To find the cells adjacent by faces, we re-use the neighborhood criterion (1). Namely, out of all neighbors of a cell found with the criterion, we take only those who have as many vertices common with the cell as the number of vertices in one face. This number is equal to 3 for tetrahedral and 4 for hexahedral cells.

For detection of discontinuous layers, we take an arbitrary cell in layer s_i and perform the breadth-first search for neighboring cells. Then, all cells from the set of accessible vertices are moved to a new sublayer, s_j^*, and the search starts again. Finally, for each discontinuous layer, we have a set of sublayers, $\{s_j^*\}$; for each continuous layer – a single sublayer s_1^*.

After layer s_i was formed, there may be hanging cells, the cells that have only one common vertex with this layer. This is the case for multiply-connected domains/meshes. To exclude such cells and to merge them into layer s_i, we check if this layer is discontinuous by faces. If the layer has more than one sublayer, we find the sublayer with the maximum number of cells. This sublayer will become a new s_i, while other sublayers will be merged with the previous layer s_{i-1}.

Accordingly to the neighborhood criterion, the number of layers n_s is bounded either by the number of cells in the direction of the diameter of the computational domain or by the maximum number of mesh cells along the coordinate directions. For example, for the hexagonal mesh of $10 \times 10 \times 100$ elements, the number of layers $n_s \leqslant 100$. For the unstructured mesh with the same number of tetrahedral cells, $n_s \leqslant 50$.

4   Combining Layers into Subdomains

In this section, we present different variants of combining the layers $s_j, j = 1, 2, \ldots, n_s$ of the unstructured mesh Ω into of the subdomains $\Omega_i,$ $i = 1, 2, \ldots, n_\Omega$. We estimate the load imbalance incurred by these partitioning schemes and compare it with the multilevel k-way graph partitioning algorithm, implemented in METIS [7]. We consider the following combinations of mesh layers: block, parity of layer indices, and their modifications.

Block combination of mesh layers can be summarized as follows. First, all layers are united: $\Omega = \bigcup_{i=1}^{n_s} s_i$. Then, domain Ω is partitioned into subdomains Ω_i so that $|\Omega_i| \approx m/n_\Omega,$ $i = 1, 2, \ldots, n_\Omega$, where n_Ω is the number of subdomains. Figure 4 shows the layers combined into two neighboring subdomains. Here, the subdomain Ω_1 (marked as A in the figure) includes seven full layers $s_1, s_2 \ldots, s_7$ and most of s_8. The uniform distribution of finite elements will balance the load between the processes of parallel FEM algorithm. However, it requires to revisit the layers to satisfy the neighborhood criterion. The block partitioning requires at least two layers in the subdomain to avoid the conflicts for shared data between parallel processes. Otherwise, it will be necessary to sort the cells, for example, at the coordinates of their centers.

Fig. 4. Block combination of the layers of the quasi-structured hexahedral mesh.

Parity combination of mesh layers is based on the block combination and can be summarized as follows. First, in subdomains $\Omega_i, i = 1, 2, \ldots, n_\Omega$ obtained after the block combination, all layers are split into the sets of odd and even indices (see Fig. 5). Then, odd- or even-numbered layers with indices $i, i + n_\Omega, i + 2n_\Omega, \ldots$ are merged as a new subdomain Ω_i. Therefore, the blocked layers are reshuffled based on the parity of their indices. Figure 6 illustrates selection of layers for one subdomain.

We compared the results of layer-by-layer and multilevel k-way graph partitioning in terms of load imbalance. Parity combination of layers was considered

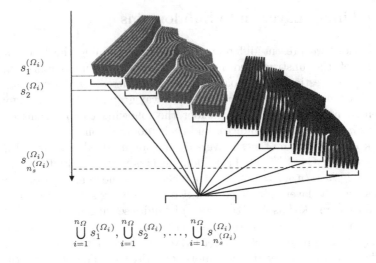

$$\bigcup_{i=1}^{n_\Omega} s_1^{(\Omega_i)}, \bigcup_{i=1}^{n_\Omega} s_2^{(\Omega_i)}, ..., \bigcup_{i=1}^{n_\Omega} s_{n_s}^{(\Omega_i)}$$

Fig. 5. Parity combination of the layers of the quasi-structured mesh: the layers with fixed indices for eight subdomains $\Omega_i, i = 1, 2, ..., n_\Omega$.

in two variants: odd/even and odd+even, which specify if the even or odd or both types of mesh layers will be processed in the OpenMP parallel regions of the FEM algorithm (see Fig. 6).

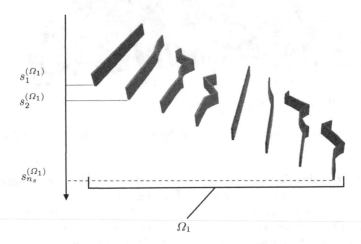

Fig. 6. Parity combination of the layers of the quasi-structured mesh into subdomain Ω_1.

Tables 2 and 3 show the maximum and minimum of numbers of cells in the subdomains obtained with the help of different partitioning algorithms. It can

Table 2. The maximum (upper) and minimum (bottom) of cells numbers m_i for quasi-structured mesh subdomains $\Omega_i, i = 1, 2, ..., n_\Omega$ at a given n_Ω.

n_Ω	METIS	block	odd	even	odd+even
8	63503	63488	31744	31744	63488
	42964	63488	31744	31744	63488
16	32343	31744	17888	17824	35712
	31507	31744	13856	13920	27776
32	16299	15872	10208	10144	20352
	15408	15872	6144	6144	12288
60	8692	8465	6112	6048	12160
	1876	8465	1568	1632	3200

be seen that the layer-by-layer algorithms based on the parity combination of layers results in the least balanced distribution of mesh cells.

Table 3. The maximum (upper) and minimum (bottom) of cells numbers m_i for unstructured mesh subdomains $\Omega_i, i = 1, 2, ..., n_\Omega$ at a given n_Ω.

n_Ω	METIS	block	odd	even	odd+even
8	61163	60731	33838	34171	68009
	60428	60726	17842	27448	55398
16	31180	30366	17781	17803	35419
	29479	30353	11477	11506	23087
32	15638	15183	12965	13334	25729
	14739	15170	4209	4469	8678
60	8340	8098	8839	9202	17605
	7861	8061	1982	1877	3859

We improved the distribution of mesh cells in the **balanced parity combination** scheme. In this scheme, first, layers $s_1, s_3, s_5, \ldots$ are combined into a subdomain $\Omega^{(odd)}$, and layers $s_2, s_4, s_6, \ldots$ – into subdomain $\Omega^{(even)}$. Then, $\Omega^{(odd)}$ is partitioned into n_Ω subdomains $\Omega_i^{(odd)}$ so that $|\Omega_i^{(odd)}| \approx |\Omega^{(odd)}|/n_\Omega$. $\Omega^{(even)}$ is partitioned in the same way. Therefore, this scheme is well-balanced by design.

5   Experimental Results

In this section, we compare the results of layer-by-layer and multilevel k-way graph partitioning in terms of memory conflicts.

Figures 7 and 8 show the unstructured tetrahedral mesh partitioned into $n_\Omega = 8$ subdomains by different algorithms. In contrast to multilevel k-way partitioning, the layer-by-layer algorithms put the subdomains in a special order. Figure 7 shows subdomains obtained by means of block combination (a) and parity combination (b). In both cases, all the subdomains have two neighbors, whereas in Fig. 8, several subdomains have three or more neighbors.

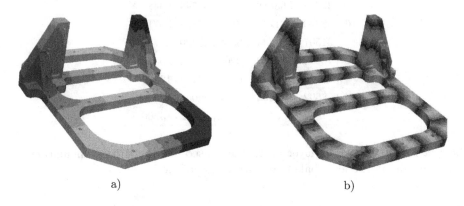

a)   b)

Fig. 7. Layer-by-layer partitioning of unstructured mesh, $n_\Omega = 8$: (a) block combination; (b) parity combination.

Note that after layer-by-layer partitioning (*Algorithm 1*), each mesh node belongs to no more than two subdomains. This is not fulfilled by the multilevel algorithm implemented in METIS, especially for large values of n_Ω on adaptive unstructured meshes as in Fig. 8.

Fig. 8. The unstructured mesh partitioned by the multilevel k-way algorithm, $n_\Omega = 8$.

When applied to multiply-connected domain, both layer-by-layer and multilevel partitioning schemes resulted in discontinuous subdomains. As mentioned above, discontinuous subdomains do not affect the computational stability of parallel element-by-element finite element schemes but slow down memory access. Usually, the discontinuous parts of subdomains are either merged with other subdomains to preserve the number of subdomains or considered as independent subdomains. The proposed ordering of layers and subdomains simplifies merging of the neighboring subdomains.

Table 4 shows the number of potential conflicts in parallel finite element assembly of the vector for several mesh partitioning algorithms. A conflict occurs when different processes simultaneously access the same node in a cell. In terms of mesh, the conflict may occur for the cells than have common nodes and the same numbers (indices) within two or more subdomains. Therefore, we estimate the number of potential conflicts by the number of such cells.

In columns of Table 4, the numerator represents the number of cells whose nodes are accessed from different subdomains (threads), the denominator – the number of cells whose nodes are accessed from different subdomains using identical local indices. The estimate presented in the denominator is more strict, based on the fact that the parallel processes (threads) start at the same time and access each cell within equal periods of time from any of the parallel processes. If this estimate is different from zero (Table 4), then the parallel assembly of finite element vectors my result in computational errors.

Table 4. The number of possible conflicts in parallel vector assembly.

n_Ω	by generation	METIS	block	odd	even	odd+even
Quasi-structured mesh, $m = 507904$						
8	0	0	0	0	0	0
16	0	384/0	0	0	0	0
32	0	24/0	0	0	0	0
60	0	60/2	0	0	0	0
Unstructured mesh, $m = 485843$						
8	284/88	7/2	0	0	0	0
16	623/198	13/4	0	0	0	0
32	1144/352	60/25	0	0	0	0
60	2425/815	169/71	167/40	0	0	4/3

In the experiments labeled "by generation", we used the cell ordering obtained from the mesh generator. The mesh was partitioned into n_Ω subdomains, with approximately the same number of cells each: $\Omega_i \approx m/n_\Omega$. Computational errors due to the conflicts in shared memory may not show up because of cell ordering (see Table 4, the second column).

When unstructured mesh is partitioned by parity of layers (odd+even), the conflicts are related to load imbalance, namely, uneven distribution of cells between subdomains. In this case, the mesh cells are accessed in a sequence of two parallel OpenMP regions, separately for the odd (or even) layers.

In the case of the block partitioning, conflicts are the result of discontinuity of the layers. Combining layers eliminates these conflicts, but results in a less balanced partitioning.

Finally, we examined the computational cost of the proposed layer-by-layer partitioning algorithms based on *Algorithm 1*. Forming layers (first step of *Algorithm 1*) is the most expensive part, which took 0.8–0.9 sec for unstructured mesh ($m = 485843$) and 0.39–0.41 s for quasi-structured mesh ($m = 507904$). Layers combining (second step of *Algorithm 1*) by parity took 0.005–0.006 s, the block combining took 0.003–0.004 s.

6 Conclusion

In this paper, we presented the approach to adapt finite element algorithms to stream processing paradigm, characteristic for hybrid architectures with massively parallel accelerators. The approach is based on partitioning unstructured meshes so that concurrent access to shared data is eliminated and computational load is balanced.

With layer-by-layer partitioning, we achieved 68% speedup of parallel finite element vector assembly on a 61-core processor Xeon Phi 7110X, and 75% speedup on two 4-core Xeon E5-2609. Due to the high efficiency of the assembly operation, we achieved nearly linear speedup of the matrix-vector product in the element-by-element scheme on unstructured meshes.

Comparison of layer-by-layer and multilevel partitioning shown that: (i) the algorithms proposed in this paper outperform METIS up to three times; (ii) in contrast to the multilevel algorithm, the execution time to form subdomains is practically independent of the number of subdomains; (iii) ordering of layers and, as a result, mesh subdomains optimizes communications in both partitioning and assembly operations.

In future, we plan to develop algorithms for parallel forming layers, algorithms for hierarchy of subdomains, and algorithms for partitioning multiply-connected unstructured meshes into connected subdomains.

Acknowledgments. This work is supported by the Russian Foundation for Basic Research (grants 14–01–00055-a, 16–01–00129-a).

References

1. Hendrickson, B., Devine, K.: Dynamic load balancing in computational mechanics. Comput. Methods Appl. Mech. Eng. **184**, 485–500 (2000)
2. Kopysov, S.P., Novikov, A.K.: Parallel algorithms of adaptive refinement and partitioning of unstructured grids. Math. Models Comput. Simul. **14**(9), 91–96 (2002)
3. Berger, M.J., Bokhari, S.H.: A partitioning strategy for nonuniform problems on multiprocessors. IEEE Trans. Comput. **36**(5), 570–580 (1987)
4. Farhat, C., Lesoinne, M.: Automatic partitioning of unstructured meshes for the parallel solution of problems in computational mechanics. Int. J. Numer. Methods Eng. **5**(36), 745–764 (1993)

5. Pilkington, J., Baden, S.: Partitioning with space-filling curves. Technical report CS94-349, Deparment of Computer Science and Engineering, University of California, San Diego, p. 49 (1994)
6. Fortmeier, O., Bucker, H.M., Auer, B.O., Bisseling, R.H.: A new metric enabling an exact hypergraph model for the communication volume in distributed-memory parallel applications. Parallel Comput. **39**, 319–335 (2013)
7. Karypis, G.: METIS: a software package for partitioning unstructured graphs, partitioning meshes, and computing fill-reducing orderings of sparse matrices, Version 5.1, p. 35. University of Minnesota, Minneapolis (2013)
8. Hendrickson, B., Leland, R.: Multidimensional spectral load balancing. Technical report SAND93-0074. Sandia National Labs, Albuquerque, pp. 1–13 (1993)
9. Golovchenko, E.N., Yakobovskiy, M.V.: Parallel partitioning tool GridSpiderPar for large mesh decomposition. In: Russian Supercomputing Days: Proceedings of the International Conference, Moscow, Russia, 28–29 September 2015, pp. 303–315. Moscow State University, Moscow (2015)
10. Kopysov, S.P., Kuz'min, I.M., Nedozhogin, N.S., Novikov, A.K., Rychkov, V.N., Sagdeeva, Y.A., Tonkov, L.E.: Parallel implementation of a finite-element algorithms on a graphics accelerator in the software package FEStudio. Comput. Res. Model. **6**(1), 79–97 (2014)
11. Fuhry, M., Giuliani, A., Krivodonova, L.: Discontinuous Galerkin methods on graphics processing units for nonlinear hyperbolic conservation laws. Int. J. Numer. Methods Fluids **76**(12), 982–1003 (2014)
12. Saad, Y.: Iterative Methods for Sparse Linear Systems. SIAM, Philadelphia (2003)
13. Pirova, A., Meerov, I.B., Kozinov, E.A., Lebedev, S.A.: A parallel multilevel nested dissection algorithm for shared-memory computing systems. Numer. Methods Program. **16**, 407–420 (2015)
14. Komatitsch, D., Micha, D., Erlebacher, G.: Porting a high-order finite-element earthquake modeling application to NVIDIA graphics cards using CUDA. J. Parallel Distrib. Comput. **69**(5), 451–460 (2009)
15. Kopysov, S., Kuzmin, I., Nedozhogin, N., Novikov, A., Sagdeeva, Y.: Hybrid multi-GPU solver based on schur complement method. In: Malyshkin, V. (ed.) PaCT 2013. LNCS, vol. 7979, pp. 65–79. Springer, Heidelberg (2013). doi:10.1007/978-3-642-39958-9_6

Multilevel Parallelization: Grid Methods for Solving Direct and Inverse Problems

Sofya S. Titarenko[1]([✉]), Igor M. Kulikov[2,3], Igor G.Chernykh[2],
Maxim A. Shishlenin[2,3,4], Olga I. Krivorot'ko[2,3], Dmitry A. Voronov[2,3],
and Mark Hildyard[1]

[1] School of Earth and Environment, University of Leeds, Leeds LS2 9JT, UK
S.Titarenko@leeds.ac.uk
[2] Institute of Computational Mathematics and Mathematical Geophysics,
SB RAS, Novosibirsk 630090, Russia
[3] Novosibirsk State University, Novosibirsk 630090, Russia
[4] Sobolev Institute of Mathematics, SB RAS, Novosibirsk 630090, Russia

Abstract. In this paper we present grid methods which we have developed for solving direct and inverse problems, and their realization with different levels of optimization. We have focused on solving systems of hyperbolic equations using finite difference and finite volume numerical methods on multicore architectures. Several levels of parallelism have been applied: geometric decomposition of the calculative domain, workload distribution over threads within OpenMP directives, and vectorization. The run-time efficiency of these methods has been investigated. These developments have been tested using the astrophysics code AstroPhi on a hybrid cluster Polytechnic RSC PetaStream (consisting of Intel Xeon Phi accelerators) and a geophysics (seismic wave) code on an Intel Core i7-3930K multicore processor. We present the results of the calculations and study MPI run-time energy efficiency.

Keywords: High performance computing · Intel Xeon Phi accelerators · Grid-based numerical methods

1 Introduction

Numerical methods have become very powerful tools for modeling problems in physics and engineering. Many such problems can be described as a set of hyperbolic equations. In the last decade, a large number of numerical methods have been developed and improved, with finite difference and finite volume methods being almost the most popular. As models become more complex and often require high accuracies of calculation, the use of modern accelerators is more desirable. When moving towards exascale computing, energy consumption increases dramatically and the run-time energy efficiency of calculations becomes very important. In the recent past, geometrical decomposition of the solution domain of a problem (through message passing interface, MPI) was the only tool in parallelization. Since the release of multicore processors (e.g.

© Springer International Publishing AG 2016
V. Voevodin and S. Sobolev (Eds.): RuSCDays 2016, CCIS 687, pp. 118–131, 2016.
DOI: 10.1007/978-3-319-55669-7_10

Graphics Processing Units and Xeon Phi processors), the combination of geometrical decomposition with multithread parallelization and vectorization of the calculations has become increasingly important.

Direct and inverse problems in geophysics are often impossible to solve analytically and hence numerical solution is the only option. One important example is forward modeling of wave propagation through an elastic medium. This problem was first solved numerically using a finite difference scheme by Alterman in 1968 [2]. Later this method was applied to generate synthetic seismograms by Kelly in 1976 [13]. A similar approach has been used to generate sound fields in acoustic problems [22, 24]. Solution of the direct wave propagation problem is widely used in full waveform inversion problems where a good initial guess is extremely important. This problem demands large computing resources and time and hence more and more scientists have optimized their codes using APIs, GPU and MPI parallelization. Examples of parallelizing large scale geophysical problems can be found in [3, 5, 7, 20, 23, 26].

When applying finite difference schemes to a problem it is necessary to calculate a derivative on a stencil type structure. Unfortunately, it is impossible to apply standard automatic vectorization techniques to a stencil type loop. In this work we present a method of memory rearrangement which allows vectorization with high level instructions only. This method is universal to any stencil type structure and its application considerably decreases the calculative time. Moreover, a derivative order has very little effect on CPU time. This allows a considerable increase in the accuracy of a scheme, without changing the CPU time. In this paper we discuss issues of using the proposed method together with OpenMP multithreading. We demonstrate the efficiency of the method on wave propagation through an elastic medium.

Another example of a complex problem which requires parallelization is the modeling of magnetic fields in astrophysics. Magnetic fields play a key role in the formation of astrophysical objects. Taking magnetic fields into account when modeling the evolution of interstellar turbulence makes a considerable difference to the results (see [18]). In recent years, modeling magnetohydrodynamic turbulence problems has helped our understanding of sub-alpha currents [19] and the rate of star formation [4]. A comparison of different codes for subsonic turbulence is presented in [14]. Classical methods for simulation of magnetohydrodynamic turbulence such as adaptive mesh refinement (AMR) and smoothed-particle hydrodynamics are still widely used, but in recent years an impressive range of new methods have been proposed (a good review of these methods can be found in [15–17]).

The inverse coefficient problem for a system of 2D hyperbolic equations has been studied in [1]. In this paper the acoustic tomography problem was reformulated as an inverse coefficient problem for a system of first order hyperbolic equations (system of acoustic equations). To solve the inverse problem the gradient method to optimize an objective functional was chosen. This method is widely used in inverse and ill-posed problem theory [8–12]. The main idea of the method is to solve direct and conjugate problems at every time step. This means a numerical method to solve the direct problem needs to be well optimized. Here we present a method of optimization which proved to be very efficient.

In the first two sections we discuss various levels of optimization for the above astrophysics problem. The third section explains difficulties of automatic vectorization when applied to finite difference schemes. A method is proposed which overcomes the difficulties and considerably improves performance. The fourth section discusses the importance of run-time energy efficiency of calculations and demonstrates impressive results for the AstroPhi code. The fifth section presents results from numerical experiments.

2 Geometric Decomposition Pattern

The use of a uniform mesh gives us a possibility to apply a generic Cartesian topology for decomposition of the calculative domain. This approach leads to potentially infinite scalability of the problem. As shown in [17], the AstroPhi code implements multilevel one-dimensional geometric decomposition of the calculative domain. The first coordinate corresponds to the MPI level of parallelization. Every MPI thread sends tasks to OpenMP threads, optimized for MIC architectures. This type of topology is related to the topology and architecture of the hybrid cluster RSC PetaSteam, which has been used for the numerical simulations.

Various levels of AstroPhi code scalability have been tested on Intel Xeon Phi 5120 D accelerators. A grid $512p \times 256 \times 256$ has been used (where p is a number of accelerators). Every accelerator has 4 logical cores. The calculative domain is divided into data chunks of equal size, then the chunks are sent to the accelerators. To study the scalability we have estimated the total run-time (in seconds) for different numbers of Intel Xeon Phi accelerators. At every time step a certain number of processes has to be completed. We calculate the total run-time as the sum of the run-times of all these processes. The scalability has been calculated according to the formula

$$T = \frac{\text{Total}_1}{\text{Total}_p}, \tag{1}$$

where Total_1 and Total_p are a run-time and a calculation time on a single processor respectively, and the problem runs on p processors. The results are presented in Table 1. From the table it is clear to see we have achieved an efficiency of 73% on 256 Intel Xeon Phi 5120 D processors.

3 Multicore Threading on Intel Xeon Phi Accelerators

Parallelization of the AstroPhi code on Intel Xeon Phi accelerators has been achieved through a standard technique:

Table 1. Scalability T of the AstroPhi code on the hybrid cluster RSC PetaStream. Time is in seconds.

MIC	Total (SPb)	Scalability (SPb)
1	55.5742	1.0000
8	56.3752	0.9857
64	64.1803	0.8659
128	68.6065	0.8101
256	76.1687	0.7296

1. decomposition of the calculative domain;
2. workload distribution amongst the available threads;

For this problem we applied decomposition to a 512×256^2 grid on a single Xeon Phi accelerator. To calculate the acceleration we have measured the calculation time of each function of the numerical simulation and then calculated its summation on one thread and on p threads. The acceleration has been calculated according to the formula

$$P = \frac{\text{Total}_1}{\text{Total}_K},\tag{2}$$

where Total_1 is the calculative time for one logical core, Total_K is the calculative time for K logical cores. The results of testing the AstroPhi code on hybrid cluster PetaStream (SPb) are presented in the Table 2.

Table 2. Acceleration P with increasing numbers of logical cores (on a single Xeon Phi accelerator). The code has been tested on a hybrid cluster RSC PetaStream (SPb). Time is in seconds.

Threads	Total (SPb)	P (SPb)
1	219.7956	1.0000
8	27.7089	7.9323
32	7.9673	27.5872
128	2.6271	83.6647
240	2.5905	84.8467

4   Vectorization

The first processor supporting a SIMD (Single Instruction Multiple Data) instruction set was designed by Intel in 1999. This Streaming SIMD (SSE) extension accelerates the calculation due to the use of larger registers. The first SIMD registers were designed to hold four 32-bit floats/two 64-bit doubles (128-bit registers). This means that 4 floats/2 doubles can be uploaded into a register and the arithmetic and logic routine can be applied to a *vector* instead of a *single* value.

This simple idea has become very popular and nowadays almost all modern architectures support SIMD operations. The capacity of SIMD registers has also been considerably increased. For example the Sandy Bridge microarchitecture includes the AVX extension with 256-bit SIMD registers and the Skylake microarchitecture includes the AVX-512 (Xeon models only) extension which operates with 512-bit SIMD registers.

To be able to take advantage of automatic vectorization either an optimization flag (/O2 and higher for Intel machines) needs to be switched on, or a microarchitecture specific flag (ex. /QxSSE4.1, /QxAVX for Intel) should be chosen. It is important to note that an automatic vectorization routine cannot be applied to any loop. The memory in the loop needs to be aligned and the vector length must be divisible by 4 (8 or 16 depending on the size of a SIMD register and the bit size of values operated with). It is important that there shouldn't be any read-write memory conflict, for example a cycle of the type

```
for(int i = 0; i < N; i++){
    a[i] = b[i] + c[i];
}
```

is automatically vectorizable (assuming the memory is aligned and N is divisible by 4). However a loop of the type

```
__assumed_aligned(a, 32);
__assumed_aligned(b, 32);
#pragma simd
for(int i = 0; i < N; i++){
    a[i] = b[i] + b[i + 2];
}
```

cannot be vectorized by high level instructions.

When calculating the derivative on a stencil structure we get a loop which is not automatically vectorizable. However, the situation changes if we *rearrange the memory* in the way described in Fig. 1. In this case the compiler will be uploading and applying arithmetic and logical operations to all 4 values simultaneously and the necessary acceleration will be achieved.

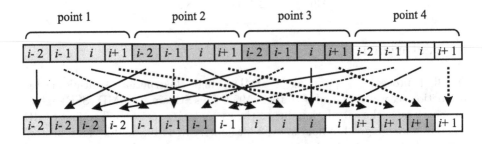

Fig. 1. Memory reorganization for high level vectorization.

In the case of larger than 128-bit registers or smaller bit size values it is always possible to rearrange the data in a such way the compiler would work with a vector of 8, 16, etc. In this case the values we use for calculation of one derivative i will be located with the step spacing 8, 16 etc. from each other.

A very important property of such rearrangement is that the speed of derivative calculation is almost unaffected by the order of the derivative. This property is especially important for problems where it is problematic to achieve the required degree of accuracy due to long run times.

We have tested this property by calculating 2nd, 4th, 6th, 8th and 10th order derivatives on the grid 8192×8192 (for 500 cycles). Note that the order of the derivative increases the size of the stencil. Table 3 shows the calculation time *without* memory rearrangement and automatic optimization (flag /0d for Intel), *without* memory rearrangement and *with* aggressive automatic vectorization (flag /03 for Intel) and *with* memory rearrangement and *with* aggressive automatic vectorization. The table shows that in the initial case, run time is 74% greater for the 10th order than the 2nd order. When automatic vectorization is used without memory rearrangement, run time is much shorter but the difference between 10th order and second order becomes 200%. Finally, when vectorization is combined with memory rearrangement, run time becomes almost independent of the order of the derivative. The acceleration for 10th order is ≈ 11.57 times, compared with ≈ 6.63 times for 2nd order.

Table 3. Applying different levels of optimization (see text for details) for derivatives of a high order. Run time is in seconds.

Order	/0d	/03, $[1 \times 1]$	/03, $[4 \times 1]$
2	311.601	63.978	40.987
4	378.303	96.804	41.502
6	433.277	122.147	42.687
8	485.380	158.753	41.350
10	543.615	192.288	46.996

5   Study of Run-Time Energy Efficiency

Nowadays, run-time energy efficiency is mostly used for commercial projects working with big data problems. However, the idea of exascale computing is becoming more and more popular and petascale computers are expected to be widely used in the near future. Exascale computing is capable of more then 10^{18} calculations per second. The first petascale computers were released in 2008 and are considered to be very promising and powerful tools for solving big data problems, for example modeling for climate, in geophysics and astrophysics. Running such a computer efficiently is of great importance. If only 10 MW of energy for running an exascale supercomputer were used inefficiently it could cancel out all the advantage of using it. Overall the definition of run-time energy efficiency includes about 20 parameters, most of them related to run time efficiency. In this work we assume the code to be efficient if

1. CPU cores and CPUs are used in the most efficient way;
2. the data exchange between the CPU cores and CPUs is minimized;
3. the code has good balance.

By minimizing data exchange between CPU cores/CPUs we reduce the waiting time for a CPU/CPU cluster (the time while the CPU/CPU cluster doesn't work, awaiting the completion of all necessary processes). Good balance allows tasks to be distributed in between cores and accelerators evenly. By applying these ideas to the AstroPhi code we have reduced the time spent on data exchange by MPI instructions to 7–8% of the total run-time. The level of imbalance has been reduced to no more then 2–3% between all the threads. This helped us to achieve 72% efficiency (scalability in the "weak" sense) in parallelizing on a 256 Intel Xeon Phi accelerator (more than 50K cores). Modern accelerators help to achieve the maximum run-time efficiency by multithreading and vectorization. By applying vectorization to the AstroPhi code we have achieved 6.5 times acceleration and by run-time energy efficiency we have approached the efficiency of libraries like MAGMA MIC [6].

6   Results

6.1   Modeling of Wave Propagation Through an Elastic Medium

To model wave propagation through two-dimensional elastic media the second order wave equation is often reduced to a system of first order partial differential equations

$$
\begin{cases}
\rho \frac{\partial u}{\partial t} = \frac{\partial \sigma_{11}}{\partial x_1} + \frac{\partial \sigma_{12}}{\partial x_2}, \quad \rho \frac{\partial v}{\partial t} = \frac{\partial \sigma_{22}}{\partial x_2} + \frac{\partial \sigma_{12}}{\partial x_1}, \\
\frac{\partial \sigma_{11}}{\partial t} = \left(\lambda + 2\mu \right) \frac{\partial u}{\partial x_1} + \lambda \frac{\partial v}{\partial x_2}, \quad \frac{\partial \sigma_{12}}{\partial t} = \mu \frac{\partial u}{\partial x_2} + \mu \frac{\partial v}{\partial x_1}, \\
\frac{\partial \sigma_{22}}{\partial t} = \lambda \frac{\partial u}{\partial x_1} + \left(\lambda + 2\mu \right) \frac{\partial v}{\partial x_2},
\end{cases}
\tag{3}
$$

where (u, v) is the velocity vector, σ_{ij} is the stress tensor, and λ and μ are Lamé parameters.

It is also usual to move from an ordinary grid to a *staggered grid*. The method was first proposed in [25] to solve elastic wave propagation problems and has been proved to have better stability and dispersion behavior for 4th order accuracy schemes [21].

Let us define

$$
\begin{aligned}
\delta u &\equiv u^{i-2} - 27u^{i-1} + 27u^i - u^{i+1}, \\
\delta v &\equiv v^{j-2} - 27v^{j-1} + 27v^j - v^{j+1}, \\
\delta \sigma_{11} &\equiv \sigma_{11}^{i-1} - 27\sigma_{11}^i + 27\sigma_{11}^{i+1} - \sigma_{11}^{i+2}, \\
\delta \sigma_{22} &\equiv \sigma_{11}^{j-1} - 27\sigma_{11}^j + 27\sigma_{11}^{j+1} - \sigma_{11}^{j+2}, \\
\delta u^+ &\equiv u^{j-1} - 27u^j + 27u^{j+1} - u^{j+2}, \\
\delta v^+ &\equiv v^{i-1} - 27v^i + 27v^{i+1} - v^{i+2}.
\end{aligned}
\tag{4}
$$

Then according to Finite Difference rules the new expression for σ_{11}, σ_{22}, σ_{12} and velocities u, v for every time step can be found as

$$\sigma_{11}^{t} = \sigma_{11}^{t-1} + \frac{(\lambda+2\mu)\Delta t}{24\Delta x_1}\delta u + \frac{\mu\Delta t}{24\Delta x_2}\delta v,$$

$$\sigma_{22}^{t} = \sigma_{22}^{t-1} + \frac{\mu\Delta t}{24\Delta x_1}\delta u + \frac{(\lambda+2\mu)\Delta t}{24\Delta x_2}\delta v,$$

$$\sigma_{12}^{t} = \sigma_{12}^{t-1} + \frac{\Delta t\mu}{24\Delta x_2}\delta u^+ + \frac{\Delta t\mu}{24\Delta x_1}\delta v^+, \tag{5}$$

$$u^{t+1/2} = u^{t-1/2} + \frac{\Delta t}{24\rho\Delta x_1}\delta\sigma_{11} + \frac{\Delta t}{24\rho\Delta x_2}\left(\sigma_{12}^{j-2} - 27\sigma_{12}^{j-1} + 27\sigma_{12}^{j} - \sigma_{12}^{j+1}\right),$$

$$v^{t+1/2} = v^{t-1/2}\frac{\Delta t}{24\rho\Delta x_1}\left(\sigma_{12}^{i-2} - 27\sigma_{12}^{i-1} + 27\sigma_{12}^{i} - \sigma_{12}^{i+1}\right) + \frac{\Delta t}{24\rho\Delta x_2}\delta\sigma_{22}.$$

Figure 2 shows the general scheme for vectorization. It is clear to see that application of boundary conditions for the problem will lead to divergence. This happens because the boundary conditions will be applied to the internal points of the problem. To eliminate this bottleneck we introduce *virtual* blocks. The boundary conditions will be applied to virtual blocks instead and at every time step we have to copy the data from internal points to the virtual ones to achieve convergence.

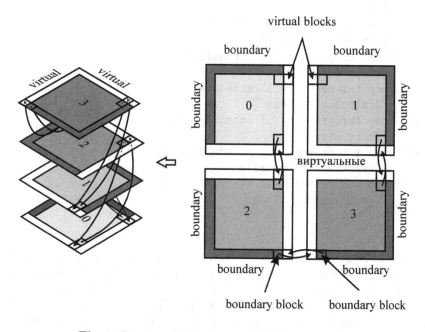

Fig. 2. Copying of data from real blocks to virtual.

Table 4 shows the acceleration of the problem for different sizes. If the flag /O1 is switched on the compiler applies automatic optimization, but not vectorization. Flag /O2 and higher enables automatic vectorization.

Table 4. Comparison of run time for problems of different sizes ($N \times N$) without optimization and with flags /O1 and /O3. The problems have been calculated on one thread. Time is in seconds.

N	/Od	/O1	/O3	Acceleration
$32 \times 32 \times 2$	216.761	54.143	26.133	8.295
$64 \times 32 \times 2$	974.981	186.679	81.356	11.984
$128 \times 32 \times 2$	3789.820	705.704	309.508	12.245

The problem can easily be parallelized on available CPU cores by applying OpenMP directives. The whole domain is divided into blocks. Each block has so-called *buffer* points. At every time step the values from the *buffer* points are updated by copying from the internal grid points of corresponding blocks. The blocks are run in a random order. For our problem we have found the following OpenMP structure gives the best acceleration

```
#pragma omp parallel
{
    for (int j = 0; j < numberOfSteps; j++){
        int i;
#pragma omp for private (i) schedule(auto)
        //calculate velocity
#pragma omp for private (i) schedule(auto)
        //copy velocities
#pragma omp for private (i) schedule(auto)
        //calculate stress
#pragma omp for private (i) schedule(auto)
        //copy stress
#pragma omp for private (i) schedule(auto)
        //calculate boundaries
#pragma omp for private (i) schedule(auto)
        //copy virtual blocks
    }
}
```

It should be mentioned, that vectorization blocks can be arranged in a different order and the order has a small influence on acceleration. For this problem we studied $[4 \times 1]$, $[1 \times 4]$, $[2 \times 2]$ structures. They are presented in Fig. 3.

Experiments and results from VTune Intel Amplifier profiling proved the best vectorization structure to be $[4 \times 1]$. This structure is used in Table 5 for OpenMP parallelization.

6.2   Modeling of Magneto-Hydrodynamics Turbulence Evolution

This numerical model is based on coupling of equations for multidimensional magneto-gas-dynamics, the ordinary differential equation for the evolution of the concentration of ionized hydrogen, and a special form for external force. External force is found from the mass conservation law and Poisson equation.

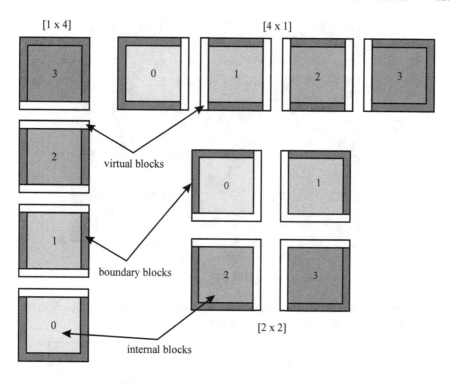

Fig. 3. Types of vectorization structure.

Table 5. Parallelization of seismic wave problem through OpenMP API. The size of the problem is [4096 × 4096], the vectorization structure is [4 × 1]. In all cases the most aggressive automatic vectorization has been applied. Time is in seconds.

N threads	Run time	Acceleration
1	82.300	1.000
4	28.157	2.923
6	25.286	3.255

Its time derivative can be described by the Cauchy–Kovalevskaya equation. By using this mathematical model it becomes possible to formulate a generalized parallelization calculation method [16], which is based on a combination of an operator-splitting method, Godunov method, and a piecewise-parabolic approximation on a regular grid cell. Figure 4 shows the result of the numerical simulations described above. The figure shows the high density area of a "palm tree" shape, which resembles the nebular NGC 6188. Figure 5 shows the correlation of $M \sim n^2$ (white line) and most of a nebular cloud $n > 10\,\mathrm{m}^{-3}$ is in the super-Alfvenic speed area. Contours of the cosine of the colinear angle between velocity vector and magnetic field have a saddle shape (see Fig. 6). This means that compression occurs along magnetic field lines.

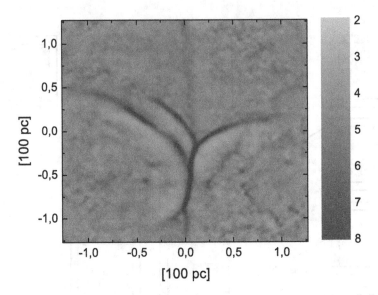

Fig. 4. Numerical simulation for magneto-gas-dynamics. Gas concentration is in cm^{-3} at a time $t = 15$ million years (back from now).

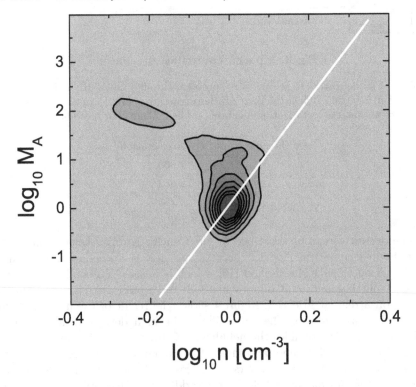

Fig. 5. Dependence of Alfven velocity (M_A) on gas density (n).

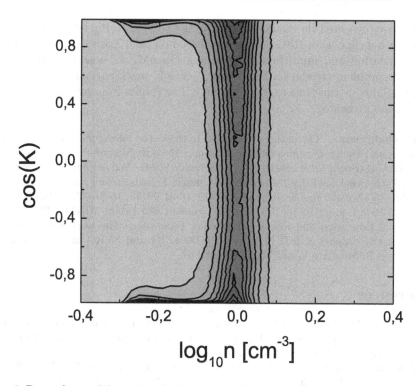

Fig. 6. Dependence of the cosine of colinear angle between velocity vector and magnetic field from gas density.

7   Conclusions

In this paper we have demonstrated the following.

1. It is important to apply all levels of parallelization to big data problems: vectorization, multithread parallelization within shared memory API (OpenMP directives in the test cases) and clusters.
2. Run-time energy efficiency is very important in parallelization as demonstrated by application to modeling of magnetic fields in astrophysics problems.
3. It is impossible to apply high level vectorization methodology to a stencil type loop for a standard memory structure. We have presented a new method of memory rearrangement which overcomes this bottleneck and allows automatic vectorization to take place. In application to finite difference schemes we have demonstrated the method to be particular efficient when using high order derivatives. We proved that if the suggested memory organization is used, the run-time for derivative calculations is almost independent of its order (which is not the case for a standard memory structure!).

4. We have presented the results of astrophysics code AstroPhi efficiency, (tested on the hybrid cluster RSC PetaStream based on Intel Xeon Phi accelerators), vectorization and multithreading withing OpenMP of seismic wave propagation problem (tested on Intel Core i7-3930K machine) and the study on AstroPhi code run-time energy efficiency. The results from numerical simulations are presented.

Acknowledgments. The authors would like to thank the colleagues from Intel (Nikolai Mester and Dmitri Petunin) and RSC Group (Alexader Moscowski, Pavel Lavrenko and Boris Gagarinov) for access to RSC PetaStream cluster and helpful instructions for its use. Authors acknowledge the support of Russian Foundation for Basic Research 14-01-00208, 15-31-20150 mol-a-ved, 15-01-00508, 16-01-00755, 16-31-00382, 16-29-15120 and 16-31-00189, grant of the President of Russian Federation MK-1445.2017.9 and Ministry of Education and Science of Russian Federation. The work has been also done with the support of RCUK grant NE/L000423/1 and NERC and Environment Agency and Radioactive Waste Management Ltd.

References

1. Kulikov, I.M., Novikov, N.S., Shishlenin, M.A.: Mathematical modeling of ultrasound wave propagation in 2D medium: direct and inverse problem. Sib. Electron. Math. Rep. **12**, 219–228 (2015). (in Russian)
2. Alterman, Z., Karal, F.C.: Propagation of elastic waves in layered media by finite-difference methods. Bull. Seismol. Soc. Am. **58**, 367–398 (1968)
3. Aochi, H., Dupros, F.: MPI-OpenMP hybrid simulations using boundary integral equation and finite difference methods for earthquake dynamics and wave propagation: application to the 2007 Niigata Chuetsu-Oki earthquake (mw6.6). Procedia Comput. Sci. **4**, 1496–1505 (2011). doi:10.1016/j.procs.2011.04.162
4. Federrath, C., Klessen, R.S.: The star formation rate of turbulent magnetized clouds: comparing theory, simulations, and observations. Astrophys. J. **761**(2), 156 (2012). doi:10.1088/0004-637X/761/2/156
5. Guyau, A., Cupillard, P., Kutsenko, A.: Accelerating a finite-difference time-domain code for the acoustic wave propagation. In: 35th Gocad Meeting – 2015 RING Meeting. ASGA (2015)
6. Haidar, A., Dongarra, J., Kabir, K., Gates, M., Luszczek, P., Tomov, S., Jia, Y.: HPC programming on Intel many-integrated-core hardware with MAGMA port to Xeon Phi. Sci. Program. **23** (2015). doi:10.3233/SPR-140404
7. Hernández, M., Imbernón, B., Navarro, J.M., García, J.M., Cebrián, J.M., Cecilia, J.M.: Evaluation of the 3-D finite difference implementation of the acoustic diffusion equation model on massively parallel architectures. Comput. Electr. Eng. **46**, 190–201 (2015). doi:10.1016/j.compeleceng.2015.07.001
8. Kabanikhin, S., Shishlenin, M.: Quasi-solution in inverse coefficient problems. J. Inverse Ill-Posed Prob. **16**(7), 705–713 (2008)
9. Kabanikhin, S., Shishlenin, M.: About the usage of the a priori information in coefficient inverse problems for hyperbolic equations. Proc. IMM UrB RAS **18**(1), 147–164 (2012)
10. Kabanikhin, S.I., Gasimov, Y.S., Nurseitov, D.B., Shishlenin, M.A., Sholpanbaev, B.B., Kasenov, S.: Regularization of the continuation problem for elliptic equations. J. Inverse Ill-Posed Probl. **21**(6), 871–884 (2013)

11. Kabanikhin, S.I., Nurseitov, D.B., Shishlenin, M.A., Sholpanbaev, B.B.: Inverse problems for the ground penetrating radar. J. Inverse Ill-Posed Probl. **21**(6), 885–892 (2013)
12. Kabanikhin, S.I., Shishlenin, M.A., Nurseitov, D.B., Nurseitova, A.T., Kasenov, S.E.: Comparative analysis of methods for regularizing an initial boundary value problem for the Helmholtz equation. J. Appl. Math. **2014** (2014). doi:10.1155/2014/786326
13. Kelly, K.R., Ward, R.W., Tritel, S., Alford, R.M.: Synthetic seismograms: a finite-difference approach. Geophysics **41**, 2–27 (1976)
14. Kritsuk, A.G., Nordlund, Å., Collins, D., Padoan, P., Norman, M.L., Abel, T., Banerjee, R., Federrath, C., Flock, M., Lee, D., Li, P.S., Muller, W.C., Teyssier, R., Ustyugov, S.D., Vogel, C., Xu, H.: Comparing numerical methods for isothermal magnetized supersonic turbulence. Astrophys. J. **737**(1), 13 (2011). doi:10.1088/0004-637X/737/1/13
15. Kulikov, I.: GPUPEGAS: a new GPU-accelerated hydrodynamic code for numerical simulations of interacting galaxies. Astrophys. J. Suppl. Ser. **214**(1), 12 (2014). doi:10.1088/0067-0049/214/1/12
16. Kulikov, I., Vorobyov, E.: Using the PPML approach for constructing a low-dissipation, operator-splitting scheme for numerical simulations of hydrodynamic flows. J. Comput. Phys. **317**, 318–346 (2016). doi:10.1016/j.jcp.2016.04.057
17. Kulikov, I.M., Chernykh, I.G., Snytnikov, A.V., Glinskiy, B.M., Tutukov, A.V.: AstroPhi: a code for complex simulation of the dynamics of astrophysical objects using hybrid supercomputers. Comput. Phys. Commun. **186**, 71–80 (2015). doi:10.1016/j.cpc.2014.09.004
18. Mason, J., Perez, J.C., Cattaneo, F., Boldyrev, S.: Extended scaling laws in numerical simulations of magnetohydrodynamic turbulence. Astrophys. J. Lett. **735**(2), L26 (2011). doi:10.1088/2041-8205/735/2/L26
19. McKee, C.F., Li, P.S., Klein, R.I.: Sub-Alfvénic non-ideal MHD turbulence simulations with ambipolar diffusion. II. Comparison with observation, clump properties, and scaling to physical units. Astrophys. J. **720**(2), 1612–1634 (2010). doi:10.1088/0004-637X/720/2/1612
20. Michéa, D., Komatitsch, D.: Accelerating a three-dimensional finite-difference wave propagation code using GPU graphics cards. Geophys. J. Int. **182**(1), 389–402 (2010). doi:10.1111/j.1365-246X.2010.04616.x
21. Moczo, P., Kristek, J., Bystrický, E.: Stability and grid dispersion of the P-SV 4th-order staggered-grid finite-difference schemes. Stud. Geophys. Geod. **44**(3), 381–402 (2000). doi:10.1023/A:1022112620994
22. Sakamoto, S., Seimiya, T., Tachibana, H.: Visualization of sound reflection and diffraction using finite difference time domain method. Acoust. Sci. Technol. **23**(1), 34–39 (2002). doi:10.1250/ast.23.34
23. Sheen, D.H., Tuncay, K., Baag, C.E., Ortoleva, P.J.: Parallel impelemntation of a velocity-stress staggered-grid finite-difference method for 2-D poroelastic wave propagation. Comput. Geosci. **32**, 1182–1191 (2006)
24. Siltanen, S., Robinson, P.W., Saarelma, J., Pätynen, J., Tervo, S., Savioja, L., Lokki, T.: Acoustic visualizations using surface mapping. J. Acoust. Soc. Am. **135**(6), EL344–EL349 (2014). doi:10.1121/1.4879670
25. Virieux, J.: P-SV wave propagation in heterogeneous media; velocity-stress finite-difference method. Geophysics **51**(4), 889–901 (1986). doi:10.1190/1.1442147
26. Zhang, Y., Gao, J.: A 3D staggered-grid finite difference scheme for poroelastic wave equation. J. Appl. Geophys. **109**, 281–291 (2014). doi:10.1016/j.jappgeo.2014.08.007

Numerical Model of Shallow Water: The Use of NVIDIA CUDA Graphics Processors

Tatyana Dyakonova[✉], Alexander Khoperskov, and Sergey Khrapov

Volgograd State University, Volgograd, Russia
{dyakonova,khoperskov,khrapov}@volsu.ru

Abstract. In the paper we discuss the main features of the software package for numerical simulations of the surface water dynamics. We consider an approximation of the shallow water equations together with the parallel technologies for NVIDIA CUDA graphics processors. The numerical hydrodynamic code is based on the combined Lagrangian-Euler method (CSPH-TVD). We focused on the features of the parallel implementation of Tesla line of graphics processors: C2070, K20, K40, K80. By using hierarchical grid systems at different spatial scales we increase the efficiency of the computing resources usage and speed up our simulations of a various flooding problems.

Keywords: Numerical simulation · Parallel technology · Graphics processors · Shallow water equations

1 Introduction

Various hydrology problems for the real terrain surface $b(x, y)$ with taking into account important physical factors for large areas can be studied by using modern computational technologies [11]. Believed that control of the hydrological regime of the floodplain landscape during the spring flood on large rivers is one of the most important problem facing the numerical simulation [13]. Solution of this problem requires a very efficient numerical method based on a parallel technology [5]. For example, to make a both ecological and economic management of the Volga-Akhtuba floodplain we have to solve the optimization problem of hydrograph for the specific conditions of each year [14]. We also need to explore the results of different modes of operation of tens hydraulic structures in the floodplain and for a new facilities projects. Each that research requires hundreds of numerical experiments on the basis of a direct hydrodynamic simulation of the shallow water dynamics on the area of 2000×20000 square kilometers.

Our practice of using large supercomputers[1] for a large number of hydrodynamic simulations arises a number of problems related to necessity to do a numerous simulations during the short time period and then following transfer

---

[1] In particular, the one is at Research Computing Center of M.V. Lomonosov Moscow State University [12].

V. Voevodin and S. Sobolev (Eds.): RuSCDays 2016, CCIS 687, pp. 132–145, 2016.
DOI: 10.1007/978-3-319-55669-7_11

of a large massive of data for later processing and analysis. Both performance of calculations and post-processing of the simulation data are an important factors in the usage of such models in practice. An additional problem is the visualization of the calculations, which seems a common difficulty for a very high-performance machines [9]. However we can partly solve these problems in case of using the computing resources such as personal supercomputers based on GPU accelerators. The paper discusses the results of the software package development for the parallel hydrodynamic simulations on the nodes C2070, K20, K40, K80.

2   Mathematical and Numerical Models

2.1   Basic Equations

Numerical simulations are based on the shallow water model (Saint-Venant equations) in the following form:

$$\frac{\partial H}{\partial t} + \frac{\partial H U_x}{\partial x} + \frac{\partial H U_y}{\partial y} = \sigma \, , \tag{1}$$

$$\frac{\partial U_x}{\partial t} + U_x \frac{\partial U_x}{\partial x} + U_y \frac{\partial U_x}{\partial y} = -g \frac{\partial \eta}{\partial x} + F_x + \frac{\sigma}{H}(V_x - U_x) \, , \tag{2}$$

$$\frac{\partial U_y}{\partial t} + U_x \frac{\partial U_y}{\partial x} + U_y \frac{\partial U_y}{\partial y} = -g \frac{\partial \eta}{\partial y} + F_y + \frac{\sigma}{H}(V_y - U_y) \, , \tag{3}$$

where H is the water depth, U_x, U_y are the horizontal components of water velocity vector U, which is averaged along the vertical direction, σ is the surface density of the water sources and drains [m/sec], g is gravitational acceleration, $\eta(x, y, t) = H(x, y, t) + b(x, y)$ is the free water surface level, V_x, V_y are the mean horizontal velocity vector components of water at the source or drain ($V = V_x e_x + V_y e_y$), F_x, F_y are the horizontal components of the external and internal forces ($F = F_x e_x + F_y e_y$) acting the water layer. The total density of the forces can be written as

$$F = F^{fric} + F^{visc} + F^{cor} + F^{wind} \, , \tag{4}$$

where $F^{fric} = -\frac{\lambda}{2} U |U|$ is the force of bottom friction, $\lambda = \frac{2gn_M^2}{H^{4/3}}$ is the value of hydraulic friction, n_M is the phenomenological Manning roughness coefficient, $F^{visc} = \nu(\frac{\partial^2 U_x}{\partial x^2} + \frac{\partial^2 U_y}{\partial y^2})$ is the viscous force of internal friction between layers of flow, ν is the kinematic turbulent viscosity, $F^{cor} = 2[U \times \Omega]$ is the Coriolis force, Ω is the angular velocity of Earth's rotation, $f^{wind} = C_a \frac{\rho_a}{\rho H}(W - U)|W - U|$ is the wind force acting on the water layer, parameter C_a determines the state of the water surface, ρ_a and ρ are the densities of air and water, respectively, W is the wind velocity vector in the horizontal direction.

The model (1)–(3) takes into account the following factors [3]: irregular, inhomogeneous terrain $b(x, y)$; flow interaction with the underlying inhomogeneous topography; Earth's rotation; interaction of water flow with wind; sources, caused by work of hydraulic structures and rainfall; filtration and evaporation; internal friction due to turbulent transport.

In work [5] proposed the so-called "Co-design" approach of the computational models construction. It takes into account the architecture of the supercomputer when one is creating the program code. The increased efficiency is based on the maximization of independent calculations and taking into account the peculiarities of the numerical algorithms for solving the equations of Saint-Venant in inhomogeneous terrain $b(x, y)$.

2.2   Grids System and Matrix of Digital Terrain Elevation

Despite the complex, irregular topography of large rivers (Volga, Akhtuba) riverbeds, numerous channels and small ducts, in the simulations with the unsteady "wet-dry" type boundaries we used an uniform Cartesian grid $\Delta x_i = \Delta y_j = \Delta x = \Delta y$ which let us increase the efficiency of CSPH-TVD method (see Sect. 2.3). Typical grid size ℓ is limited by the depth of the fluid H. Since our problem is strongly non-steady, the fluid depth in computational cells can vary from 10 cm in the flooded areas of land up to 30 m in riverbed of the Volga. Therefore, for the shallow water model we use a large-scale grid for the simulation of deep channel areas and a small-scale grid for the calculations of flooded land areas.

Figure 1 shows a grid structure which allows us to use efficiently computational resources because the fluid flow occurs only in a small number of cells. We simulate the dynamics of the surface water on hierarchical grid system (HGS) sequentially from the smaller to the largest scales with taking into account not smooth source distribution. Zoom-in technology is used only for mission-critical areas [2]. This type of simulation is based on usage of two (or even more) grids with different spatial resolution depending on physical parameters of the water flow. In this case the simulated flows on the small-scale grid affects on the simulation on the larger grid. Zoom-in models without this feedback are less accurate but faster in computational sense.

2.3   The Numerical Hydrodynamic Scheme

Figure 2 shows the calculation scheme of the CSPH-TVD method [1,6–8], where $h = \Delta x = \Delta y$ is the spatial resolution, $0 < K < 1$ is the Courant number which determines the stability of our numerical scheme, $U_p = \max[|U_x^n + \text{sign}(F_x)\sqrt{hF_x}|, |U_y^n + \text{sign}(F_y)\sqrt{hF_y}|]$, $U_s = \max(|U_x^n| + \sqrt{gH^n}, |U_y^n| + \sqrt{gH^n}$. There are four main stages of computations at a given time step t_n. Lagrangian approach is applied to the I and II stages, and the third and fourth stages are based on Euler approach. Both source terms Q and σ (see Fig. 2 and (1)) are determined by external and internal forces respectively, and they have to be calculated firstly. Time step τ_n is also calculated at this stage. Then, at the second stage, we calculate the changes of variables q by using the results we obtained at the first stage. Thus, we find the displacement of the Lagrangian particles Δr inside the cells. At this stage the predictor-corrector scheme gives the second

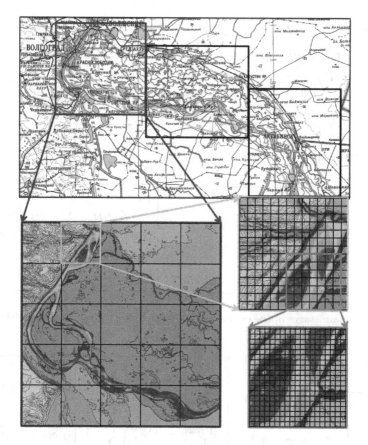

Fig. 1. The hierarchical system of grids for the flooded area between the Volga and Akhtuba Rivers.

order accuracy for time integration. In the third stage, the fluxes of mass and momentum through the boundaries of Euler cells are calculated by using the approximate solution of the Riemann problem. In the last stage we update the values of q on the next time step t_{n+1} and here we also put back the particles to the centers of the grid cells (x_i, y_j).

There are several advantages of the numerical scheme described above, such as the second order of accuracy for the both time and spatial integration, the conservativeness, the well-balanced property and the straightforward simulation of a "water – dry bottom" dynamical boundaries without any regularization [4].

3   Parallel Realization of Numerical Model

CUDA technology was used to parallelize the CSPH-TVD numerical scheme which in turn let us to use efficiently the hierarchical grid system (HGS, see Sect. 2.1). This is due to the fact that HGS blocks are a kind of analogue of

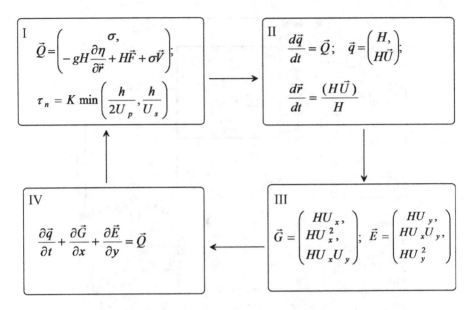

Fig. 2. The main stages of the computational scheme for the solution of shallow water equations.

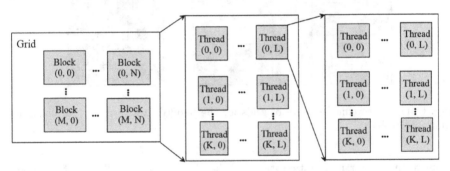

Fig. 3. The hierarchy of threads on the GPU devices adjusted for the dynamic parallelization.

CUDA stream blocks, which provide the execution of CUDA-cores (Fig. 3). Using CUDA dynamic parallelism is a feature of implemented approach that allows detailed calculations of hydrodynamic flows in the small-scale grids with additional threads for the most important spatial zones associated with irregular topography.

The computational algorithm described in Sect. 2 is parallelized by using the hybrid OpenMP-CUDA parallel programming model. Activity diagram of the main stages of the numerical algorithm is shown in Fig. 4. We use the following notations for computing CUDA cores: K1—kernel_Index_block is the determination of water-filled blocks; K2—kernel_forces_predictor calculates the

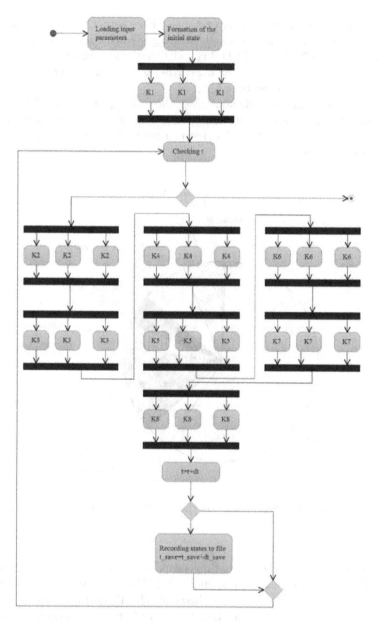

Fig. 4. Activity diagram for the calculation module.

forces at the time t_n on the Lagrangian stage; K3—kernel_dt calculates the time step Δt_{n+1}, depending on the flow parameters on the layer n; K4—kernel_SPH_predictor calculates the new provisions of the particles and the integral characteristics at time $t_{n+1/2}$; K5—kernel_forces_corrector determines the forces on the intermediate time layer $t_{n+1/2}$; K6—kernel_SPH_corrector calculates the positions of the particles and the integrated characteristics for the next time layer t_{n+1}; K7—kernel_TVD_flux calculates the flux physical quantities through the cell boundaries at time t_{n+1}; K8—kernel_Final determines the final hydrodynamic parameters at time t_{n+1}.

The diagram in the Fig. 4 demonstrates the features of CSPH–TVD method. We emphasize that the computational algorithm separation is optimal usage of GPU resources of eight CUDA-cores in case of shallow water flows on the irregular topography.

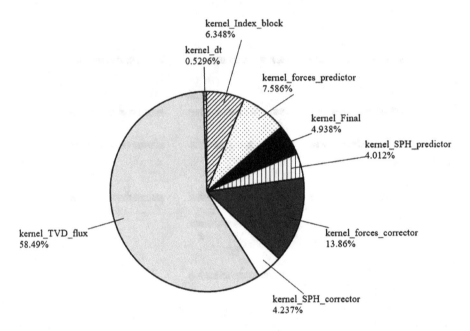

Fig. 5. Contributions of different stages of the CSPH-TVD numerical scheme at a given time step.

Figure. 5 shows the execution time proportions of the main stages of the numerical algorithm for the corresponding CUDA-cores. We spend almost 60% of the time on the calculation of the fluxes for the TVD-stage kernel_TVD_flux (see Fig. 5). This is because at this stage we solve the Riemann problem. Taking into account the calculation of forces kernel_forces_predictor and kernel_forces_corrector the total contribution from TVD-stage increases and it is dominant. We have just over 8% from the Lagrangian stage (sum of kernel_SPH_predictor and kernel_SPH_corrector). Thus, the SPH-stage significantly improves the properties of the numerical scheme, but an additional stage has little effect on the final calculation time.

Fragment of the program (See below) implements the final stage of the calculation (CUDA-core K8) with checking of the presence of liquid in the CUDA-block with size of 16×16 cells. If liquid is absent in the started block (parameter Index_block is zero), the computations are skipped for all threads in the block. Similar approach we apply for other CUDA-cores as well (K2, K4, K5, K6 and K7).

The code fragment for CUDA-cores K8

```
__global__ void kernel_Final(double *H, double2 *HV, double *Ht,
        double2 *HVt, double *Fh, double2 *Fv, int2 *Index_block,
        double tau){
    int ib=blockIdx.x+blockIdx.y*gridDim.x;
    int x = threadIdx.x + blockIdx.x * blockDim.x, y = threadIdx.y +
        blockIdx.y * blockDim.y;
    int ind = x + y * blockDim.x * gridDim.x;
 if(Index_block[ib].x > 0 || Index_block[ib].y > 0){
    double dt_h = tau/dd.hp, Eps=dd.Eps, ht, Flux_h;
    double2 hv=make_double2(0,0);
    ht = Ht[ind]; Flux_h=Fh[ind];
    if( ht>Eps || fabs(Flux_h)>Eps){
        ht = dev_h(ht + dt_h*Flux_h);
        hv.x = dev_hv(HVt[ind].x + dt_h*Fv[ind].x,ht);
        hv.y = dev_hv(HVt[ind].y + dt_h*Fv[ind].y,ht);
    }
    H[ind]=ht; HV[ind]=hv; Ht[ind]=0; HVt[ind]=make_double2(0,0);
 }else {Ht[ind]=0; HVt[ind] = make_double2(0,0);}
}
```

Parameter Index_block is determined in the CUDA-core K1 (See below). The variable Index_block is the structure of type int2 containing two integer fields Index_block.x and Index_block.y. The condition Index_block.x > 0 indicates the presence of water at least in one cell of this CUDA-block. If the condition Index_block.y > 0 is satisfied, there is a water at least in one of the CUDA-block surrounding cells. Since there are liquid fluxes through the CUDA-block boundaries on the Euler stage in the CUDA-core K7, we have to check the availability of water in the boundary cells of the surrounding CUDA-blocks.

The code fragment for CUDA-core K1

```
__global__ void kernel_Index_block(int2 *Index_block, int *Index_Q,
    double *H){
    __shared__ int2 Sij[ithbx*ithby];
    int ind_thb = threadIdx.x + ithbx*threadIdx.y;
    int ib=blockIdx.x+blockIdx.y*gridDim.x;
    int x = threadIdx.x + blockIdx.x*blockDim.x, y = threadIdx.y +
        blockIdx.y*blockDim.y;
    int tid, xx, yy, i, j, si, sj, Ni, Nj, isi, jsj, m;
    double2 D=make_double2(0,0); int2 iD=make_int2(0,0);
    if(threadIdx.x == 0){Ni=1; si=-1;}
    else if(threadIdx.x == ithbx-1){Ni=1; si=1;}
```

```
else {Ni=0; si=0;}
if(threadIdx.y == 0){Nj=1; sj=-1;}
else if(threadIdx.y == ithby-1){Nj=1; sj=1;}
else {Nj=0; sj=0;}
for(i=0; i<=Ni; i++){
  isi = i*si;
  if(x==0 && isi<=-1) xx = x;
  else if(x==dd.Nx-1 && isi>=1) xx = x;
  else xx = x + isi;
  if(i==0) m=Nj; else m=0;
  for(j=0; j<=m; j++){
    jsj = j*sj;
    if(y==0 && jsj<=-1) yy = y;
    else if(y==dd.Ny-1 && jsj>=1) yy = y;
    else yy = y + jsj;
    tid = xx + yy*dd.Nx;
    if(i>0 || j>0)  {D.y += H[tid]; iD.y += Index_Q[tid]; }
    if(i==0 & j==0) {D.x += H[tid]; iD.x += Index_Q[tid]; }
  }
}
if(D.x>dd.Eps || iD.x>0) Sij[ind_thb].x=1;
else Sij[ind_thb].x=0;
if(D.y>dd.Eps || iD.y>0) Sij[ind_thb].y=1;
else Sij[ind_thb].y=0;
__syncthreads();
int k = ithbx*ithby/2;
while(k != 0){
  if(ind_thb < k) {Sij[ind_thb].x += Sij[ind_thb+k].x;
                   Sij[ind_thb].y += Sij[ind_thb+k].y; }
  __syncthreads();
  k /= 2; }
if(ind_thb == 0) {Index_block[ib].x = Sij[0].x;
                  Index_block[ib].y = Sij[0].y; }
}
```

The two-level parallelization scheme (Fig. 6a) is more suitable for hybrid systems type CPU + $n\times$ GPU. Direct Access technology provides the fast data exchange at different GPU. This technology is applicable only for the GPUs, which are connected to the PCI Express buses under the control of one CPU (Fig. 6b).

4 Comparison of the Effectiveness for Different GPU

The Fig. 7a shows the diagram of Software Productivity for four Tesla graphics cards. The capture time executing CUDA kernels on the GPU (or GPU utilization) are quoted in percentages. We used the NVIDIA Parallel Nsight for profiling the program. In the transition to more efficient GPU with a large number of scalar cores we have a decrease in the percentage of GPU utilization and it is related to

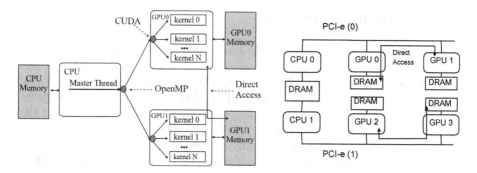

Fig. 6. (a) The two-level scheme of parallelization with OpenMP–CUDA. (b) Architecture 2×CPU+4×GPU.

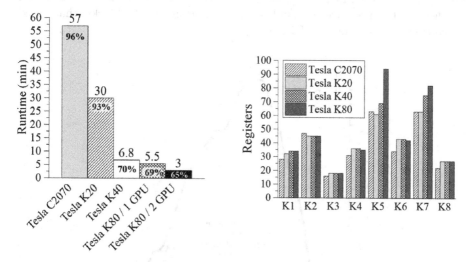

Fig. 7. (a) Time calculation of flooding dynamics for the northern part of the floodplain for a period of 20 h on different GPU using the grid 1024×1024. (b) Distribution of memory registers of GPU-multiprocessors on CUDA-cores for different GPU.

the number of computational cells 1024×1024 (with $\Delta x = \Delta y = 50$ m). GPU utilization increases in the case of $10 - 25$ m spatial resolution. It is important to emphasize that the use of the spatial resolution < 10 m may violate the approximation of the shallow water equations. Thus, the use of personal supercomputers with multi-GPU is the most suitable for the simulation of flooding over an area of about 10^4 km^2 considering the factor of GPU Utilization.

The Fig. 7 b shows a diagram of the distribution of memory registers on a stream for CUDA-cores. We selected the parameters of the program for our graphic accelerators to avoid the spilling of registers.

5   The Simulation Results

We have a good agreement between the results of our numerical simulations and observation data for the dynamics of water level on the gauging stations and for the flooding area in May 2011 (Fig. 8, See details in the [6,7,11]). As an example, consider the problem of emergency water discharge from the dam for the rate of $100000\,m^3/s$. Simulations were performed for the north part of the Volga-Akhtuba floodplain of the area $51200\,m \times 51200\,m$. Breaking wave formed due to the emergency discharge leads to the complete flooding of the studied area of the floodplain for 20 h. The mean flow velocity in the floodplain is $5\,m/s$ and the average depth equals to $6\,m$. The ratio between the water and land in the entire field of modeling is $\sim 35\%$, and its maximum is $\sim 60\%$. For such problems, it is recommended to use the described approach (See Sect. 3) based on check of the presence of liquid in the computing CUDA-blocks, that speed up calculations by a factor of $1.5 - 2$.

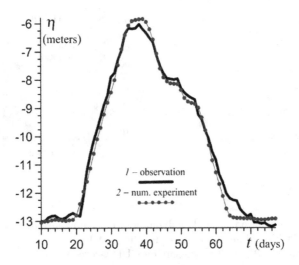

Fig. 8. A comparison of water levels η from observations (line 1) at gauging station "Svetlyj Jar" with numerical simulation result (line 2).

The Fig. 9 shows the results of numerical modeling of flooding territories with taking into account of the zoom-in approach. Calculations were performed for the two grids:
— the global (main) grid which covers the entire area of simulations with the number of cells equal to 1024×1024 ($\Delta x = \Delta y = 50\,m$);
— the local grid for a critical region where we adopted a much higher spatial resolution $\Delta x = \Delta y = 12.5\,m$ (size of calculation domain is 1024×1024, in the vicinity of village or complex terrain).

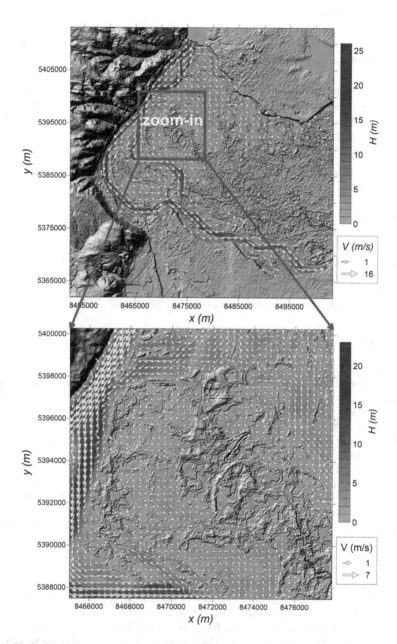

Fig. 9. Hydrological state in the floodplain at time $t = 10\,\mathrm{h}$ using the zoom-in approach. Top frame shows the flow structure on the global grid. ($\Delta x = \Delta y = 50\,\mathrm{m}$), bottom frame shows the flow structure on the small grid ($\Delta x = \Delta y = 12.5\,\mathrm{m}$). The water depth distributions and the velocity field are shown by the color and arrowheads respectively. (Color figure online)

6 Conclusion

We have investigated some features of the parallel implementation of numerical models for the Saint-Venant equations in the case when the flooded area changes in a wide range over the time. For example, the area under the water level may increase by a factor of tens or even hundreds in the period of spring floods for the Volga-Akhtuba floodplain. To improve the efficiency of these calculations we used a hybrid OpenMP-CUDA parallelization approach and developed the method for choice of the CUDA-blocks with the aim to control the presence of liquid in the computational cells. Our parallel implementation reduced the computation time by a factor of $100 - 1200$ for different GPU in comparison to sequential code.

Acknowledgments. The numerical simulations have been performed at the Research Computing Center (Moscow State University). AVK has been supported by the Russian Scientific Foundation (grant 15-02-06204), SSK is thankful to the RFBR (grant 16-07-01037). The simulation results part was developed under support from the RFBR and the Administration of Volgograd region grant 15–45-02655 (TAD). The study is carried out within the framework of government task by the Ministry of Education and Science of the Russian Federation (research work title No. 2.852.2017/PCH).

References

1. Belousov, A.V., Khrapov, S.S.: Gasdynamic modeling on the basis of the Lagrangian and Euler scheme LES-ASG. J. Vestnik Volgogradskogo gosudarstvennogo universiteta. Seriya 1: Matematika. Fizika (Sci. J. VolSU. Math. Phys.) **5**(30), 36–51 (2015)
2. Bonoli, S., Mayer, L., Kazantzidis, S., Madau, P., Bellovary, J., Governato, F.: Black hole starvation and bulge evolution in a Milky Way-like galaxy. J. Mon. Not. R. Astron. Soc. **459**, 2603–2617 (2016)
3. Dyakonova, T.A., Pisarev, A.V., Khoperskov, A.V., Khrapov, S.S.:: Mathematical model of surface water dynamics. J. Vestnik Volgogradskogo gosudarstvennogo universiteta. Seriya 1: Matematika. Fizika (Sci. J. VolSU. Math. Phys.) **1**(20), 35–44 (2014)
4. Dyakonova, T.A., Khrapov, S.S., Khoperskov, A.V.: The problem of boundary conditions for the shallow water equations. Vestnik Udmurtskogo Universiteta. Mekhanika. Komp'yuternye nauki (Bull. Udmurt Univ. Math. Mech.) **26**(3), 401–417 (2016)
5. Glinskiy, B., Kulikov, I., Snytnikov, A., Romanenko, A., Chernykh, I., Vshivkov, V.: Co-design of parallel numerical methods for plasma physis and astrophysics. J. Supercomput. Front. Innov. **1**, 88–98 (2014)
6. Khrapov, S.S., Khoperskov, A.V., Kuz'min, N.M., Pisarev, A.V., Kobelev, I.A.: A numerical scheme for simulating the dynamics of surface water on the basis of the combined SPH-TVD approach. J. Vychislitel'nye Metody i Programmirovanie (Numer. Methods Program.) **12**(2), 282–297 (2011)
7. Khrapov, S., Pisarev, A., Kobelev, I., Zhumaliev, A., Agafonnikova, E., Losev, A., Khoperskov, A.: The numerical simulation of shallow water: estimation of the roughness coefficient on the flood stage. J. Adv. Mech. Eng. **5**(2013), 1–11 (2013). doi:10.1155/2013/787016

8. Kuz'min, N.M., Belousov, A.V., Shushkevich, T.S., Khrapov, S.S.: Numerical scheme CSPH-TVD: investigation of influence slope limiters. J. Vestnik Volgogradskogo Gosudarstvennogo Universiteta. Seriya 1: Matematika. Fizika (Sci. J. VolSU. Math. Phys.) 1(20), 22–34 (2014)
9. Moreland, K., Larsen, M., Childs, H.: Visualization for exascale: portable performance is critical. J. Supercomput. Front. Innov. 2(3), 67–75 (2015)
10. Pisarev, A.V., Khrapov, S.S., Voronin, A.A., Dyakonova, T.A., Tsyrkova, E.A.: The role of infiltration and evaporation in the flooding dynamics of the Volga-Akhtuba floodplain. J. Vestnik Volgogradskogo Gosudarstvennogo Universiteta. Seriya 1, Matematika. Fizika (Sci. J. VolSU. Math. Phys.) 1(16), 36–41 (2012)
11. Pisarev, A.V., Khrapov, S.S., Agafonnikova, E.O., Khoperskov, A.V.: Numerical model of shallow water dynamics in the channel of the Volga: estimation of roughness. J. Vestnik Udmurtskogo Universiteta. Matematika. Mekhanika. Komp'yuternye nauki (Bull. Udmurt Univ. Math. Mech.) 1, 114–130 (2013)
12. Sadovnichy, V., Tikhonravov, A., Voevodin, V., Opanasenko, V.: Lomonosov: supercomputing at Moscow state university. In: Contemporary High Performance Computing: From Petascale toward Exascale, pp. 283–307. Chapman & Hall/CRC Computational Science, Boca Raton/CRC Press (2013)
13. Voronin, A.A., Eliseeva, M.V., Khrapov, S.S., Pisarev, A.V., Khoperskov, A.V.: The Regimen control task in the eco-economic system Volzhskaya hydroelectric power station - the Volga-Akhtuba floodplain. II. Synthesis of Control System. J. Problemy Upravleniya (Control Sci.) 6, 19–25 (2012)
14. Voronin, A., Vasilchenko, A., Pisareva, M., Pisarev, A., Khoperskov, A., Khrapov, S., Podschipkova, J.: Designing a system for ecological-economical management of the Volga-Akhtuba floodplain on basis of hydrodynamic and geoinformational simulation. J. Upravlenie Bol'shimi Sistemami (Large-Scale Syst. Control) 55, 79–102 (2015)

Parallel Algorithm for Simulation of Fragmentation and Formation of Filamentous Structures in Molecular Clouds

Boris Rybakin[1]([⊠]), Nikolai Smirnov[1], and Valery Goryachev[2]

[1] Department of Gas and Wave Dynamics, Moscow State University,
Moscow, Russia
rybakin@vip.niisi.ru

[2] Department of Mathematics, Tver State Technical University, Tver, Russia
gdv.vdg@yandex.ru

Abstract. The report is devoted to numerical simulation of interaction between the post-shock wave frontal of supernova blast remnants and the gas of two molecular clouds (MC). The dynamical formation of MC structures associated with Kelvin-Helmholtz and Richtmayer-Meshkov instabilities occurring in the cloud and interstellar medium interaction zone is simulated. The MC gas flow evolution is derived from the time dependent equations of mass, momentum, and energy conservation. High resolution computational meshes (more than two billion nodes) were used in parallel computing on multiprocessor hybrid computers. In the model two initially spatially separated clouds with different gas density distribution fields interact with the post-shock medium. The peculiarities of clump and shell fragmentation of clouds and formation of filamentous rudiment structures are considered.

Keywords: Parallel computing · Supersonic turbulence · Shock waves · Supernova blast remnants · Small molecular clouds

1 Introduction

The development of parallel algorithms intended for the solution of astrophysics problems of sizeable spatial and time scales is a necessary step to further exaflops calculations. The paper presents the results of using new computational codes to simulate the interaction of shock waves of different nature and the molecular clouds (MC) unevenly distributed in galaxies. There are many sources of shock wave occurrence. They are formed by supernova explosions, collisions of giant molecular clouds, and so forth [1]. Propagation and collision of shock waves and their interaction with MCs are instrumental in the event chain resulting in a star formation via self-gravity of gas in dynamically varying molecular clouds.

The concept of long-lived MCs has given place to the recognition that in terms of galaxy age standards MCs are low-lifetime objects. Such clouds are quickly generated from the matter of interstellar medium (ISM), a part of which falls under strong shock wave compression in extended filaments and globules that eventually collapse and partially convert into stars [2]. The ISM matter concentrates and condenses in a spatial

V. Voevodin and S. Sobolev (Eds.): RuSCDays 2016, CCIS 687, pp. 146–157, 2016.
DOI: 10.1007/978-3-319-55669-7_12

network of filaments. Suitable conditions for the gravity force commencement are engendering during the formation of the abovementioned structures. MCs are cold and dense enough, so gravity force is a major contributor to the emergence of very dense areas. At sufficiently small scales (about one parsec) the gravity begins dominating over dynamic forces what can be accompanied by supersonic fluctuations of a matter. This gives rise to numerous yet denser gas bulges, specific cores of protostars and their clusters. Simulation of such astrophysical processes based on the gas-dynamic description of turbulent mediums enables us to speak uniformly when describing the form alteration of matter in galaxies [3].

Processing of astrophysical objects images obtained by means of orbiting and Earth-based telescopes shows that the filament structures are widespread [4]. Star formation is believed to occur within their limits. Current models of the MC formation are associated with shock influence on the process of gravitational instability and turbulence [5, 6]. The interstellar medium is heavily fragmented. Even at the largest scales, small molecular clouds (SMCs) are seen as fragments inside the giant molecular clouds (GMCs). These extended regions are caused by turbulence and have similar fractal structures in larger formation [7].

In the paper the authors simulate and analyze the influence of shock waves and generated turbulence on forming primary filamentary structures of evolving SMCs. The gas density contrast $\chi = \frac{\rho_{cl}}{\rho_{ism}}$ varies in the range of 100–5000. The merge of SMCs into GMCs with more powerful hydrodynamic and magnetic effects are required in order to form dense structures. In some ways, observed in calculations the turbulent transformations of denser layers of clouds are initiating the further filamentous structures of galactic scale. The analysis of filamentary structures is possible at primary stages of their formation. This approach is common in a number of studies devoted to the simulation of supersonic turbulent flows when solving the problems of astrophysics of different spatial and temporal scales differing by several orders of magnitude [8].

In recent times, the concept that large-scale magnetic instabilities initiate the formation of MCs, with their lifetime being $\sim 2 \times 10^7$ years, has got widespread use [9]. Ambipolar diffusion of magnetic flux [10] and turbulence transition of MCs [11] are proposed as mechanisms causing the delay of star formation in MCs. The recent studies have shown that only a small proportion of MCs are dense enough to be sources of protostellar cores. Therefore, the observed low rate of star formation is caused by small volumetric fraction of MCs that are sufficiently dense to contain protostellar cores [12].

Current models of the MC formation are associated with the gravitational instability and turbulence influence on these processes [5]. In terms of galaxy, gravity can lead to the highly dynamic collapse of MC matter ($\sim 10^7$ years). However, turbulence processes caused by the gravity, the effects of shock waves from supernova explosions, the collisions with the remnants of supernova shells, the density drops of galaxy spiral arms, the MC collisions with other clouds may accelerate the collapse of MC matters. Viewed in this way, a MC state is an intermediate redistribution of turbulent energy cascade when it enters a MC at small dissipative scales [13].

2   Simulation of Shock Wave Collision with Two MCs

The objects studied in the paper have a wide range of sizes and densities. The inter-stellar medium that fills in our and other galaxies has a density of about 10^{-25} g · cm^{-3}; the density of MCs which are generated by interstellar matter is two or three orders of magnitude greater. After the shock wave interaction with MCs their density further increases by several orders of magnitude. Such a wide range of values for reference distances and densities imposes narrow constraints on numerical mesh size, which are necessary to achieve a sufficient spatial resolution of emerging flows in simulation. Small-size mesh calculations lead to loss of resolution of fast-fluctuating variables and important flow details. As shown in [9], the acceleration and mixing in MCs occur up to five times as fast on a low resolution mesh.

In calculations, we used meshes with resolution from $512 \times 256 \times 256$ to $2048 \times 1024 \times 1024$ nodes. To make hydrodynamic calculations of the shock wave interaction with a MC in adiabatic approximation it is necessary to use at least about 100 mesh nodes per MC initial radius. The chosen resolution affects the visualization of filament formation areas. The calculations made suggest that it is difficult to distinguish the details of formation of vortices and strains of filamentous structures in low reso-lution meshes.

Numerical simulation is conducted for the case of shock wave interaction with two MCs, with Mach number being equal to seven, which are at various compression stages and therefore have different radial density distribution over their volumes. Main gas-dynamic characteristics are put in correspondence with the accepted values of a pioneer work [3] and of recent studies [2, 14–16], where the shock wave interaction with a single cloud was simulated.

2.1   Initial Conditions

Before disturbance the interstellar medium contains substance in a plasma state ($T \sim 10000$ K). Cold molecular clouds ($T \sim 100$ K) of high density are nonuniformly distributed in the medium. Initially the clouds are in dynamic equilibrium with back-ground gas. The model uses the ideal gas law with $\gamma = 5/3$. The density of interstellar medium is based on $\rho_{ism} = 2.15 \cdot 10^{-25}$ g · cm^{-3}, the temperature is $T_{ism} = 10^4$ K, $u_{ism} = 0.0$. The cloud density is $\rho_{cl} = 1.075 \cdot 10^{-22}$ g · cm^{-3}, the temperature is $T_{cl} = 100$ K, $u_{cl} = 0.0$. Gas parameters behind the shock wave are determined by the Rankine–Hugoniot equations. Mach number for a shock wave is equal to $M = 7$, the density is $\rho_{sw} = 8.6 \cdot 10^{-25}$ g · cm^{-3}, the temperature is $T_{sw} = 1.5 \cdot 10^5$ K, the velocity is $u_{sh} = 104$ km/s The shock front thickness is rather large and is $\sim 2-5$ pc which is much more than a cloud radius. Cloud dimensions are of 0.1 pc, the period of time the shock wave propagates the upper cloud diameter $-t_{swoc}$ is about 2000 years ("swoc" – shock wave over cloud).

Figure 1 shows the initial location of clouds. At the initial moment the shock wave impacting the conventional boundary of the upper cloud is approaching to spherical MCs. The assumed layout of lower C_1 and upper C_2 clouds is provided with various laws of density distribution, each one emphasizing the individual character of their gravity fields.

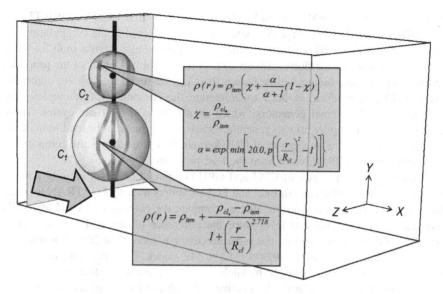

Fig. 1. Computational domain layout

The density profiles shown are determined in accordance with the recommendations of smoothing boundary density distributions in [14, 17] where collisions of shock waves with single MCs were simulated. The initial mass of clouds C_1 and C_2 are assumed as $0.005M_\odot$ and $0.007M_\odot$, respectively (in solar mass fractions). The contrast density ratio in the globular formation center is assumed as $\chi = \frac{\rho_{cl}}{\rho_{ism}} = 500$. The computation domain is a parallelepiped of $3.2 \times 1.6 \times 1.6$ pc. The MC standard radius corresponds to 128 mesh nodes. Boundary conditions for primary variables at the lateral boundaries of the computation region are taken as open.

2.2 Parallelization and Performance Optimization

To compute the MC 3D motion and evolution, a set of Euler equations written in a conservative form are used. To solve the equations a numerical simulation with difference scheme of high resolution like TVD is used. The difference scheme has second-order accuracy and allows high-accuracy computations to be done for the zones close to shock waves and contact discontinuities, and nonphysical oscillations to be prevented. The problem is solved with an Euler mesh. The mathematical statement of the problem and its numerical simulation is detailed in [18].

A computation procedure is used to compute on multiprocessor hybrid computers with OpenMP and CUDA. OpenMP was set up with Intel VTune Amplifier XE. The amplifier can profile applications in the cluster node directly. Lightweight Hotspots analysis is used. Xeon E2630 and Xeon E5 2650 Ivy Bridge processors are set up.

The parallel part of computations is performed with either GPU or Intel Xeon processors. The computations are performed on cluster nodes having 24- or 40-core processors and with OpenMP. When CUDA is applied, NVIDIA GeForce 980 TI with 6 Gb DDR5 is used. 70% of the computation time is devoted to the computations of hydrodynamic flows at 3D mesh cell boundaries. A parallel algorithm is built for faster execution; it computes the flows with all CPU cores available. Details of the proposed algorithm and its fine-tuning with software tools can be found in [16]. An algorithm based on CUDA platform is built to accelerate the computations. It employs the technology of CUF kernel generation which is usually used for basic cycles. These directives tell the compiler how to generate kernels for the basic embedded loop of the host program. Thus, the generation of CUF kernels for cycles allows the computation to be performed in GPU directly from the host program.

Comparing the efficiency of CPU and GPU computations is not entirely correct. The thing is that the efficiency of Intel and PGI compilers used by FORTRAN is pretty much different. For this reason the authors compare the efficiency of CPU and GPU computations at the estimated level since the difference of one and the same code (OpenMP) efficiency on one and the same processor is more than 30%. For not very large mesh dimensions, up to $1024 \times 512 \times 512$ nodes, GPU parallelization gives better results in time than OpenMP parallelization. With the mesh dimension increase GPU computation speed reduces. It is connected with the larger volume of the information transmitted via rather low bus PCI-E.

The computations performed with various computation meshes shows qualitatively identical results but in different degree of detail. The chosen resolution limits the identification of flow instability details and fluctuation magnitudes, the monitoring of the compression of shell boundary layers and the formation of emerging filaments.

2.3   Analysis of MC Forming after Shock Impact

At initial time, when a bow shock wave rounds clouds, a wave is formed behind its front. The wave moves towards the flow and forms a primary disturbance. Because of the sudden gas density change at MC boundaries the Richtmyer–Meshkov instability occurs. Simultaneously the flow velocity slope increases between the cloud boundary layers in the region of mixing with surrounding gas, with the Kelvin–Helmholtz instability occurring. There occur convective acceleration and whirling of the boundary layers of the conventional border between a MC and a surrounding matter, zones with large and small gas density. The mixing of C_1 and C_2 cloud spatial tracks looks like the formation of Kármán vortex streets.

Figure 2 shows the dynamics of shaping with numerical schlierens in a central cross-section of the computational domain, morphology of cloud mixing, and density gradient distribution with shock wave wakes. The figures show how shock waves compress the selected regions of gas flows and sharply increase their density. The interference of reflected shock waves and the intensive fluctuations of supersonic velocity fields in gradient zones lead to sharp differentiation of gas density, up to the contrast density ratio $\chi \sim 2000$. Gas compression zones concentrate along film shells of a conventional cylinder-conical form and elongated in the direction of the shock wave propagation.

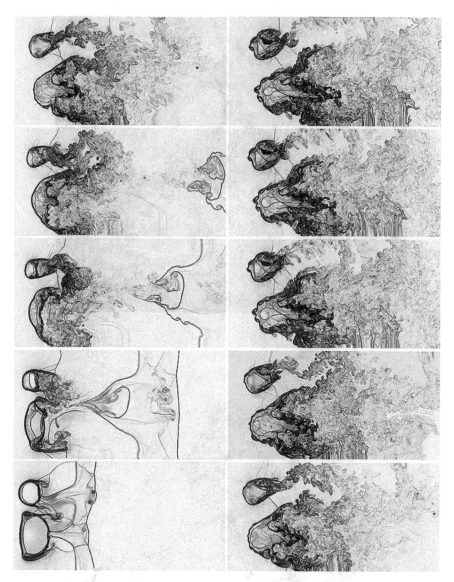

Fig. 2. Change of MC density structure from $t = 40 \cdot t_{swoc}$ to $t = 600 \cdot t_{swoc}$

The global circulation of a gas flow in the mixing zone begins to appear after cloud C_1 being rounded by a shock wave and finds its source in two vortex lines born inside the cloud at the back side. The flow swirl occurs in accordance with the scheme of spatial twin vortex, see Fig. 3. Positive and negative boundary values of the longitudinal component variable of the vortex field rotation are shown in light and dark tones, respectively. The variable distribution is shown on the Q-criterion field isosurface. Initially, a vortex wake of the upper cloud is sucked inside the rear part of the lower one, and then the transformation of helixes with reverse twist and turn in the direction

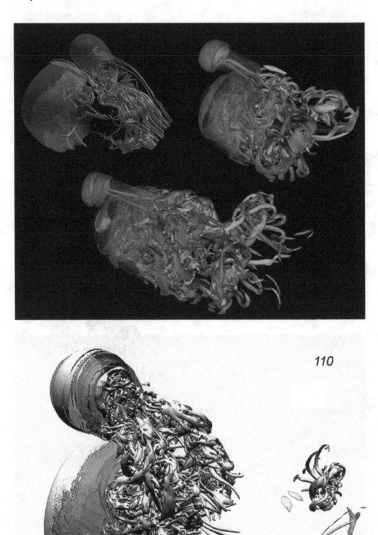

Fig. 3. General flow circulation inside MCs at initial stages from $t = 80 \cdot t_{swoc}$ to $t = 120 \cdot t_{swoc}$

of cloud drift occurs. The wake shifts and their upstream turn happen. With the flow development the vortex lines elongate, kink, take the form of hairpins, and expand in the bend region.

Figure 4 shows typical vortex formation with torus-like structures, elongated loops, and helical deformations.

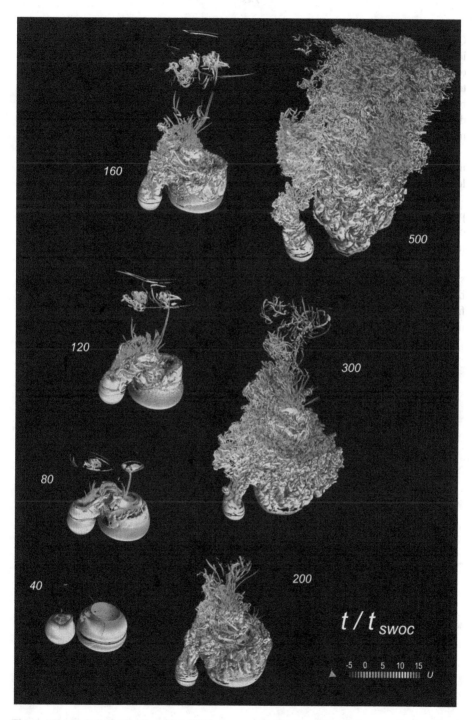

Fig. 4. Q-criterion distribution $(Q = 100)$ – vortex indicator with velocity contours over and inside a MCs for different time stages t/t_{swoc}

To differentiate some peculiarities of the flow vortex structure, the authors compute Q-criterion fields, the second invariant of a velocity gradient tensor being used to identify the regions of small-scale vortex concentrations. The grey color palette maps of local velocity shown on Q-criterion surfaces are illustrative of gas flow intermittency. Vortices have smaller distribution density within the mixing region, at the boundaries and surfaces of elongated filamentous film rudiments the vortex distribution densities are significantly higher and show a local velocity slope in different MC regions.

The gas flows in the mixing region show high turbulence. The fluctuation intensity in boundary zones increases by an order, and the fluctuations take on values exceeding local supersonic velocity by more than one third at the measuring points.

Figure 5 shows the distribution fields of the peculiarities for one of MC evolution periods. The spatial location analysis of intensive fluctuations shows that their sharp change zones are associated with their following the conventional boundaries of filamentous formations.

Figure 6 illustrates a MC fragmentation. To provide visual convenience the authors show only half of 3D presentation of density visibility isosurfaces ($\chi = 10; 100; 500$), bounded χ–surfaces are shown as translucent ($\chi = 1000; 2000$). The right part of the figure shows breaking cloud fragmentation with a spatial division of forming filamentous regions with $\chi = 600$, condensations formed at longer evolution periods.

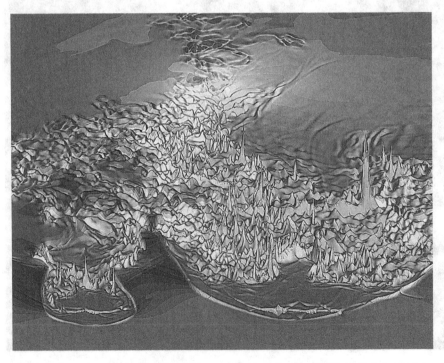

Fig. 5. High relief of turbulent velocity fluctuation intensity for $u'/|U|$, built in mid-plane of MC mixing region, at $t = 300 \cdot t_{swoc}$

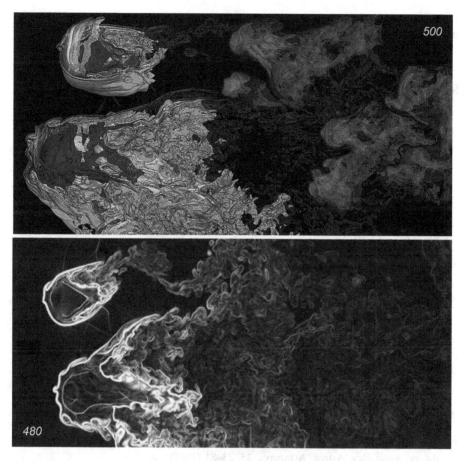

Fig. 6. Gas density fragmentation at $t = 500 \cdot t_{swoc}$ and filamentous structure evolution. The bottom fragment shows the numerical schlieren for density gradient at $t = 480 \cdot t_{swoc}$

The analysis of forming 'clump and shell' filamentous structures elongated by shock wave strengths shows that the fragmentation of a two cloud system in time generally follows time pictures of breaking described in [2, 14–17] for single MCs. Under more general simulation conditions it is more evident and natural when shock waves interact with initially spatially distributed clouds and various gravity field distributions. The continuation of collision simulation for molecular cloud systems allows the intermediate filamentous rudiment formations to be revealed [19].

Tracking changes in a gas size distribution in an actual time of numerical experiments shows that longer time of deforming cloud drift leads to the fact that the median of MC density fraction distributions shifts to their density visibility increase χ up to the magnitudes of order of 700. To clarify possible variations of this factor value, probably oscillating in time, the additional numerical research is necessary. The research should be predicted till the estimated time of about $10000 \cdot t_{swoc}$, the model being added with magnetic fields with computational meshes two-three times greater than the present work provides.

Acknowledgements. The work has been funded by the Russian Foundation for Basic Research grants Nos. 14-29-06055 and 14-07-00065.

References

1. Elmegreen, B.G.: Gravitational collapse in dust lanes and the appearance of spiral structure in galaxies. Astrophys. J. **231**, 372–383 (1979)
2. Banda-Barragan, W.E., Parkin, E.R., Federrath, C., Crocker, R.M., Bicknell, G.V.: Filament formation in wind-cloud interactions. I. Spherical clouds in uniform magnetic fields. MNRAS **455**, 1309–1333 (2016)
3. Klein, R.I., McKee, C.F., Collela, P.: On the hydrodynamical interaction of the shock waves with interstellar clouds. Astrophys. J. **420**, 213–236 (1994)
4. Schneider, S., Elmegreen, B.G.: A catalog of dark globular filaments. Astrophys. J. Suppl. Ser. **41**, 87–95 (1979)
5. Vázquez-Semandeni, E., Ostriker, E.C., Passot, T., Gammie, C.F., Stone, J.M.: Compressible MHD turbulence: implications for molecular cloud and star formation. In: Mannings, V., et al. (eds.) Protostars and Planets IV, pp. 3–28. University of Arizona, Tucson (2000)
6. Wooden, D.H., Charnley, S.B., Ehrenfreund, P.: Composition and Evolution of Interstellar Clouds: Comets II. University of Arizona Press, Tucson (2005)
7. Falgarone, E., Phillips, T.G., Walker, C.K.: The edges of molecular clouds: fractal boundaries and density structure. Astrophys. J. **378**, 186–201 (1991)
8. Kritsuk, A.G., Wagner, R., Norman, M.L., Padoan, P.: High resolution simulations of supersonic turbulence in molecular clouds. In: Pogorelov, N., Raeder, J., Yee, H.C. Zank, G. (eds.) Numerical Modeling of Space Plasma Flows, ASP Conference Series, vol. XXX, p. 7 (2006)
9. Blitz, L., Shu, F.H.: The origin and lifetime of giant molecular cloud complexes. Astrophys. J. **238**, 148–157 (1980)
10. Shu, F.H., Adams, F.C., Lizano, S.: Star formation in molecular clouds - observation and theory. Annu. Rev. Astron. Astrophys. **25**, 23–81 (1987)
11. Nakano, T.: Star formation in magnetic clouds. Astrophys. J. **494**, 587–604 (1998)
12. Boss, A.P.: From molecular clouds to circumstellar disks. In: Festou, M.C. et al. (eds.) Comets II, p. 745. University of Arizona, Tucson (2004)
13. Vázquez-Semandeni, E.: The turbulent star formation model. Outline and tests. In: Burton, M. et al. (eds.) IAU Symposium 221: Star Formation at High Angular Resolution, p. 512. San Francisco Publishing of Astronomical Society of the Pacific (2004)
14. Pittard, J.M., Falle, S.A.E.G., Hartquist, T.W., Dyson, J.E.: The turbulent destruction of clouds. MNRAS **394**, 1351–1378 (2009)
15. Rybakin, B.P., Goryachev, V.D., Stamov, L.I., Michalchenko, E.V.: Parallel algorithm for mathematical modeling of interaction of a strong shock wave with a molecular cloud. In: Voevodin, V. (ed.) Proceedings of the 1st Russian Conference on Supercomputing (RuSCDays 2015), Moscow, Russia. CEUR Workshop Proceedings, vol. 1482, pp. 324–331, Published on CEUR-WS.org (2015). URN: urn:nbn:de:0074-1482-7, ISSN 1613-0073. http://ceur-ws.org
16. Rybakin, B., Goryachev, V.: Treatment and visual analysis of numerical simulation of supersonic flows with extensive output after parallel calculation. In: Cardone, J. (ed.) PSFVIP 10-121, E-Book Proceedings of the 10th Pacific Symposium on Flow Visualization and Image Processing, Naples, Italy, p. ID-121-1-8 (2015). http://www.psfvip10.unina.it/Ebook/PSFVIP-10.html

17. Johansson, E.P.G., Ziegler, U.: Radiative interaction of shocks with small interstellar clouds as a pre-stage to star formation. Astrophys. J. **766**, 1–20 (2011)
18. Rybakin, B.P., Stamov, L.I., Egorova, E.V.: Accelerated solution of problems of combustion gas dynamics on GPUs. Comput. Fluids **90**, 164–171 (2014)
19. Rybakin, B., Goryachev, V.: Coherent instabilities leading to fragmentation of molecular clouds interacted with shock wave of supernova blast remnants. In: Sajjadi, S.G., Fernando, H.J.S. (eds.) Turbulence, Waves and Mixing: In Honour of Lord Julian Hunt's 75th Birthday, King's College, Cambridge, pp. 45–48. Institute of Mathematics & its Applications, British Library Cataloguing in Publication Data, Printed in Great Britain by Modern Graphic Art Limited (2016)

Parallel Algorithms for a 3D Photochemical Model of Pollutant Transport in the Atmosphere

Alexander Starchenko$^{(\boxtimes)}$, Evgeniy Danilkin,
Anastasiya Semenova, and Andrey Bart

National Research Tomsk State University, Tomsk, Russian Federation
{starch, ugin, semyonova.a, bart}@math.tsu.ru

Abstract. In this paper, a numerical scheme for solving a system of convection-diffusion-kinetics equations of a mathematical model of transport of small pollutant components with their chemical interactions in the atmospheric boundary layer is presented. A new monotonized high-accuracy spline scheme is proposed to approximate the convective terms. Various approaches to parallelization of the computational algorithm are developed and tested. These are based on a two-dimensional decomposition of the calculation domain with synchronous or asynchronous interprocessor data communications for distributed memory computer systems.

Keywords: Pollutant transport in the atmosphere · Parallel computing · Two-dimensional decomposition · High-accuracy numerical scheme

1 Introduction

Mathematical modeling is presently an efficient tool for monitoring and prediction of atmospheric air pollution on regional and urban scales under increased emissions of anthropogenic pollutants into the atmosphere due to industrial development, heat and electrical power engineering, combined with ever larger vehicular emissions.

The methods of air pollution estimation were greatly improved when the empirical-statistical analysis of air quality by Gaussian-type models was replaced by atmospheric diffusion theories [1]. The description of pollutant transport by turbulent diffusion equations is more versatile, since it allows studying pollutant dispersion from sources of various types and taking into account precipitation, chemical reactions, and other processes under variable weather conditions by using meteorological parameters calculated with mesoscale meteorological models [2].

A detailed simulation of complex atmospheric processes on regional scales involves many calculations. The computational workload will only increase when such models involve finer spatial resolution and incorporate a wider range of atmospheric phenomena [3].

When implementing numerically the atmospheric boundary layer and pollutant transport models, it is important to choose a good-quality difference scheme with minimum artificial viscosity for the approximation of the convective terms of the

© Springer International Publishing AG 2016
V. Voevodin and S. Sobolev (Eds.): RuSCDays 2016, CCIS 687, pp. 158–171, 2016.
DOI: 10.1007/978-3-319-55669-7_13

equation. In the present paper, we propose a high-accuracy monotonized scheme that has an advantage over such schemes as MUSCL, ENO, and SUPERBEE [4–6], which are widely used for solving transport problems.

The purpose of this study is to develop a difference scheme based on local weight splines to approximate the convective terms of the transport equation and test some parallel algorithms based on a two-dimensional decomposition of the grid domain for solving numerically the equations of a three-dimensional predictive model of pollutant transport with chemical reactions.

2 Three-Dimensional Predictive Model of Pollutant Transport with Chemical and Photochemical Reactions

To calculate the concentrations of pollution components with chemical interactions between them, we use an Eulerian turbulent diffusion model with transport equations of advection, turbulent diffusion, and chemical reactions [7]:

$$\frac{\partial C_i}{\partial t} + \frac{\partial U C_i}{\partial x} + \frac{\partial V C_i}{\partial y} + \frac{\partial W C_i}{\partial z} = -\frac{\partial}{\partial x}\langle c_i u\rangle - \frac{\partial}{\partial y}\langle c_i v\rangle - \frac{\partial}{\partial z}\langle c_i w\rangle$$
$$-\sigma_i C_i + S_i + R_i, \quad i = 1,..,n_s. \quad (1)$$

Here C_i and c_i are the mean and pulsational components of the ith pollutant component concentration, respectively; U, V, u, and v are the mean and pulsational components of the horizontal wind velocity vector; W and w are the mean and pulsational components of the vertical pollutant velocity component; $\langle\ \rangle$ denotes Reynolds averaging; S_i is a source term of emission of pollutant components into the atmosphere; R_i is the rate of formation and transformation of the pollutant by chemical and photochemical reactions with participation of the pollutant components; σ_i is the wet deposition rate due to precipitation; n_s is the number of the pollutant chemical components with concentrations to be determined; x and y are the horizontal coordinates, Ox-axis is directed to the east and Oy-axis, to the north; z is the vertical coordinate; t is the time and T is the simulation time period. The calculation domain is a parallelepiped, L_x and L_y are its horizontal dimensions and h is its height, $-L_x/2 \le x \le L_x/2$, $-L_y/2 \le y \le L_y/2$, $0 \le z \le h, 0 \le t \le T$.

Equation (1) are underdetermined, since, in addition to the concentrations C_i to be determined, they have some unknown functions, the correlations $\langle c_i u\rangle$, $\langle c_i v\rangle$, and $\langle c_i w\rangle$, which simulate turbulent diffusion of the pollutant. In this paper, these are determined from some closure relations obtained by equilibrium approximations for the differential equations of turbulent mass fluxes under the conditions of local homogeneity of the atmospheric boundary layer [7]:

$$\langle c_i u\rangle = -\frac{\tau}{C_{1\theta}}\left((1 - C_{2\theta})\langle c_i w\rangle\frac{\partial U}{\partial z} + \langle u^2\rangle\frac{\partial C_i}{\partial x} + \langle vu\rangle\frac{\partial C_i}{\partial y} + \langle wu\rangle\frac{\partial C_i}{\partial z}\right); \quad (2)$$

$$\langle c_i v \rangle = -\frac{\tau}{C_{1\theta}} \left((1 - C_{2\theta}) \langle c_i w \rangle \frac{\partial V}{\partial z} + \langle uv \rangle \frac{\partial C_i}{\partial x} + \langle v^2 \rangle \frac{\partial C_i}{\partial y} + \langle wv \rangle \frac{\partial C_i}{\partial z} \right); \quad (3)$$

$$\langle c_i w \rangle = -\frac{\tau}{C_{1\theta} + D_{1C} F} \left(-(1 - C_{3\theta}) \frac{g}{\Theta} \langle c_i \theta \rangle + \langle uw \rangle \frac{\partial C_i}{\partial x} + \langle vw \rangle \frac{\partial C_i}{\partial y} + \langle w^2 \rangle \frac{\partial C_i}{\partial z} \right); \quad (4)$$

$$\langle c_i \theta \rangle = -\tau \frac{c_x}{2} \left(\langle c_i w \rangle \frac{\partial \Theta}{\partial z} + \langle \theta u \rangle \frac{\partial C_i}{\partial x} + \langle \theta v \rangle \frac{\partial C_i}{\partial y} + \langle \theta w \rangle \frac{\partial C_i}{\partial z} \right). \quad (5)$$

Here τ is the time scale of turbulent pulsations, g is the acceleration due to gravity, $C_{1\theta}$, $C_{2\theta}$, $C_{3\theta}$, D_{1C}, and c_x are empirical constants, F is a function determining the influence of the surface on the turbulent structure of the flow, Θ and θ are mean and pulsational components of the potential temperature: $\Theta = T(P_0/P)^{R/c_p}$, P is the pressure, P_0 is the pressure on the surface, c_p is the specific heat capacity of air at constant pressure, T is the absolute temperature, and R is the gas constant.

In this system, the Reynolds stresses $\langle uv \rangle$, $\langle uw \rangle$, $\langle vw \rangle$ and the turbulent heat fluxes $\langle u\theta \rangle$, $\langle v\theta \rangle$, $\langle w\theta \rangle$ are unknown. Some relations presented in paper [8] are used to specify these correlations.

A simplified photochemical scheme of Danish Meteorological Institute (DMI) is currently incorporated into the model [9].

3   Initial and Boundary Conditions. Deposition and Emission

Boundary conditions of dry deposition of the pollution components in the form of a simple model of resistance and of the pollutants coming from ground-based sources S_i^g are specified on the lower boundary [10]:

$$-\langle c_i w \rangle = V d_i C_i - S_i^g, \; i = 1, .., n_s; \quad (6)$$

$$V d_i = \frac{1}{r_a + r_b + r_c}; \; r_a = \frac{\Psi(z/z_0, z/L)}{v_*}; \; r_b = \frac{2(Sc_i/0.72)^{2/3}}{\kappa v_*}; \quad (7)$$

$$\Psi(z/z_0, z/L) = \begin{cases} \ln(z/z_0) - 2 \ln \left(\frac{1 + \sqrt{1 - 9z/L}}{1 + \sqrt{1 - 9z_0/L}} \right), \; z/L \leq 0 \\ \ln(z/z_0) + 6, 34 \left(\frac{z}{L} - \frac{z_0}{L} \right), \; z/L > 0. \end{cases} \quad (8)$$

Here $V d_i$ is the rate of dry deposition of the ith pollutant component, r_a is the aerodynamic resistance of turbulent atmosphere by the topographic elements of the surface, r_b is the laminar sublayer resistance by the roughness elements of the surface, r_c is the vegetation-caused resistance, z_0 is the roughness height, L is the Obukhov scale, Sc_i is the Schmidt number for the ith pollutant component, $\kappa = 0.41$ is the von Karman constant, and v_* is the dynamic velocity.

Simple gradient conditions are specified on the upper boundary for the concentrations C_i:

$$\frac{\partial C_i}{\partial z} = 0, \ i = 1, \ldots, n_s. \tag{9}$$

On the lateral boundaries of the calculation domain, we use "radiation"-type conditions allowing perturbations generated in the domain (errors of the method and rounding errors) to leave it without reflection. These are formulated as follows (index i is omitted):

$$\frac{\partial C}{\partial t} + V_{ph}\frac{\partial C}{\partial n} = \frac{\partial C_0}{\partial t} + V_{ph}\frac{\partial C_0}{\partial n}, \tag{10}$$

where C is a concentration, C_0 is a basic concentration of the pollutant, n is the normal to the surface, and V_{ph} is the phase velocity.

The basic values are taken as initial conditions for the pollutant concentration distribution in the atmosphere. The values of the pollutant components in the atmosphere after some preliminary simulation period (several dozens of hours) are close to those in real conditions. The basic values of the pollutant components are calculated separately in the absence of anthropogenic sources.

The fields of meteorological characteristics of the atmospheric boundary layer are calculated by a mesoscale meteorological model [11].

4 Numerical Calculation Method

To describe the numerical method, we restrict our consideration to the one-dimensional nonsteady convection diffusion equation ($u = $ const, $D = $ const > 0):

$$\frac{\partial \Phi}{\partial t} + u\frac{\partial \Phi}{\partial x} = D\frac{\partial^2 \Phi}{\partial x^2} + S_\Phi, \ 0 < x < X, \ 0 < t \leq T. \tag{11}$$

Initial condition: $t = 0$, $\Phi(0, x) = \Phi_{00}(x)$.
Boundary conditions: $x = 0$, $\Phi(t, 0) = \Phi_0(t)$; $x = X$, $\frac{\partial \Phi}{\partial x} = 0$.

For the domain $0 \leq x \leq X$, we construct a grid (Fig. 1) with non-overlapping finite volumes, which is uniform in time and space:

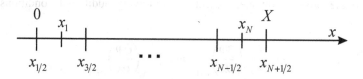

Fig. 1. Uniform grid along the Ox-axis.

$$\bar{\omega}_{h,\tau} = \{(t^n, x_m), \ t^n = \tau \cdot n, \ x_m = h \cdot m + h/2, \ n = 0, \ldots, N, \ m = 0, \ldots, M-1,$$
$$h = X/M, \ \tau = T/N\}. \tag{12}$$

Here $x_{m+1/2}$, $m = 0, \ldots, M$ are the locations of finite volume edges; and x_m, $m = 1, \ldots, M$ are the locations of finite volume nodes. Let $\Phi_m^n \approx \Phi(t^n, x_m)$ be a grid function.

Then, according to the second step of the finite volume method, all terms of Eq. (11) are integrated over the mth finite volume on the time interval $[t^n, t^{n+1}]$. Some interpolation formulas are used to calculate the integrals, and finite difference ones to calculate the derivatives. Thus, we obtain an explicit difference scheme of the following form:

$$\frac{\Phi_m^{n+1} - \Phi_m^n}{\tau} + u \frac{\Phi_{m+1/2}^n - \Phi_{m-1/2}^n}{h} = D \frac{\Phi_{m+1}^n - 2\Phi_m^n + \Phi_{m-1}^n}{h^2} + S_\Phi. \tag{13}$$

Each term of Eq. (11) is expressed in terms of discrete values of Φ at some neighboring nodes, except for the convective term, in which the dependent variable values at the finite volume edges are used. To determine these values in terms of the grid function values at the grid nodes, schemes of various orders of accuracy are currently used. The upstream and central difference schemes are most popular [12]. The former has considerable scheme viscosity, and the latter is non-monotonic. Therefore, to numerically describe the processes with dominant convection, high-order monotonized schemes [4–6, 13] (each of them has certain advantages and shortcomings) are preferable.

In this paper, a new monotonized scheme of high order accuracy (up to the fourth order where the grid function is monotonic) is proposed to approximate the convective term of the equation under consideration. This scheme is constructed with local weight cubic splines capable of reproducing a monotonic distribution of the dependent variable at the next time level.

Consider a local weight cubic interpolation spline constructed using "slopes".

Let us introduce a local grid $\bar{\omega}_h = \{x_{j-1}; x_j; x_{j+1}; x_{j+2}\}$ on the interval $[0, X]$. Here, h denotes the distance between the grid nodes. On each elementary interval $[x_i, x_{i+1}]$ we introduce a third-order polynomial $S_i(x)$ whose coefficients are to be determined. For convenience, $S_i(x)$ is written in the following form:

$$S_i(x) = a_{i0} + a_{i1}(x - x_i) + a_{i2}(x - x_i)^2 + a_{i3}(x - x_i)^3,$$
$$x \in [x_i, x_{i+1}], \ i = j-1, j, j+1; \ S_i(x_i) = \Phi_i, \ i = j-1, j, j+1, j+2; \tag{14}$$
$$S_i'(x_i) = S_{i-1}'(x_i), \ i = j, j+1; \ w_{i-1}S_{i-1}''(x_i) = w_i S_i''(x_i), \ i = j, j+1.$$

Since it is planned to construct (determine the coefficients of) the weight cubic spline using "slopes", $(S_i'(x_i) = m_i, \ S_i'(x_{i+1}) = m_{i+1}, \ i = j-1, j, j+1)$, and make the solution to the problem unique, we set the following additional conditions at the boundaries:

$$m_{j-1} = \left(\frac{\partial \Phi}{\partial x}\right)_{j-1}; \ m_{j+2} = \left(\frac{\partial \Phi}{\partial x}\right)_{j+2}. \tag{15}$$

To determine the spline coefficients, we use the continuity of the first derivative, the conditions of interpolation, and $w_{i-1}S''_{i-1}(x_i) = w_iS''_i(x_i)$, $i = j, j+1$. Finally, we obtain the following system:

$$\begin{cases} m_{j-1} = \left(\dfrac{\partial \Phi}{\partial x}\right)_{j-1} \\[2mm] \dfrac{w_{i-1}}{h}m_{i-1} + 2\left(\dfrac{w_{i-1}}{h} + \dfrac{w_i}{h}\right)m_i + \dfrac{w_i}{h}m_{i+1} = \\[2mm] \quad 3\left(\dfrac{\Phi_i - \Phi_{i-1}}{h^2}w_{i-1} + \dfrac{\Phi_{i+1} - \Phi_i}{h^2}w_i\right), \quad i = j, j+1 \\[2mm] m_{j+2} = \left(\dfrac{\partial \Phi}{\partial x}\right)_{j+2}. \end{cases} \tag{16}$$

The derivatives in this system are approximated by second-order finite differences. This system has strict diagonal dominance, which provides the existence and uniqueness of the weight cubic spline.

Now we obtain the following formula for the construction of splines using "slopes":

$$S_i(x) = \frac{m_{i+1} + m_i}{h^2}(x - x_i)^3 - 2\frac{\Phi_{i+1} - \Phi_i}{h^3}(x - x_i)^3 + 3\frac{\Phi_{i+1} - \Phi_i}{h^2}(x - x_i)^2$$
$$- \frac{2m_i + m_{i+1}}{h}(x - x_i)^2 + m_i(x - x_i) + \Phi_i. \tag{17}$$

The weights w_i may be specified in various ways. For a spline without oscillations to interpolate monotonic data, B.I. Kvasov's theorem presented below can be used [14].

Theorem 1. *Let a weight cubic spline $S \in C^1[a, b]$ with boundary conditions $S'(x_0) = f'_0$ and $S'(x_{n+1}) = f'_{n+1}$ interpolate monotonic data $\{f_i\}$, $i = 0, \ldots, n+1$. If the inequalities $0 \le f'_0 \le 3f[x_0, x_1]$, $0, le f'_{N+1}, le 3f[x_N, x_{N+1}]$,*

$$\frac{w_{i-1}}{w_i}\frac{h_i}{h_{i-1}} \ge \frac{f[x_i, x_{i+1}]}{f[x_{i-1}, x_i]} - 2, \quad and \quad \frac{w_i}{w_{i-1}}\frac{h_{i-1}}{h_i} \ge \frac{f[x_{i-1}, x_i]}{f[x_i, x_{i+1}]} - 2,$$

are valid, $S'(x) \ge 0$ for all $x \in [a, b]$, that is, the spline S is monotonic on $[a, b]$.

An analog of this theorem can be formulated for monotonically decreasing data.

The following algorithm may be used in calculating the weight coefficients [14]. Let a parameter w_{i-1} be given. Assume that $w_i = w_{i-1}$. Check the inequalities of the theorem to be used. If some condition is violated, express w_i from this and change the inequality into an equality. Start the algorithm with $w_0 = 1$, and find all the parameters $\{w_i\}$ providing monotonicity of the weight cubic spline for arbitrary monotonic data. If at some step $w_i < \varepsilon$, $(w_i > 1/\varepsilon)$, set $w_i = \varepsilon$, $(w_i = 1/\varepsilon)$, where ε is a sufficiently small positive number that makes it possible to avoid nulling (overflow).

Returning to the construction of the difference scheme, consider ae method of approximating $\Phi^n_{m+1/2}$ in the convective term based on the thus obtained local cubic spline:

$$\Phi^n_{m+1/2} = S_m(x_{m+1/2}).\tag{18}$$

The spline construction algorithm allows a monotonic approximation of monotonic data. However, if the data are not monotonic, there arises the problem of monotonicity. To avoid this problem, a limiter is further applied to the spline [15].

In this case, the spline approximation is as follows:

$$\Phi^n_{m+1/2} = \begin{cases} \Phi^n_m + \frac{1}{2} \cdot \max[0, \min(2\Theta, \Psi, 2)] \cdot (\Phi^n_{m+1} - \Phi^n_m), & u > 0 \\ \Phi^n_{m+1} - \frac{1}{2} \cdot \max\left[0, \min\left(2\hat\Theta, \hat\Psi, 2\right)\right] \cdot (\Phi^n_{m+1} - \Phi^n_m), & u \le 0, \end{cases}\tag{19}$$

where

$$\Theta = \frac{\Phi^n_m - \Phi^n_{m-1}}{\Phi^n_{m+1} - \Phi^n_m}, \quad \Psi = \frac{S_m(x_{m+1/2}) - \Phi^n_m}{0.5(\Phi^n_{m+1} - \Phi^n_m)},\tag{20}$$

$$\hat\Theta = \frac{\Phi^n_{m+2} - \Phi^n_{m+1}}{\Phi^n_{m+1} - \Phi^n_m}, \quad \hat\Psi = \frac{\Phi^n_{m+1} - S_m(x_{m+1/2})}{0.5(\Phi^n_{m+1} - \Phi^n_m)}.\tag{21}$$

To test the feasibility of the constructed scheme, consider the problem (11) at $S_\Phi(t, x) = 0$, $X = 2$ without diffusion and at the following initial conditions:

$$\Phi_{00}(x) = \begin{cases} 1, & x \in (0.75; \ 1.25) \\ 0, & x \notin (0.75; \ 1.25). \end{cases}\tag{22}$$

The exact solution of this problem may be written as $\Phi_*(t, x) = \Phi_{00}(x - ut)$.

This problem has been calculated under the following conditions: $M = 100$, $u = 1$, $D = 0$, $h = 0.02$, $\tau = 0.004$, $T = 50\ \tau$.

The problem has also been solved by the other above-listed [4–6] methods. The error norm values are presented in Table 1.

Table 1. Comparison of the methods accuracy

	Method	$\max\limits_{m=0,\dots,M} \left|\Phi^N_m - \Phi_*(T, x_m)\right|$	$\sqrt{\dfrac{\sum\limits_{m=0,\dots,M}\left|\Phi^N_m - \Phi_*(T, x_m)\right|^2}{M}}$
1.	Upwind	0.44374	0.01129
2.	MLU	0.28130	0.00522
3.	MUSCL	0.27333	0.00513
4.	ENO	0.31731	0.00618
5.	Spline	0.21682	0.00401
6.	Harten scheme	0.36789	0.00810
7.	SuperBee	0.32330	0.00607

On the basis of these data one can conclude that the use of the spline function has an advantage over the other schemes.

5   Parallelization of the Numerical Algorithm

The calculation domain is a $50 \times 50 \times 2$ km parallelepiped located in the atmospheric surface layer such that its square base has a part of the underlying surface with a large population area in its center. In the calculations the pollutant emission from some area sources on the surface is taken into account. The domain is covered with a calculation grid of $N_x \times N_y \times N_z = 100 \times 100 \times 30$ nodes. The prediction simulation period is typically 48 h. The time step is $\Delta t_c = 6$ s for the calculations of chemical transformations and $\Delta t = 60$ s for the convection-diffusion ones. The fields of meteorological characteristics needed for the calculations are prepared before the calculation of the concentration fields using a mesoscale model [11].

This model is mostly used for short-term predictions of air quality over urban territories. Therefore, it is important for the calculation time to be as small as possible. The currently available workstations and servers with multicore architecture cannot provide the needed speedup when using the typical sequential or shared memory version of the program. However, the usage of the distributed memory computers can decrease the overall calculation time.

In parallel implementation of the algorithm, choosing an optimal method of distributing the calculations among the processors is very important. This is determined by the peculiarities of the algorithm itself, the computer system architecture, the number of processors available for the calculations, as well as by some physical and computational considerations. Let us consider some peculiarities in the development of parallel programs and assess the speedup and efficiency at a two-dimensional grid domain decomposition. The scheme of the calculation domain is presented in Fig. 2.

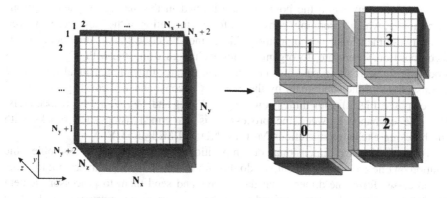

Fig. 2. Schematic diagram of the calculation domain and two-dimensional decomposition at parallel implementation of the explicit-implicit method using four processors as an example ($p = 4$, $p_x = 2$, $p_y = 2$).

The $(2 \leq i_x \leq N_x + 1) \times (2 \leq i_y \leq N_y + 1)$ calculation domain is bordered by two rows of fictitious cells along the perimeter to implement the calculation template being used and provide second or higher order of approximation.

6 Implementation of the Two-Dimensional Decomposition

The two-dimensional decomposition implies better scalability of the problem solved by finite difference or finite element methods on multi-processor computer systems, since this approach allows using a larger number of processors. For example, under some constraint on the width of the grid domain being processed p^2 parallel processes can be used at the two-dimensional decomposition.

The two-dimensional implementation using the MPI library [16] differs considerably from the one-dimensional one. First, it is necessary to organize a two-dimensional topology of the simultaneous processes. In the present paper, a Cartesian topology version is chosen. After the MPI is initialized and the number of processors to be used is defined, the MPI_Dims_Create procedure choosing optimal parameters $p_x \times p_y = p$ for the two-dimensional decomposition is called. With these parameters the MPI_Cart_Create procedure creates a Cartesian topology with renumbering of the processes (p_x and p_y are the numbers of one-dimensional decompositions performed independently along the Ox and Oy coordinate directions).

Subsequent calls of the MPI_Comm_Rank and MPI_Cart_Shift procedures allow determining the new number of the process and the new numbers of neighboring processes in the Cartesian topology. The dimensions of the subdomains for each of the processes are calculated simultaneously. This approach makes it possible to organize calculations at any number of processors. However, for a balanced distribution of the load it is recommended to choose an value for p satisfying the condition $p_x = p_y = \sqrt{p}$.

Second, the two-dimensional decomposition implementation of the interprocessor data exchanges is more complicated. In the one-dimensional case, the direction of decomposition can always be chosen so that the elements of the three-dimensional concentration array along the boundary are located in the memory successively one after another, which is certainly convenient for the formation of the message. However, at the two-dimensional decomposition this cannot be done for all directions of the decomposition. Therefore, to organize data exchanges using the MPI_Type_Hvector procedure, new data types are created. These are three-dimensional arrays of cross-sections of the subdomain by the planes $X = \text{const}$ and $Y = \text{const}$ (g_vector and h_vector, see Fig. 2) with dimensions $(N_z, 1, N_y/p_y)$ and $(N_z, N_x/p_x, 1)$, respectively. The data exchange between the processors is performed using such blocks with non-blocking exchange operations MPI_ISEND and MPI_IRECV.

Third, at the two-dimensional decomposition it is not easy to "assemble" the solution on one processor element. To do this, for instance, before delivering it to a file, the processes form one-dimensional data arrays and send them to processor element "0", which collects the data received and rewrites them in an appropriate order to a three-dimensional array.

7 Speedup and Efficiency

In this paper, the speedup of the parallel algorithm in comparison to the serial one and its efficiency are studied experimentally. For this aim a series of calculations has been performed on a TSU cluster called Cyberia for a simulation period of one hour, a $100 \times 100 \times 30$ calculation grid, and a DMI kinetic scheme of chemical and photo-chemical reactions [9]. To provide load balancing and, thereby, achieve the best efficiency, calculations are made with a number of processes such that the number of grid nodes processed by each processor element being the same (weak speedup estimation). 1, 4, 16, 25, and 100 processors of the calculation cluster are used. The speedups are presented in Fig. 3 with two variants of data package exchange between the processors:

- Configuration 1: two-dimensional decomposition, synchronous exchanges (send/recv);
- Configuration 2: two-dimensional decomposition, asynchronous exchanges, "advanced transmitions" (isend/irecv).

In the case of a small number of processes (up to 16) the use of advanced transmitions does not provide any advantage. However, the situation changes when the number of processes increases. With 16 and more processors the number of inter-processor exchanges increases, which increases the idle time of the processors waiting for completion of the message transfer with blocking operations. The efficiency of configuration 1 decreases, which is especially demonstrated by the high performance of the program with asynchronous operations (Configuration 2) (Fig. 3).

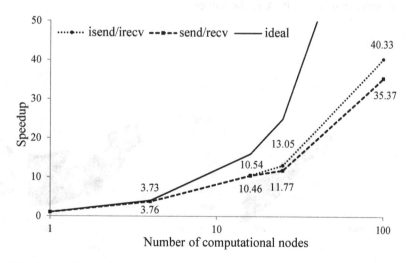

Fig. 3. Parallel program speedup for various calculation domain decompositions.

Thus, the results of the calculations have shown that when more than 16 processes are used the technology of advanced sending makes it possible to increase the efficiency of the parallel program operation by 8–12% in comparison to the use of blocking procedures of message exchange. This is a rather impressive figure when a large volume of calculations is performed.

8   Numerical Experiment

The problem to be solved is as follows: there is a permanent pollutant emission in an urban region at one grid cell of the zone being considered. The horizontal dimensions of the zone are 50×50 km. The grid has 100 nodes in each horizontal direction and 30 height levels. A two-day wind field predicted by a mesoscale model is also used in the calculations [11].

In the center of the zone there is an area from which a gaseous pollutant is emitted at a rate of 10^4 mg/s. The problem is to predict the concentration with the wind field for the subsequent time. To solve this problem, three methods of approximating the convective terms of the transport equations are considered MLU, MUSCL [4, 15], and the spline interpolation being proposed. The figures below show the pollutant concentration propagation and the wind field for 29 and 41 h of physical time after the start of the simulation.

It follows from the wind fields shown in Fig. 4 that in the first case (a) the surface wind direction is North-West. This situation leads to a considerable scheme diffusion, since the wind is directed at an angle of $45°$ to the grid lines [12]. The other case (b) is also of interest, with restructuring of the surface wind field and "returning" of the pollutant being transported back to the source.

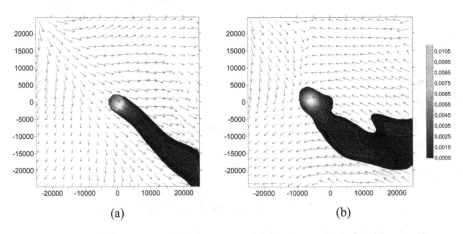

(a)   (b)

Fig. 4. Surface pollutant concentrations and wind field after (a) 29 h and (b) 41 h.

Figures 5 and 6 show the pollutant concentration distributions at the same times when the MLU schemes [4] and MUSCL [15], as well as the spline scheme, are used. The comparison of the results is made using the local concentration maxima in the right lower corner of the figures. The schemes MLU and MUSCL show similar results. The results of the spline scheme provide a larger size of the zone with higher pollutant concentration. That is, according to Fig. 5 in this zone the surface pollutant concentration is 0.002 mg/m^3 over the larger part of the territory, and with the two other schemes a concentration of 0.0015 mg/m^3 is predicted. The spline scheme more accurately predicts the concentration level with less smoothing of the numerical solution. Similar conclusions can be made from Fig. 6.

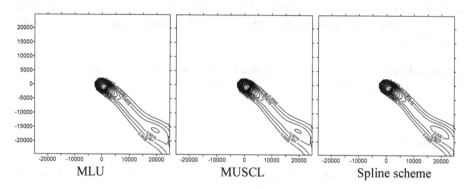

Fig. 5. Pollutant concentration after 29 h.

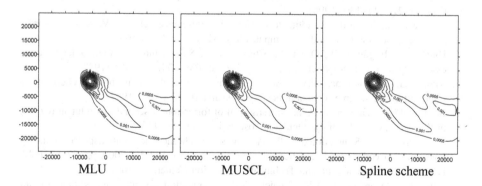

Fig. 6. Pollutant concentration after 41 h.

The above results make it possible to conclude that the spline scheme constructed to approximate the convective terms of the transport equation has less scheme diffusion. This allows one to make better prediction of local maxima in the numerical solution of pollutant transport problems in the atmosphere.

9 Conclusions

A numerical scheme for solving a system of convection-diffusion-kinetics equations of the mathematical model for transport of small pollutant components with chemical interactions in the atmospheric boundary layer is presented. A new monotonized high-accuracy spline scheme is proposed to approximate the convective terms. It has been shown that this scheme has an advantage over the monotonized schemes of second or third order used to solve such problems. Various approaches to construct the parallel computational algorithm are developed and tested. These approaches are based on a two-dimensional decomposition of the calculation domain with synchronous or asynchronous methods of interprocessor data transmission for distributed-memory computer systems. It has been shown that advanced transmissions of the calculated boundary values of the grid function with asynchronous data exchanges speeds up the calculations.

Acknowledgements. This work was supported by the Ministry of Education and Science of the Russian Federation, Public Contract no. 5.628.2014/K.

References

1. Berlyand, M.E.: Prediction and Regulation of Air Pollution. Gidrometeoizdat, Leningrad (1985)
2. Penenko, V.V., Aloyan, A.E.: Models and Methods for Environmental Protection Problems. Nauka, Novosibirsk (1985)
3. Dabdub, D., Seinfeld, J.H.: Parallel computation in atmospheric chemical modeling. Parallel Comput. **22**, 111–130 (1996)
4. van Leer, B.: Towards the ultimate conservative difference scheme. V. A second-order sequel to Godunov's method. J. Comput. Phys. **32**, 101–136 (1979)
5. Harten, A., Engquist, B., Osher, S., Chakravarthy, S.R.: Uniformly high order accurate essentially non-oscillatory schemes, III. J. Comput. Phys. **71**, 231–303 (1987)
6. Rivin, G.S., Voronina, P.V.: Aerosol transport in the atmosphere: choosing a finite-difference scheme. Optika atmosfery i okeana **10**, 623–633 (1997)
7. Belikov, D.A., Starchenko, A.V.: Investigation of formation of secondary pollutant (ozone) in the atmosphere of Tomsk. Optika atmosfery i okeana **18**, 435–443 (2005)
8. Starchenko, A.V.: Simulation of pollutant transport in a homogeneous atmospheric boundary layer. In: Proceedings of International Conference on ENVIROMIS-2000, pp. 77–82. Publ. House of Tomsk Scientific and Technical Information Center, Tomsk (2000)
9. Bart, A.A., Starchenko, A.V., Fazliev, A.Z.: Information-computational system for air quality short-range prognosis over territory of Tomsk. Optika atmosfery i okeana **25**, 594–601 (2012)
10. Hurley, P.J.: The air pollution model (TAPM) version 2. Part 1: technical description. CSIRO Atmospheric Research Technical Paper, No. 55 (2002)
11. Starchenko, A.V., Bart, A.A., Bogoslovskiy, N.N., Danilkin, E.A., Terenteva, M.V.: Mathematical modelling of atmospheric processes above an industrial centre. In: Proceedings of SPIE 9292, 20th International Symposium on Atmospheric and Ocean Optics: Atmospheric Physics, pp. 929249/1-9 (2014)

12. Patankar, S.V.: Numerical Methods for Solving Heat Transfer and Fluid Dynamics Problems. Energoatomizdat, Moscow (1984). Transl. from English, Ed.: V.D. Violensky
13. van Leer, B.: Towards the ultimate conservative difference scheme. IV. A new approach to numerical convection. J. Comput. Phys. **23**, 276–299 (1977)
14. Kvasov, B.I.: Methods of Isogeometric Spline Approximation. FIZMATLIT, Moscow (2006)
15. Čada, M., Torrilhon, M.: Compact third-order limiter functions for finite volume methods. J. Comput. Phys. **228**, 4118–4145 (2009)
16. Starchenko, A.V., Danilkin, E.A., Laeva, V.I., Prokhanov, S.A.: A Practical Course on Parallel Computation Methods. Publ. House of MSU, Moscow (2010)

Parallel Computation of Normalized Legendre Polynomials Using Graphics Processors

Konstantin Isupov$^{(\boxtimes)}$, Vladimir Knyazkov, Alexander Kuvaev,
and Mikhail Popov

Department of Electronic Computing Machines,
Vyatka State University, Kirov 610000, Russia
{ks_isupov,knyazkov}@vyatsu.ru

Abstract. To carry out some calculations in physics and Earth sciences, for example, to determine spherical harmonics in geodesy or angular momentum in quantum mechanics, it is necessary to compute normalized Legendre polynomials. We consider the solution to this problem on modern graphics processing units, whose massively parallel architectures allow to perform calculations for many arguments, orders and degrees of polynomials simultaneously. For higher degrees of a polynomial, computations are characterized by a considerable spread in numerical values and lead to overflow and/or underflow problems. In order to avoid such problems, support for extended-range arithmetic has been implemented.

Keywords: Normalized Legendre polynomials · Extended-range arithmetic · GPU · CUDA

1 Introduction

Associated Legendre polynomials are solutions of the differential equation

$$(1 - x^2)\frac{d^2}{dx^2}P_n^m(x) - 2x\frac{d}{dx}P_n^m(x) + \left[n(n+1) - \frac{m^2}{1-x^2}\right]P_n^m(x) = 0, \quad (1)$$

where degree n and order m are integers satisfying $0 \le n$, $0 \le m \le n$, and x is a real variable in the interval $[-1,1]$ which is usually expressed as $\cos\theta$, where θ represents the colatitude [1, Chap. 15].

These polynomials are important when defining geopotential of the Earth's surface [2,3], spherical functions in molecular dynamics [4], angular momentum in quantum mechanics [5], as well as in a number of other physical applications. The accuracy and scale of numerical simulations directly depend on the maximum degree of a polynomial which can be correctly computed. Modern applications typically operate upon first-kind ($m = 0$) polynomials at a degree of 10^3 or higher. The functions $P_n^m(x)$ grow combinatorially with n and can overflow for n larger than about 150. Therefore, for large n, instead of $P_n^m(x)$, *normalized* associated Legendre polynomials are computed. There are a number

© Springer International Publishing AG 2016
V. Voevodin and S. Sobolev (Eds.): RuSCDays 2016, CCIS 687, pp. 172–184, 2016.
DOI: 10.1007/978-3-319-55669-7_14

of different normalization methods [6, Chap. 7]. We consider the computation of fully normalized polynomials

$$\bar{P}_n^m(x) = \sqrt{\frac{2n+1}{2}\frac{(n-m)!}{(n+m)!}}(1-x^2)^{m/2}\frac{\partial^m}{\partial x^m}P_n(x), \tag{2}$$

which satisfy the following equation:

$$\int_{-1}^{1}\{\bar{P}_n^m(x)\}^2 dx = 1. \tag{3}$$

Mathematical properties and numerical tables of $\bar{P}_n^m(x)$ are given in [7]. A number of recursive algorithms are suggested to evaluate $\bar{P}_n^m(x)$ [8–10]. One of the most common ones is based on the following relation [8]:

$$\bar{P}_n^{m-1}(x) = \frac{2mx}{\sqrt{(1-x^2)(n+m)(n-m+1)}}\bar{P}_n^m(x) - \sqrt{\frac{(n-m)(n+m+1)}{(n+m)(n-m+1)}}\bar{P}_n^{m+1}(x). \tag{4}$$

The starting points for recursion (4) are the values $\bar{P}_n^{n+1}(x) = 0$ and

$$\bar{P}_n^n(x) = \sqrt{\frac{1}{2}\frac{3\cdot5\cdots(2n+1)}{2\cdot4\cdots2n}}(1-x^2)^{n/2}. \tag{5}$$

Equations (4) and (5) are asymptotically stable at any admissible parameters x, m and n, so if we consider them in terms of pure mathematics, they are appropriate for computing polynomials of an arbitrarily high degree. In practical computation, however, there are difficulties in computing (4) and (5) when n becomes large [3]. This is due to the following reasons:

- computations take an unacceptable long time;
- overflow or underflow exceptions may occur.

The first of these problems stems from the fact that during the numerical simulation it is required to compute many polynomials of different degrees at a fixed angle, or many fixed degree polynomials for a variety of angles. An effective solution to this problem has been made possible thanks to the intensive development of new generation massively parallel computing architectures, such as graphics processing units (GPUs).

The second problem is related to the limited dynamic range of real numbers which are represented in computers [11]. As a result, if x is about ±1, the computation of the starting value $\bar{P}_n^n(x)$ leads to underflow, even though the desired value $\bar{P}_n^m(x)$ is within an acceptable dynamic range. For example, if $x = 0.984808$, which corresponds to angle $\theta \approx 10°$, then $\bar{P}_{5000}^{5000}(x) \approx 1.42 \times 10^{-3801}$ while $\bar{P}_{5000}^0(x) \approx 3.32 \times 10^{-1}$. The smallest normal value in IEEE-754 double-precision format is approximately equal to 10^{-308}. Thus, to evaluate $\bar{P}_{5000}^{5000}(x)$, it is necessary to extend the dynamic range by more than an order of magnitude. The value of $\bar{P}_{5000}^{5000}(x)$ is not of independent practical interest, however, it is impossible to start recursion for calculating $\bar{P}^0_{5000}(x)$ without it being correctly computed, because if, due to underflow, $\bar{P}^{5000}_{5000}(x) = 0$, then all following values $\bar{P}^{4999}_{5000}(x)$, $\bar{P}^{4998}_{5000}(x)$, etc. will also become zero. On the other hand, calculating the fraction in (5) in a conventional way (first the numerator, and then the denominator) may lead to overflow exception. Some information about the range of angles and limitations to polynomial degrees at which calculations in IEEE-754 arithmetic do not result in exceptions is given in [2].

To avoid overflow or underflow problems, methods using global scaling coefficients are suggested [9]. However, as noted in [3], this solves the problem only for limited ranges of arguments and degrees. The general solution to the overflow and/or underflow problem when computing the normalized Legendre polynomials is suggested in [8] and involves the use of extended-range arithmetic.

In this paper we consider parallel computation of normalized polynomials $\bar{P}^m_n(x)$ of high degrees in extended-range arithmetic using CUDA-compatible GPUs. Due to a high level of task parallelism, the transfer of computations to the GPU has allowed to achieve significant performance improvement, as compared with the CPU implementation.

2   Extended-Range Arithmetic

2.1   Basic Algorithms

Currently, IEEE-754 standard is the main standard for binary floating-point arithmetic [12]. It defines two most widely used formats: a single-precision format (binary32) and a double-precision format (binary64). These formats are supported, to some extent, at both the hardware level and the level of programming language. In 2008, a revision to IEEE-754 standard was published, which further describes the quadruple-precision binary format—binary128, and two decimal formats—decimal64 and decimal128 [13]. However, support for these new formats is currently implemented in quite rare cases. The properties of single- and double-precision binary formats are presented in Table 1. In this table, the number of digits of the significand, p, defines precision of the format; integers e_{min} and e_{max} are the extremal exponents; n_{max} is the largest positive finite number, n_{min} is the smallest positive normal number, and s_{min} is the smallest positive subnormal number; the segment $[n_{min}, n_{max}]$ specifies the range of positive normal numbers, and the segment $[s_{min}, n_{max}]$ specifies the total range of positive finite numbers.

Table 1. The properties of IEEE-754 single-precision and double-precision formats

	p	e_{min}	e_{max}	n_{min}	n_{max}	s_{min}
binary32	24	-126	$+127$	2^{-126}	$(2 - 2^{-23}) \times 2^{127}$	2^{-149}
binary64	53	-1022	$+1023$	2^{-1022}	$(2 - 2^{-52}) \times 2^{1023}$	2^{-1074}

The situation when the intermediate result of an arithmetic operation or function exceeds in magnitude the largest finite floating-point number $n_{max} = (2 - 2^{1-p}) \times 2^{e_{max}}$ in IEEE-754 standard is defined as *overflow*. When there is overflow, the result, depending on the used rounding mode, is replaced with infinity ($\pm\infty$) or the largest finite floating-point number. The situation when the intermediate result of an arithmetic operation is too close to zero, i.e. in magnitude it is strictly less than the smallest positive normal number $n_{min} = 2^{e_{min}}$ is defined as *underflow* [13,14]. When there is underflow, the result is replaced with zero, subnormal number, or the smallest positive normal number. In all cases, the sign of the rounded result coincides with the sign of the intermediate result. The exceptions examined are presented in Fig. 1.

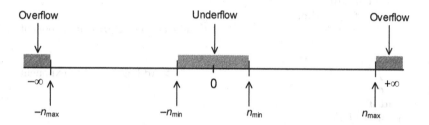

Fig. 1. Overflow and underflow in floating-point arithmetic

One of the ways to eliminate overflow or underflow is scaling. This method requires estimating the source operands and multiplying them by factor K chosen so that all intermediate results are within the normal range. After the computation of the final result, scaling is carried out by dividing it by K [14]. In terms of computing speed, this technique is evidently the best one. However, it requires a detailed analysis of the whole computing process and is not applicable in many cases. A more common approach is emulation of extended-range arithmetic. To do this, the integer e is paired with a conventional floating-point number f, and this pair is considered as a number

$$f \times B^e, \tag{6}$$

where B is a predetermined constant that is a power of the floating-point base [8, 11]. Significand f can take values in the interval $(1/B, B)$. Given this, B must be such as for any arithmetic operation performed with f, no underflow or overflow occurs. It is advisable that B is a natural power of two (for a binary computer).
For instance, if

- f is a double-precision number (binary64),
- e is a 32-bit signed integer ($-2147483648 \leq e \leq 2147483647$) and
- $B = 2^{256}$,

then the range of the represented numbers will exceed $10^{\pm 165492990270}$.

The algorithms for basic extended-range operations are considered in [8,11], and therefore, we will focus only on some of them. In the following, we will assume that the base of exponent B is uniquely determined and the extended-range number is represented by a pair (f, e). Algorithm 1 performs the "adjustment" of the number. It is one of the basic extended-range arithmetic algorithms. It provides control of the value range of significand f, as well as its correction in case the input is incorrect encoding of zero.

Algorithm 1. Adjustment of the extended-range representation

1: **procedure** ADJUST(f, e)
2:    **if** $f = 0$ **then**
3:        **return** $(0, 0)$
4:    **else if** $|f| \geq B$ **then**
5:        $f \leftarrow f/2^{\log_2 B}$ ▷ Subtracting $\log_2 B$ from exponent of f
6:        $e \leftarrow e + 1$
7:    **else if** $|f| \leq 1/B$ **then**
8:        $f \leftarrow f \times 2^{\log_2 B}$ ▷ Adding $\log_2 B$ to exponent of f
9:        $e \leftarrow e - 1$
10:   **end if**
11:   **return** (f, e)
12: **end procedure**

It is important to note that it is not always enough to carry out the ADJUST procedure only one time. This can take place at least in the following two cases: (a) when the number is converted from the machine format or a format with different from the current exponent base; (b) when subtraction of almost equal numbers (or addition with different signs) is performed. In any of these cases, it is possible that, after the ADJUST procedure has been performed, significand f is less than $1/B$. If it is ignored and the computation process is continued, gradual "zeroing" of the result is likely to take place. To avoid this, it is possible to use a cyclic adjustment procedure which is implemented by Algorithm 2.

Algorithm 2. Cyclic adjustment of the extended-range representation. Procedure should be used in conversion, signed addition and subtraction algorithms.

1: **procedure** CYCLICADJUST(f, e)
2:    $(f_1, e_1) \leftarrow$ ADJUST(f, e)
3:    **while** $e \neq e_1$ **or** $f \neq f_1$ **do**
4:        $(f, e) \leftarrow (f_1, e_1)$
5:        $(f_1, e_1) \leftarrow$ ADJUST(f, e)
6:    **end while**
7:    **return** (f, e)
8: **end procedure**

Algorithm 3 performs addition of extended-range representations. Algorithms for subtraction and comparison are quite similar to the addition algorithm, so their description seems to be unnecessary.

Algorithm 3. Adding extended-range numbers, $(f_z, e_z) \leftarrow (f_x, e_x) + (f_y, e_y)$

1: **procedure** ADD(f_x, e_x, f_y, e_y)
2:     **if** $f_x = 0$ **and** $e_x = 0$ **then return** (f_y, e_y)
3:     **else if** $f_y = 0$ **and** $e_y = 0$ **then return** (f_x, e_x)
4:     **end if**
5:     $\Delta e = |e_x - e_y|$
6:     **if** $e_x > e_y$ **then**
7:         $f_z \leftarrow f_x + f_y \times 2^{-\Delta e \times \log_2 B}$
8:         $e_z \leftarrow e_x$
9:     **else if** $e_y > e_x$ **then**
10:         $f_z \leftarrow f_y + f_x \times 2^{-\Delta e \times \log_2 B}$
11:         $e_z \leftarrow e_y$
12:     **else if** $e_y = e_x$ **then**
13:         $f_z \leftarrow f_x + f_y$
14:         $e_z \leftarrow e_x$
15:     **end if**
16:     **return** CYCLICADJUST(f_z, e_z)
17: **end procedure**

2.2   Implementation of Extended-Range Arithmetic

We have implemented all basic algorithms of extended-range arithmetic, and a number of mathematical functions for CPUs and NVIDIA CUDA-compatible GPUs. Do to it, data types shown in Fig. 2 were declared.

```
typedef struct {
  er_frac_t frac;     //significand
  er_exp_t exp;       //exponent
} __extended_range_struct;

//single number:
typedef __extended_range_struct *er_t;
//arrays:
typedef __extended_range_struct *er_arr_t;
//for device side code:
typedef __extended_range_struct er_static_t;
```

Fig. 2. Extended-range data types: er_frac_t—standard floating-point number (**double** by default), er_exp_t—machine integer (**int64_t** by default)

The list of implemented CPU- and CUDA-functions includes the following:

- memory management and constants initialization;
- addition, subtraction, multiplication and division, supporting four IEEE-754 rounding modes, as well as comparison functions;
- integer floor and ceiling functions, computation of the fractional part;
- functions of converting numbers from the double-precision IEEE-754 data type to extended-range data type, and vice versa;
- factorial, power, square root, and a number of other mathematical functions.

The exponent base B is defined in parameters. By default $B = 2$. It is quite enough for the computation carried out. The declaration of CPU- and GPU-functions is identical (cuda namespace is used for GPU-functions). Pointers are used for effective passing of parameters. All functions are thread-safe.

Efficiency of extended-range arithmetic is largely determined by the speed of converting numbers from the machine floating-point representation to extended-range representation, and vice versa. To implement these procedures, we used bitwise operations. In particular, Fig. 3 shows the subroutine er_set_d that converts a conventional IEEE-754 double-precision number into the extended-range representation. This subroutine uses the DoubleIntUnion structure, which allows storing double and integer data types in the same memory location.

```
union DoubleIntUnion {
    double   dvalue;
    uint64_t ivalue;
}

void er_set_d(er_t res, const double x) {
    DoubleIntUnion u;
    if (x == 0) {
        res->exp = res->frac = 0;
        return;
    }
    u.dvalue = x;
    uint8_t sign = (uint8_t) (u.ivalue >> SIGN_OFFSET);
    res->exp = ((u.ivalue & ~((uint64_t) 1 << SIGN_OFFSET))
            >> EXP_OFFSET) - EXP_BIAS;
    u.ivalue = u.ivalue & (((uint64_t) 1 << EXP_OFFSET) - 1)
            | ((uint64_t) EXP_BIAS << EXP_OFFSET);
    res->frac = u.dvalue;
    if (sign)
        res->frac = -res->frac;
    cyclic_adjust(res);
}
```

Fig. 3. Conversion of a double-precision floating-point number into the extended-range representation. For the double data type, SIGN_OFFSET = 63, EXP_OFFSET = 52 and EXP_BIAS = 1023.

3 Computation of Normalized Legendre Polynomials on CPU and GPU

3.1 Computation of Starting Point of Recursion

Our implementation of normalized Legendre polynomials computation is based on the recursion (4), which, in turn, requires computation of relation (5). In case of high degree n of polynomial, direct computation of (5) is time-consuming since it requires computing two double factorials in the extended-range arithmetic, $(2n+1)!! = 3 \cdot 5 \cdots (2n+1)$ and $(2n)!! = 2 \cdot 4 \cdots 2n$. When one polynomial is computed, it is not critical. However, the problem becomes urgent when many polynomials of various degrees are computed sequentially. In addition, in case of direct computation of factorials in the machine-precision floating-point arithmetic, significant rounding errors can accumulate. To partially solve the problem, the ROM lookup table (LUT) can be used which stores values $\frac{(i \cdot 2h+1)!!}{(i \cdot 2h)!!}$ for $i = 1, 2, \ldots, N$, where h and N are some integers. Then, for computing $\frac{(2n+1)!!}{(2n)!!}$, where n is the polynomial degree, one has to take the $\lfloor n/h \rfloor$-th value from LUT, and compute $\prod_{i=1}^{n-q} \frac{2(q+i)+1}{2(q+i)}$, where $q = h\lfloor n/h \rfloor$, and multiply these two values. The size of LUT is determined by step h and the maximum degree of polynomial n we want to compute. For instance, if $n = 50000$ and $h = 100$, LUT will contain $N = 500$ values. LUT content is computed in advance with high precision, after which it is converted into the extended-range format.

3.2 Developed Software for Computing Legendre Polynomials

Based on the implemented extended-range arithmetic functions (Subsect. 2.2), CPU- and CUDA-subroutines were developed, which allow computing $\bar{P}_n^m(x)$ for large $n \geq 0$ and at any m, $0 \leq m \leq n$. They are shown in Table 2.

For implementation on the GPU, the direct paralleling scheme was chosen, according to which i-th GPU thread computes a polynomial of degree $n[i]$ and order $m[i]$ for argument $x[i]$. The result is written with a corresponding offset to res array. The number of the required thread blocks is defined by the following:

$$N = \left\lceil \frac{vector\_size}{max\_threads\_per\_block} \right\rceil, \tag{7}$$

where $vector\_size$ is the size of the input vectors, $max\_threads\_per\_block$ is the maximum number of threads in a block. If N does not exceed the maximum number of blocks for the device, fully parallel computation of all polynomials is possible. Otherwise, some threads compute more than one polynomial. Listing of CUDA kernel legendre_1st is given in Fig. 4.

4 Experimental Results

The evaluation of correctness and efficiency of the developed subroutines was carried out within HP SL390+NVIDIA Tesla M2090 stand of UniHUB.ru platform at the Institute for System Programming of the Russian Academy of Sciences [15]. Three software implementations of the recursive algorithm (4) have

Table 2. Subroutines to compute normalized Legendre polynomials

Subroutine	Parameters	Description
legendre_eqls	er_t res er_t x uint32_t const n	Computation of $\bar{P}_n^n(x)$ in accordance with (5) with optimization from Subsect. 3.1. The result is a pointer **res**.
legendre_recur	er_t res er_t x er_t p1 er_t p2 uint32_t const n uint32_t const m	One iteration of recursion (4). For the given $\bar{P}_n^m(x)$ (parameter **p1**) and $\bar{P}_n^{m+1}(x)$ (parameter **p2**) $\bar{P}_n^{m-1}(x)$ is computed. The result is a pointer **res**.
legendre	er_t res double const x uint32_t n uint32_t m	Computation of normalized Legendre polynomial $\bar{P}_n^m(x)$ of degree n and order m. The result is a pointer **res**.
legendre_lst	er_arr_t res double const *x uint32_t const *n uint32_t const *m uint32_t size	Computation of the vector of normalized Legendre polynomials for given vector of arguments x, vector of degrees n, and vector of orders m. The result is a pointer **res** to an array.

been examined: CPU- and GPU-implementations based on extended-range arithmetic, and calculations using the GNU MPFR Library.

In the first experiment, we examined dependence of computation time for the first-kind polynomial on n. The value $\cos(179°) \approx -0.999848$ was taken as an argument. The degree n varied in the range of 100 to 53200, and it was doubled at each stage of the testing procedure. The results are presented in Fig. 5(a).

In the second experiment, vectors of polynomials were calculated at fixed $m = 0$ and $n = 20000$, whose size ranged from 32 to 8192. The arguments were defined by the formula $x_i = \cos\left(i \times \frac{180}{vector\_size}\right)$, which allowed calculations for each vector in the angular range $[0°, 180°]$, having a uniform step determined by the size of the vector ($vector\_size$). The experiment results are shown in Fig. 5(b). Longer GPU computation time, observed at the vector size greater than 2048, is explained by the limited resources of the used device.

It is worth noting that the subroutines for computation of normalized and non-normalized associated Legendre polynomials are implemented in a number of well-known software packages, such as The GNU Scientific Library, Boost, ALGLIB. However, they allow calculations only for rather small degrees (up to several thousand). Therefore, these implementations were not analysed in the experiments.

```
__global__ void legendre_1st(er_arr_t res,
                             double const *x,
                             uint32_t const *n,
                             uint32_t const *m,
                             uint32_t size){
const uint32_t id = threadIdx.x + blockIdx.x * blockDim.x;
  if (id < size) {
    uint32_t thread_n = n[id];
    uint32_t thread_m = m[id];
    er_static_t thread_x;
    cuda::er_set_d(&thread_x, x[id]);
    legendre_eqls(&res[id], &thread_x, thread_n);
    if (thread_n > thread_m) {
      er_static_t p0, p1, p2;
      cuda::er_set(&p1, &res[id]);
      cuda::er_set_d(&p2, 0.0);
      uint32_t current_m = thread_n;
      legendre_recur(&p0,&thread_x,&p1,&p2,thread_n,current_m);
      uint32_t iter_n = thread_n - thread_m - 1;
      for (uint32_t i = 0; i < iter_n; i++) {
        current_m = current_m - 1;
        cuda::er_set(&p2, &p1);
        cuda::er_set(&p1, &p0);
        legendre_recur(&p0,&thread_x,&p1,&p2,thread_n,current_m);
      }
      cuda::er_set(&res[id], &p0);
    }
  }
}
```

Fig. 4. CUDA kernel for computing normalized Legendre polynomials

The computed polynomials $\bar{P}_n^m(\cos\theta)$ for $n = 1000, 5000, 15000, 20000$, $m = 0$ with intervals of θ equal to $1°$ are shown in Fig. 6, and the logarithms of the starting values (5) for recursion (4) are shown in Fig. 7.

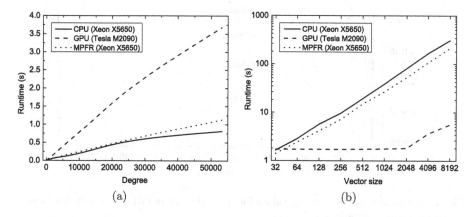

(a)   (b)

Fig. 5. Experimental results: computation time of $\bar{P}_n^m(x)$ at fixed $m = 0$ and $x = \cos(179°)$ versus degree n (a); computation time of the vector of $\bar{P}_n^m(x)$ at fixed $m = 0$ and $n = 20000$ versus the vector size (b)

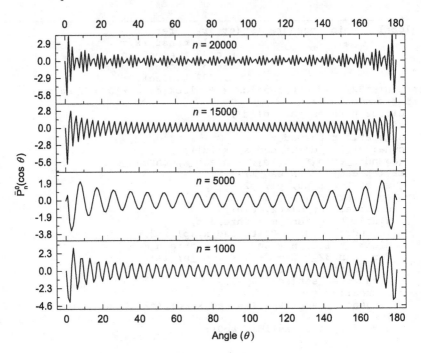

Fig. 6. Normalized associated Legendre polynomials

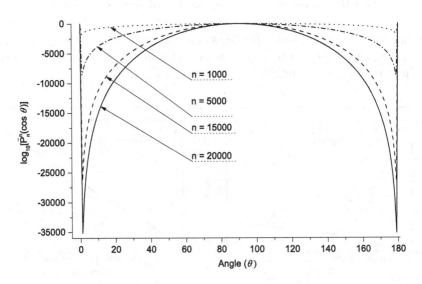

Fig. 7. Logarithms of the starting values for recursive computation of normalized associated Legendre polynomials

5   Conclusion

The paper considers the problem of GPU-based calculation of normalized associated Legendre polynomials. At high degrees n and orders m, this polynomials are characterized by a large spread of numerical values, which greatly limits the possibility of their correct computation in IEEE-754 arithmetic. In particular, when $n = 20000$, obtaining correct results in the double-precision format is only possible for angles ranging from 75° to 105°. Calculations for angles beyond this range result in underflow. To overcome this limitation, extended-range arithmetic is implemented on GPU. The experimental evaluation shows that, thanks to the natural task parallelism and a simple computation scheme, the use of GPU is effective even with small length vectors. With increasing problem size the speedup becomes more significant.

When parallel computation of polynomials of the same degree for a set of different arguments is made, the computation scheme is balanced, since time complexity of the extended-range arithmetic operations does not depend significantly on the magnitude of the arguments. If vectors of polynomials of different (high) degrees are computed, then, to improve the GPU-implementation performance, a more complicated computation scheme involving load balancing can be applied.

Acknowledgement. This work was supported by the Russian Foundation for Basic Research, project No. 16-37-60003 mol_a_dk.

References

1. Arfken, G.B., Weber, H.J., Harris, F.E.: Mathematical Methods for Physicists, 7th edn. Academic Press, Boston (2013)
2. Wittwer, T., Klees, R., Seitz, K., Heck, B.: Ultra-high degree spherical harmonic analysis and synthesis using extended-range arithmetic. J. Geodesy **82**(4), 223–229 (2008). doi:10.1007/s00190-007-0172-y
3. Fukushima, T.: Numerical computation of spherical harmonics of arbitrary degree and order by extending exponent of floating point numbers. J. Geodesy **86**(4), 271–285 (2012). doi:10.1007/s00190-011-0519-2
4. Morris, R.J., Najmanovich, R.J., Kahraman, A., Thornton, J.M.: Real spherical harmonic expansion coefficients as 3D shape descriptors for protein binding pocket and ligand comparisons. Bioinformatics **21**(10), 2347–2355 (2005). doi:10.1093/bioinformatics/bti337
5. Rose, M.: Elementary Theory of Angular Momentum. Wiley, New York (1957)
6. Galassi, M., Davies, J., Theiler, J., Gough, B., Jungman, G., Alken, P., Booth, M., Rossi, F., Ulerich, R.: GNU Scientific Library (2016). https://www.gnu.org/software/gsl/manual/gsl-ref.pdf
7. Belousov, S.: Tables of Normalized Associated Legendre Polynomials. Pergamon Press, New York (1962)
8. Smith, J.M., Olver, F.W.J., Lozier, D.W.: Extended-range arithmetic and normalized Legendre polynomials. ACM Trans. Math. Softw. **7**(1), 93–105 (1981). doi:10.1145/355934.355940

184 K. Isupov et al.

9. Wenzel, G.: Ultra-high degree geopotential models GPM98A, B, and C to degree 1800. In: Joint Meeting of the International Gravity Commission and International Geoid Commission, Trieste (1998)
10. Holmes, S.A., Featherstone, W.E.: A unified approach to the Clenshaw summation and the recursive computation of very high degree and order normalised associated Legendre functions. J. Geodesy **76**(5), 279–299 (2002). doi:10.1007/s00190-002-0216-2
11. Hauser, J.R.: Handling floating-point exceptions in numeric programs. ACM Trans. Program. Lang. Syst. **18**(2), 139–174 (1996). doi:10.1145/227699.227701
12. IEEE Standard for Binary Floating-Point Arithmetic. ANSI/IEEE Std 754-1985, pp. 1–20 (1985). doi:10.1109/IEEESTD.1985.82928
13. IEEE Standard for Floating-Point Arithmetic. IEEE Std 754-2008, pp. 1–70 (2008). doi:10.1109/IEEESTD.2008.4610935
14. Muller, J.M., Brisebarre, N., de Dinechin, F., Jeannerod, C.P., Lefèvre, V., Melquiond, G., Revol, N., Stehlé, D., Torres, S.: Handbook of Floating-Point Arithmetic. Birkhäuser Boston, New York (2010)
15. The "University cluster" program's technological platform. https://unihub.ru/resources/sl390m2090

Parallel Software for Simulation of Nonlinear Processes in Technical Microsystems

Sergey Polyakov$^{(\boxtimes)}$, Viktoriia Podryga, Dmitry Puzyrkov, and Tatiana Kudryashova

Keldysh Institute of Applied Mathematics of RAS,
Miusskaya Square 4, 125047 Moscow, Russia
polyakov@imamod.ru

Abstract. The modern stage of the industry evolution is characterized by introduction of nanotechnologies in production. Therefore, scientific researches of various technological processes and facilities at different levels of detailing up to atomic are become actual. Multiscale modeling of microsystems, which combines the methods of continuum mechanics and molecular dynamics, has become one of the possible approaches. This report presents elements of supercomputer technology and software, which enable us to solve some problems of nanotechnologies within the chosen approach.

Keywords: Mathematical simulation · Multiscale approaches · Continuum mechanics · Mesh methods · Molecular dynamics · Parallel algorithms and programs · Supercomputer computations

1 Introduction

At present time the world and domestic industry develop their production by implementing the perspective nanotechnologies. In this connection, the scientific research of engineering processes and complex technical systems at different scale levels down to atomic has become particularly relevant. The multiscale modeling of microsystems combining methods of continuum mechanics and molecular dynamics became one of widely used mathematical approaches. However the realization of multiscale computer researches demands development of new steady numerical methods, parallel algorithms and software tools for ultra-high performance calculators.

In this work the supercomputer technology details and the software allowing to solve some urgent problems of nanotechnologies within the chosen approach are represented. Micro- and nanosystems are the main object of the research. These systems are used in the solution of practically important problems related to the application of modern micro- and nanoelectronics. Creation of new nanocoverings and nanomaterials by the methods of a gasdynamic spraying the nanoparticles on a substrate [1,2], and also the analysis of chips life

V. Voevodin and S. Sobolev (Eds.): RuSCDays 2016, CCIS 687, pp. 185–198, 2016.
DOI: 10.1007/978-3-319-55669-7_15

cycle, including the problems of interconnections degradation in electronic circuits [3–6] have been chosen as applications. In this a little heterogeneous applied subject the general fundamental component was selected. It is connected to simulation of nonlinear processes in metal–metal, metal–gas, gas–gas, and also metal–semiconductor, metal–dielectric microsystems.

The chosen microsystems differ in the fact that at relatively small sizes of their active elements or layers (about tens or several hundred nanometers) the systems themselves can have quite impressive sizes (from few microns to few centimeters) in one or several dimensions. Therefore, modeling of such systems is necessary to produce at least two scales. The first of them refers to the system in general (macrolevel), the second – to the separate small parts of system down to the molecules and atoms (microlevel).

As basic models at the macrolevel the different variants of hydrodynamic system of equations were chosen. They included the elastodynamics (QGD) [7,8] system of equations supplemented by the electrodynamics equations. As basic model at the microlevel Newton's dynamic equations were used. Statements of specific mathematical problems are the boundary or initial-boundary value problems for these equations supplemented by material conditions including equations of state and connections. Numerical methods for solving problems at macrolevel are the grid and are based on the finite volume method [9]. Numerical methods for solving Newton's equations are based on the Verlet scheme [10]. The general numerical procedure represents a method of splitting into physical processes [11]. The parallelism principles on coordinates and functions [12] and the domain decomposition technology [13] are used in parallel realization of the specified numerical algorithms. Program implementation of approach is executed on a hybrid technology [14] and is oriented on application of MPI [15], OpenMP [16] and CUDA Toolkit [17] software. Computer calculations were performed on systems with central, vector and graphic processors (CPU, VPU and GPU).

2   Statements of Mathematical Problems

In this section, examples of specific problems relevant to the subject areas mentioned above are given.

2.1   Mathematical Model for the Problem of Gasdynamic Spray

The central problem of gasdynamic spray modeling is the calculation of gas mixture flow near the solid surface of installation. As mentioned above, for this purposes it is proposed to use a multiscale approach, where research is carried out on at least two scale levels – the basic one (macroscale, ranges from a few tens of microns to tens of millimeters) and the additional one (microscale, of the order of a micron or less). The mean gas flow in the channel is computed at the basic level. At the additional level, calculation is performed for the interactions: (1) between the gas molecules (determining the flow characteristics); (2) between the gas molecules and the solid surface atoms (describing boundary-layer phenomena). The implementation is based on splitting into physical processes.

At macroscales it is offered to use quasigasdynamic system of equations in relaxation approximation [7]. Motivation for this is the fact that the QGD system belongs to the class of kinetically-consistent approaches and significantly expands the possibilities of Navier-Stokes model. Their main difference of QGD from the Navier-Stokes equations is the use of a spatiotemporal averaging procedure for determining the basic gasdynamic parameters (density, momentum, and energy). Moreover, the QGD equations have terms that implement additional smoothing in time and are effective regularizers. The influence of these terms is exhibited in the case of strongly unsteady flows at Knudsen numbers close to unity. One more factor in favor of choosing the QGD equations is the nondimensionalization parameter size on space coinciding with the mean free path of gas molecules. The mean free path in temperature range 100 – 1200 K (characteristic for spraying tasks) is between tens and hundreds of nanometers to tens of microns. At last, explicit versions of sampled QGD equations allow use both structured and unstructured grids, and build effective parallel algorithms, that are easy to implement on modern high-performance computing systems.

As an example, we consider a problem about binary gas mixture flow in a microchannel of spraying installation. The length and diameter can vary in a wide range, so that the Knudsen number can range from 0.001 to 1. In the case of a gas mixture, the QGD system is written for each gas separately, but moment and energy equations include exchange terms, which are responsible for the agreement between the mixture parameters as a whole.

We write down QGD system of equations for the case of binary gas mixture in a form invariant under coordinate system, together with the equations of state:

$$\frac{\partial \rho_l}{\partial t} + \operatorname{div} \mathbf{W}_l^{(\rho)} = 0, \quad l = a, b, \tag{1}$$

$$\frac{\partial}{\partial t}\left(\rho_l u_{lk}\right) + \operatorname{div} \mathbf{W}_l^{(\rho u_k)} = S_l^{(\rho u_k)}, \quad l = a, b, \quad k = x, y, z, \tag{2}$$

$$\frac{\partial E_l}{\partial t} + \operatorname{div} \mathbf{W}_l^{(E)} = S_l^{(E)}, \quad l = a, b, \tag{3}$$

$$E_l = \rho_l\left(\frac{1}{2}|\mathbf{u}|^2 + \varepsilon_l\right), \quad p_l = Z_l \rho_l R_l T_l, \quad \varepsilon_l = c_{V,l} T_l, \quad l = a, b'. \tag{4}$$

Here a and b are labels of gases of which mixture consists. Each gas has its own mass density $\rho_l = m_l n_l$, mass of molecule m_l, number density (concentration) n_l, temperature T_l and macroscopic velocity $\mathbf{u}_l$. Other parameters: p_l are partial pressures, E_l are total energy densities, ε_l are internal energies, Z_l are compressibility factors, $c_{V,l}$ are heat capacities at constant volume, $R_l = k_B/m_l$ are gas constants (k_B is the Boltzmann's constant). Vectors $\mathbf{W}_l^{(\rho)}$, $\mathbf{W}_l^{(\rho u_k)}$, $\mathbf{W}_l^{(E)}$ coincide, up to the sign, with the density fluxes, fluxes of the corresponding components of the momentum density and energy density. They include QGD corrections proportional to Maxwell relaxation time, and depend on the coefficients of viscosity μ_l and thermal conductivity χ_l. The exchange terms $S_l^{(\rho u_k)}$

and $S_l^{(E)}$ take into account the momentum and energy redistribution between the gas mixture components depending on the molecular collision frequency.

At microscales the method of molecular dynamics (MD) [18–21] is used. In this case at the microlevel far from the channel walls the particles of two kinds (molecules of two gases) are considered, near the walls the particles making the wall material (usually metal) are added. The behavior of particles is described by the following Newton's equations:

$$m_l \frac{\mathbf{v}_{l,i}}{dt} = \mathbf{F}_{l,i}, \quad \mathbf{v}_{l,i} = \frac{d\mathbf{r}_{l,i}}{dt}, \quad i = 1, ..., N_l, \quad l = a, b, c. \tag{5}$$

Here i is particle's index, l is particle's kind (a and b designate a molecules of the first and second gases, c designates a metal atoms), N_l is total particle number of kind l. The particle of kind l with index i has mass m_l, its own position vector $\mathbf{r}_l, i$, velocity vector $\mathbf{v}_{l,i}$ and total force $\mathbf{F}_{l,i}$ acting on this particle. The interaction of particles is described using the potentials depending on the particles coordinates. The choice of interaction potential is based on comparing the mechanical properties of computer model of a potential and actual material (more details in [18–21]).

The initial conditions at the microlevel are determined by the equilibrium or quasi equilibrium thermodynamic state of particle system at given temperature, pressure, and mean momentum. The boundary conditions at the molecular level depend on the situation to be simulated. To determine the general properties of the medium, it is sufficient to consider a distinguished three-dimensional volume of it with periodic boundary conditions on all coordinates. In the study of actual geometry microsystems, such as a microchannel, one or several directions are of finite size and the shape of the object is preserved by choosing a suitable potential or fixing the system. The temperature and total momentum of the system are controlled with the help of thermostats algorithms.

The system of QGD equations is closed by initial and boundary conditions. The initial conditions correspond to an equilibrium state of the gas medium without interactions with external factors. It is possible to consider the case of a quiescent gas medium in the entire computational domain. The densities, velocities, and total energies of the gas components are set at the channel inlet. "Soft" boundary conditions [7] are specified on free surfaces. A special microscopic system consisting of gas molecules and metal atoms is introduced near the channel walls. The boundary conditions on the wall are set as third-kind conditions describing the exchange of mass, momentum components, and energy between the gas mixture in the flow and in the near-wall layer. These conditions involve accommodation coefficients determined by tabulated physical data (which is possible for limited ranges of temperatures and pressures) or computed using the MD method. Another variant of formulating the boundary conditions is the direct MD computation of density, momentum, and energy fluxes through the boundary of the near-wall layer.

To conclude this section we note that for the metal surfaces of the channel the equation of heat conduction and/or the equation of thermoelasticity should be written on a macrolevel. However, in this case they can be replaced with conditions of thermostatting and reset moment which are easily joined in a system of Eq. (5) by introducing the thermo- and the barostats adjusting the particle velocities [19].

2.2 Mathematical Model of Interconnects Degradation

The problem of interconnects degradation in electric circuits of chips is one of the reasons of rates deceleration in development of modern electronics. Physical process of interconnects degradation is associated with formation of structural defects, and then pores and breaks in the electrical lines supplying the chip [3–6]. Formation of structural defects in metal is connected in its turn with the electronic wind phenomenon when in case of current passage the freely moving electrons of a conduction band begin to pull out metal ions from its crystalline grid. This process significantly amplifies at reducing the cross section of conductors. In many previous studies it was shown that the pore formation process is not normal diffusion process. Its driving force is the free energy of disordering the atoms in a metal grid generated by at least four factors: an electromagnetic field, heat, mechanical stresses and chemical interactions at the medium boundary. It is necessary to add capillary phenomena to this also.

The description of the specified physical processes of pore formation in case of the modern interconnects sizes of hundreds and even tens of nanometers can't be carried out only on the basis of the hydro- and electrodynamics equations. It is also necessary to use multiscale approach. In this paper in addition to the macromodel used in [3] it is offered to consider the grain structure of conductors and their surroundings materials calculated at the molecular level. As a result, kinetic coefficients in the macroequations and boundary effects will be calculated directly by methods of molecular dynamics that will make calculation as realistic as possible.

Here is an example of such a complex model. First, we write the hydro- and electrodynamics equation in quasi stationary quasi neutral case:

$$
\begin{cases}
\operatorname{div} \mathbf{j} = 0, & \mathbf{j} = \sigma_e \mathbf{E}, \quad \mathbf{r} \in \Omega_1 ; \\
\operatorname{div} \mathbf{D} = 0, & \mathbf{D} = \varepsilon_e \mathbf{E}, \quad \mathbf{r} \in \Omega_2 ;
\end{cases}
\tag{6}
$$

$$
\mathbf{E} = -\nabla \varphi, \quad \mathbf{r} \in \Omega = \Omega_1 \cup \Omega_2 ;
\tag{7}
$$

$$
\operatorname{div} \mathbf{q} = \gamma_0 \left(\mathbf{E} \cdot \mathbf{j} \right), \quad \mathbf{q} = -k_T \nabla T, \quad \mathbf{r} \in \Omega ;
\tag{8}
$$

$$
\operatorname{div} \sigma_i = 0, \quad i = 1, 2, 3, \quad \mathbf{r} \in \Omega ;
\tag{9}
$$

$$
\begin{aligned}
\sigma_{ij} &= 2\mu \varepsilon_{ij} + \delta_{ij} \left(\lambda \operatorname{div} \mathbf{u} - \left(\lambda + \frac{2}{3}\mu \right) \alpha \right), \\
\varepsilon_{ij} &= \frac{1}{2} \left(\frac{\partial u_i}{\partial x_j} + \frac{\partial u_j}{\partial x_i} \right), \quad i, j = 1, 2, 3, \quad \mathbf{r} \in \Omega ;
\end{aligned}
\tag{10}
$$

$$
\frac{\partial C}{\partial t} = -\operatorname{div} \mathbf{W}, \quad \mathbf{W} = -D_C \nabla C - \bar{D}_C \nabla \Phi, \quad \mathbf{r} \in \Omega_1, \quad t > 0.
\tag{11}
$$

Here $\mathbf{r} = (x_1, x_2, x_3)^T$ is position-vector in a space, $\mathbf{j}$, $\mathbf{D}$, σ_e, ε_e, $\mathbf{E}$ and φ are vector of electric current density, electric induction vector, nonlinear conductivity of conductors, dielectric permittivity of conductors surrounding, tension and potential of electric field respectively; div and ∇ are operators of divergence and gradient in Cartesian coordinates with position-vector $\mathbf{r}$, Ω is computational domain comprising conductors Ω_1 and insulators Ω_2; $\mathbf{q}$, $\mathbf{k}_T$, T are heat flux vector, nonlinear tensor of temperature conductivity, temperature, γ_0 is dimensionless parameter; $\sigma_i = (\sigma_{i1}, \sigma_{i3}, \sigma_{i3})^T$ are column vectors constituting the tensor of thermoelastic stresses, $\mathbf{u} = (u_1, u_2, u_3)^T$ is displacement vector, $\mu = \mu(T)$, $\lambda = \lambda(T)$ are dimensionless nonlinear coefficients of the Lame, $\alpha = \alpha(T, C)$ is nonlinear function of load arising during thermal expansion of the metal and the change of its mass composition; C is mass fraction (normalized concentration) of metal in conductors, t is time, $\mathbf{W}$ is total diffusion flux, D_C and $\bar{D}_C$ nonlinear diffusion coefficients depending on C and T, $\Phi = \Phi(\varphi, C, T, H)$ is generalized thermodynamic potential (H trace of thermoelastic stresses tensor).

The boundary and initial conditions are as follows:

$$\mathbf{j} \cdot \mathbf{n} = \begin{cases} j_{in}, & \mathbf{r} \in \partial\Omega_1^{(1)}, \\ j_{out}, & \mathbf{r} \in \partial\Omega_1^{(2)}, \quad \mathbf{D} \cdot \mathbf{n} = 0, \quad \mathbf{r} \in \partial\Omega_2; \\ 0, & \mathbf{r} \in \partial\Omega_1^{(3)}, \end{cases} \tag{12}$$

$$T = T_0, \quad \mathbf{r} \in \partial\Omega; \tag{13}$$

$$\begin{cases} \dfrac{\partial\sigma_{ij}}{\partial x_j} = 0, & \mathbf{u} = \mathbf{u}_0, \quad \mathbf{r} \in \partial\Omega^{(1)}, \\ \sigma_i \cdot \mathbf{n} = 0, & \mathbf{r} \in \partial\Omega^{(2)}; \end{cases} \tag{14}$$

$$C|_{t=0} = C_{i1}, \quad \mathbf{r} \in \Omega_1; \tag{15}$$

$$\begin{cases} C = C_{i1}, & \mathbf{r} \in \partial\Omega_1^{(1)}, \\ \nabla C \cdot \mathbf{n} = 0, & \mathbf{r} \in \partial\Omega_1^{(2)}, \\ \mathbf{W} \cdot \mathbf{n} = 0, & \mathbf{r} \in \partial\Omega_1^{(3)}. \end{cases} \tag{16}$$

Here $\partial\Omega_1^{(1)}$, $\partial\Omega_1^{(2)}$, $\partial\Omega_1^{(3)}$ are boundary regions of contact of the conductor with the feeding elements (on them input currents with density j_{rmin} are set), boundary regions of the contacts which are leading out current (on them output currents with density j_{rmout} are set), the boundary regions adjoining to insulators, $\partial\Omega$ is boundary of computational domain, $\partial\Omega^{(1)}$, $\partial\Omega^{(2)}$ are two parts of the common boundary, where various conditions to thermomechanical stresses are set (on the first the displacements are set, on the second the free boundary conditions are set), $\mathbf{n}$ is external normal to boundaries, T_0 is the chip temperature, C_{i1} is equilibrium mass fraction of metal in the conductor. Note that at the internal borders between the conductors and dielectrics the standard conjugation conditions are given.

The coefficients of the Eqs. (6)–(11) are discontinuous and in a nonlinear manner depend on temperature and mass fraction of metal. The specific form of the used dependences is very diverse and is determined by conditions of the modeled physical experiment and parameters of an electric circuit. One way to determine these dependencies is to apply the methods of molecular dynamics. Therefore, the second part of the problem is to solve Newton's equations in the whole calculation domain or only at boundaries between the different materials (because information about properties of materials on these boundaries is available least of all). Actually the system of Newton's equations in this case has the same form as the system (5) and is added by different initial and boundary conditions depending on physical conditions. The quantity of particle kinds depends on composition of the used materials. For the modern types of chips as conductors the compounds of copper (the main composition of the conductor) and tantalum (the hardening shell of the conductor) are used. As insulators the silicon oxides and carbides are used.

3   Numerical Methods, Parallel Algorithms and Programs

The proposed multiscale approach can be used in two ways:

(1) using MD simulation it is possible to create the database on material properties and to use it in macromodels for determination of kinetic coefficients, the equations of state parameters and boundary conditions;
(2) it is possible to carry out MD computation during macroscopic calculations within a method of splitting into physical processes and to use them as the subgrid algorithm adjusting medium macroparameters.

The first way of calculations leads to large computational cost at the stage of calculation preparation and to large volumes of the disk space required to store the calculated initial data. The second way of calculations (direct macroscopic and microscopic simulation) is more justified if the database on material properties is incomplete or at all is absent. In this situation the second way provides a complete simulation cycle by all necessary data and allows to coordinate processes on micro- and macrolevels, and also to make calculations for a certain specific set of conditions and at the same time to accumulate the database for a case of repeated computing experiments.

Let us consider some details of developed approach in relation to the problems discussed above in the case of use the full version of computing.

3.1   Numerical Algorithm for Solving the Problems of Spraying

In solving gasdynamic spraying tasks in the full version (QGD + MD), the QGD system of the equations is sampled by the finite volume method [9] on the suitable structured or unstructured grids and is solved on the basis of explicit or implicit schemes on time (in implicit schemes case a suitable iterative process is applied). MD system of equations is used as a sub-grid algorithm and solved by means of Verlet scheme [10]. At the microlevel all calculations are made independently in each control volume of a spatial grid, except for boundary cells. In boundary cells the nonlocal MD scheme of computations can be used. This is especially relevant in a case of large Knudsen numbers (about 1 and more) when QGD system loses accuracy owing to violation of the gaseous medium continuity hypothesis.

As a result, the overall algorithm consists of four basic steps:

(1) calculating the macroparameters of gas components according to grid analogues of QGD equations excluding exchange terms in grid cells where approximation of a continuous medium is valid;
(2) MD calculation of kinetic coefficients, equations of state parameters and exchange terms in average flow field based on local algorithms;
(3) MD calculation of kinetic coefficients, equations of state parameters, exchange terms and parameters of boundary conditions near the computational domain boundaries and the boundaries between the different material types on the basis of nonlocal algorithm;
(4) return to a macrolevel and correction of gases moment and total energy densities, and also of a metal surface state.

Criterion for stopping the MD calculations is either the achievement of characteristic time of molecular system evolution (maxwellization time), or small change (for 1–2%) one or more macroparameters of molecular system (average momentum, averages kinetic and/or potential energies). A detailed presentation of the algorithm is considered in [21].

Emphasize once more that the difference of the proposed approach from other approaches in literature is the possibility of simulating the complex gas flows. It is provided by means of the MD methods determining the kinetic coefficients and exchange terms necessary for computation of the QGD system of equations, the compressibility coefficients and specific heat capacities underlying the equations of state and also boundary conditions.

3.2   Numerical Algorithm for Solving the Problems of Interconnects Degradation

The total numerical algorithm for solving the problems related to modeling of degradation processes in electronic circuits interconnects is similar to discussed in Sect. 3.1. Difference is that in this case on a macrolevel more difficult system of electrohydrodynamics (EHD) equations is considered. Also in this case it is possible to consider truncated (EHD + the database of molecular simulation) and full (EHD + MD) calculation ways. The quantity of calculation components increases both in connection with a large number of the macroequations, and with a large amount of the considered materials. In addition, at macro- and at microlevels it is necessary to take into account the formation of the defects, pores and vacuum layers appearing in investigated materials, and also the phenomena of penetration the atoms and molecules of one material in thickness of another. In general, the method of calculation remains the same:

(1) calculating the macroparameters according to grid analogues of EHD equations in grid cells where approximation of a continuous medium is valid;

(2) MD calculation of separate and mixed materials macroparameters away from the boundaries based on local algorithms;

(3) MD calculation of macroparameters near the computational domain boundaries and the boundaries between the different material types on the basis of nonlocal algorithm;

(4) return to a macrolevel and correction of macroparameters in all grid points.

3.3   Parallel Implementation of Proposed Approach

Parallel implementation of proposed multiscale approach assumes use a cluster (or a supercomputer) with the central or hybrid architecture having several multi-core central processors (CPU), and also several vector or graphic processors (VPU or GPU) on each node. Parallelization of algorithm is performed on the principles of coordinate parallelism and domain decomposition. The main calculation on a macrolevel is made by the discrete QGD or EHD equations on the grid distributed between cluster nodes by means of domain decomposition technique. Inside node the computations on a macrolevel are distributed between CPU threads. Subgrid MD computations are made by VPU or GPU in case of their existence. Multisequencing of computations on a macrolevel between the CPU threads is also made geometrically. Multisequencing of MD computations is made by dividing the entire set of particles belonging to one grid cell on groups of identical power. Each unit of VPU or GPU threads processes one or several molecular groups belonging to one or several grid nodes. More parallelization details were discussed in [19–22]. A brief study of the parallelization quality for the proposed approach is discussed in Sect. 4.

3.4   Program Realization

As mentioned above, program implementation of approach is executed on a hybrid technology [14] used MPI [15], OpenMP [16] and CUDA [17]. Development was carried out in the languages of ANSI C/C++ with templates of hybrid parallel applications developed for GIMM_NANO software tool [23], created within the state contract No. 07.524.12.4019 of the Ministry of Education and Science of the Russian Federation. As a result, following program variants have been established and registered in Rospatent:

- GIMM_APP_QGD_CPU is a program of calculation on CPU the two-dimensional and three-dimensional gas mixture flows in microchannels;
- GIMM_APP_MD_CPU_Gas_Metal is a program of calculation on CPU the gas-metal microsystems;
- GIMM_APP_MD_GPU_Gas_Metal is a program of calculation on GPU the gas-metal microsystems;
- GIMM_APP_QGD_MD_CPU is a program of calculation on CPU the two-dimensional and three-dimensional gas mixture flows in microchannels based on QGD + MD approach.

The GIMM_APP_QGD_GPU and GIMM_APP_QGD_MD_GPU programs are at the stage of testing and registration, they are oriented on more intensive use of graphics accelerators possibilities.

At a stage of development there are programs of calculation of degradation processes in electronic circuits interconnects named GIMM_APP_VOID_EGD _HYB, GIMM_APP_VOID_MD_HYB and GIMM_APP_VOID_EGD_MD_HYB. These programs are oriented on joint use of the CPU and GPU. Also the editor of microstructures GIMM_MICRO_STRUCT_EDITOR allowing to create model microstructures of the size, unlimited on data volume, is developed.

In addition the specialized software [24,25] for processing and visualization of the distributed results of molecular simulation was developed. This software will be a basis of visualization system in the cloud version of the GIMM_NANO program tool.

4   Results

In this section some of the results obtained by the developed numerical approach and software will be discussed.

4.1   Simulation of Gas Flows in Microchannels

As an illustration of using the developed parallel tool for problems of multiscale gasdynamic processes modeling following information is presented here. The gas mixture flow in a metal microchannel of square cross section was considered. The calculations were performed on the mixed algorithm represented the alternation of QGD and MD computing in all cells of the grid, including the boundary cells. Calculations of gas mixture flows were provided on K100 hybrid cluster with CPU and GPU with the help of parallel programs GIMM_APP_QGD_MD_CPU and GIMM_APP_QGD_MD_CPU. The K100 has 64 nodes, each node of the cluster has two CPU Intel Xeon X5670 (6 cores and 12 threads per one CPU) and three GPU NVidia Tesla C2050 (448 CUDA cores per one GPU). We tested our software on 3D Cartesian grid with sizes 240240720 cells. The calculations were connected with evolution of nitrogen and hydrogen jet in nickel microchannel with sizes $30 \times 30 \times 90$ microns. Each grid cell in dependence on physical conditions can consist of 500–50000 gas particles. The results of testing are shown on Fig. 1. These results confirm that using of CPU systems is more effective. Nevertheless using of GPU allows reducing calculation time by 6–10 times. The above conclusions characterize exactly the K100 architecture in which the CPUs are relatively slow and GPUs have a small graphics memory. The using of more modern GPU devices (NVidia Tesla K40 or K80) will decrease calculation times by 20–40 times and increase the efficiency of GPU computations [26].

Concurrently with development of final versions of a numerical technique and program tools the simulation of separate model subtasks was carried out. In particular, subsonic, transonic and supersonic flows of pure gases and a binary gas mixture on the example of hydrogen and nitrogen were numerically analyzed.

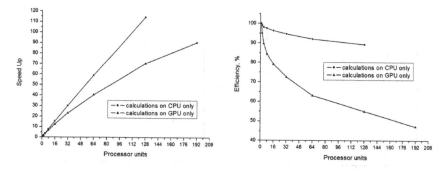

Fig. 1. Speed up (on a left) and efficiency (on a right) of QGD-MD algorithm obtained by separate calculations on CPUs and GPUs.

The results obtained in preliminary calculations in good agreement with the known tabular data and experimental results. In [27] the nitrogen and hydrogen mixture flow was calculated on an output from a micronozzle in the semi-open microchannel and further to the free space. The calculation results were close to data of experiments. In [27] combining the micro- and macromodels of the gas medium in a uniform research object was offered. In [28] interaction of a gas flow with microchannel walls on the example of nitrogen-nickel system was considered. This calculation represents a technique of receiving boundary conditions by direct MD computation. In [19–21,29,30] methods for calculating the thermodynamic equilibriums in metal–metal, gas–gas and gas–metal systems were developed. In particular, nickel–nickel, aluminum–aluminum, argon–argon, nitrogen–nitrogen, nitrogen–nickel systems were considered. In [20] on the example of nitrogen the technique of specification of the equations of state was considered. Without giving specific results of calculations here, we would like to emphasize that the developed multiscale numerical procedure has proved its efficiency on this class of tasks and leads to great prospects when using computing systems of the PFLOPS and EFLOPS performances.

4.2 Modeling the Processes of Void Formation

Simulation of degradation processes in interconnects can be observed in a full-scale experiment by means of the roentgenogram analysis where the regions of voids (pores) formation in the electrical line material are well visible. However the roentgenogram is made quite rarely and after there was a line interruption. And it is quite difficult to track an interruption dynamics. These circumstances induce to use mathematical simulation as the analysis tool.

By means of the technique provided in paper [3] the effect of pore formation was obtained (see Figs. 2 and 3) and the research of its dynamics was conducted. However calculations were carried out in two-dimensional statement and didn't consider a lot of physical factors. In this work with the help of new approach it was succeeded to repeat the received result in a three-dimensional case. Even

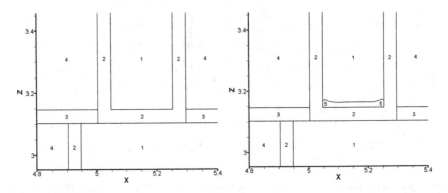

Fig. 2. Distribution of the copper mass fraction in interconnect before of pore birth (left) and in the final of its evolution (right). The digits 1, 2, 3, 4 and 5 are designate the copper, tantalum, isolator, thermal isolator and pore

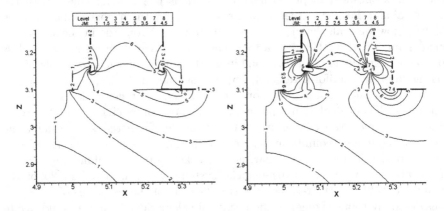

Fig. 3. Isolines of the current modulus in interconnect before of pore birth (on a left) and in the final of its evolution (on a right)

more exact calculation shall take into account the grain structure of metals and the dielectrics isolating them and the phenomenon of interpenetration of materials each other. Considering this circumstance, we note that in general the developed approach and its program implementations allow conducting such kinds of research. There is an open question of computing resources necessary for this purpose. Nevertheless, the maximum use of hybrid computing systems is the major direction in carrying out numerical experiments.

Acknowledgments. This work was supported by Russian Foundation for Basic Research (projects Nos. 14-01-00663-a, 15-07-06082-a, 15-29-07090-ofi-m, 16-07-00206-a, 16-37-00417-mol-a).

References

1. Papyrin, A., Kosarev, V., Klinkov, S., Alkhimov, A., Fomin, V.: Cold Spray Technology. Elsevier Science, Amsterdam (2007)
2. Resnick, D.: Nanoimprint lithography. In: Martin Feldman (ed.) Nanolithography. The Art of Fabricating Nanoelectronic and Nanophotonic Devices and Systems. Woodhead Publishing Limited (2014)
3. Karamzin, Y.N., Polyakov, S.V., Popov, I.V., Kobelkov, G.M., Kobelkov, S.G., Choy, J.H.: Numerical simulation of nucleation and migration voids in interconnects of electrical circuits (in Russian). Math. Simul. **19**(10), 29–43 (2007)
4. Zhu, X., Kotadia, H., Xu, S., Lu, H., Mannan, S.H., Chan, Y.-C., Bailey, C.: Multi-physics computer simulation of the electromigration phenomenon. In: 12th International Conference on Electronic Packaging Technology and High Density Packaging, pp. 448–452. ICEPT-HDP, Shanghai, China (2011)
5. Zhu, X., Kotadia, H., Xu, S., Lu, H., Mannan, S.H., Bailey, C., Chan, Y.-C.: Modeling electromigration for microelectronics design. J. Comput. Sci. Technol. **7**(2), 1–14 (2013)
6. Zhu, X., Kotadia, H., Xu, S., Mannan, S., Lu, H., Bailey, C., Chan, Y.-C.: Electromigration in Sn-Ag solder thin films under high current density. Thin Solid Films **565**, 193–201 (2014)
7. Elizarova, T.G.: Quasi-Gas Dynamic Equations. Springer, Heidelberg (2009)
8. Elizarova, T.G., Zlotnik, A.A., Chetverushkin, B.N.: On Quasi-gasdynamic and Quasi-hydrodynamic equations for binary mixtures of gases. Dokl. Math. **90**(3), 1–5 (2014)
9. Eymard, R., Gallouet, T.R., Herbin, R.: The finite volume method. In: Ciarlet, P.G., Lions, J.L. (eds.) Handbook of Numerical Analysis, vol. 7, pp. 713–1020. North-Holland, Amsterdam (2000)
10. Verlet, L.: Computer experiments on classical fluids. I. thermodynamical properties of Lennard-Jones molecules. Phys. Rev. **159**, 98–103 (1967)
11. Marchuk, G.I.: Splitting-up Methods (in Russian). Science, Moscow (1988)
12. Voevodin, V.V., Voevodin, V.V.: Parallel Computing (in Russian). BHV-Petersburg Publ., Saint-Petersburg (2002)
13. Quarteroni, A., Valli, A.: Domain Decomposition Methods for Partial Differential Equations. Clarendon press, Oxford (1999)
14. Polyakov, S.V., Karamzin, Y.N., Kosolapov, O.A., Kudryashova, T.A., Sukov, S.A.: Hybrid supercomputer platform and applications programming for the solution of continuous mechanics problems by grid methods (in Russian). Izvestiya SFedU. Eng. Sci. **6**(131), 105–115 (2012)
15. Official Documentation and Manuals on MPI. http://www.mcs.anl.gov/research/projects/mpi/
16. Official Documentation and Manuals on OpenMP. http://www.openmp.org, http://www.llnl.gov/computing/tutorials/openMP
17. CUDA Toolkit Documentation v. 7.5. Santa Clara (CA, USA): NVIDIA Corporation. http://docs.nvidia.com/cuda/index.html
18. Frenkel, D., Smit, B.: Understanding Molecular Simulation From Algorithm to Applications. Academic Press, New-York (2002)
19. Podryga, V.O., Polyakov, S.V.: Molecular dynamics simulation of thermodynamic equilibrium establishment in nickel. Math. Models Comput. Simul. **7**(5), 456–466 (2015)

20. Podryga, V.O.: Determination of real gas macroparameters by molecular dynamics (in Russian). Math. Simul. **27**(7), 80–90 (2015)

21. Podryga, V.O., Polyakov, S.V., Puzyrkov, D.V.: Supercomputer molecular modeling of thermodynamic equilibrium in gas-metal microsystems (in Russian). Numer. Methods Program. **16**(1), 123–138 (2015)

22. Podryga, V.O., Polyakov, S.V.: Parallel implementation of multiscale approach to the numerical study of gas microflows (in Russian). Numer. Methods Program. **17**(3), 138–147 (2016)

23. Bondarenko, A.A., Polyakov, S.V., Yakobovskiy, M.V., Kosolapov, O.A., Kononov, E.M.: Software Package GIMM_NANO (in Russian). In: International Supercomputer Conference: Scientific Service on the Internet: All Facets of Parallelism, pp. 1–5. CD-proceedings, Novorossiysk (2013)

24. Polyakov, S.V., Vyrodov, A.V., Puzyrkov, D.V., Yakobovskiy, M.V.: Cloud service for decision of multiscale nanotechnology problems on supercomputer systems (in Russian). Proc. Inst. Syst. Program. RAS **27**(6), 409–420 (2015)

25. Puzyrkov, D.V., Podryga, V.O., Polyakov, S.V.: Distributed data processing in application to the molecular dynamics simulation of equilibrium state in the gas-metal microsystems (in Russian). Sci. J. USATU.20 **1**(71), 175–186 (2016)

26. NVIDIA Tesla K80 GPU Test Drive: NVIDIA Corporation, November 2016. URL: http://www.nvidia.co.uk/object/k80-gpu-test-drive-uk.html. (Accessed 22.11.2016)

27. Kudryashova, T.A., Podryga, V.O., Polyakov, S.V.: Simulation of gas mixture flows in microchannels (in Russian). Bulletin of people's friendship university of Russia. Series: Mathematics. Comput. Sci. Phys. **3**, 154–163 (2014)

28. Podryga, V.O., Polyakov, S.V., Zhakhovskii, V.V.: Atomistic calculation of the nitrogen transitions in thermodynamic equilibrium over the nickel surface (in Russian). Math. Simul. **27**(7), 91–96 (2015)

29. Podryga, V.O.: Molecular dynamics method for simulation of thermodynamic equilibrium. Math. Models Comput. Simul. **3**(3), 381–388 (2011)

30. Podryga, V.O.: Molecular dynamics method for heated metal's simulation of thermodynamic equilibrium (in Russian). Math. Simul. **23**(9), 105–119 (2011)

Performance of MD-Algorithms on Hybrid Systems-on-Chip Nvidia Tegra K1 & X1

Vsevolod Nikolskii[1,3], Vyacheslav Vecher[1,2], and Vladimir Stegailov[1(✉)]

[1] Joint Institute for High Temperatures of RAS, Moscow, Russia
thevsevak@gmail.com, vecher@phystech.edu, v.stegailov@hse.ru
[2] Moscow Institute of Physics and Technology (State University),
Dolgoprudny, Russia
[3] National Research University Higher School of Economics, Moscow, Russia

Abstract. In this paper we consider the efficiency of hybrid systems-on-a-chip for high-performance calculations. Firstly, we build Roofline performance models for the systems considered using Empirical Roofline Toolkit and compare the results with the theoretical estimates. Secondly, we use LAMMPS as an example of the molecular dynamic package to demonstrate its performance and efficiency in various configurations running on Nvidia Tegra K1 & X1. Following the Roofline approach, we attempt to distinguish compute-bound and memory-bound conditions for the MD algorithm using the Lennard-Jones liquid model. The results are discussed in the context of the LAMMPS performance on Intel Xeon CPUs and the Nvidia Tesla K80 GPU.

Keywords: ARM · GPU · Maxwell · Kepler · Roofline · LAMMPS

1 Intoduction

Today, microcircuits technology is close to reaching the physical limits beyond which the increase in the density of transistors on a chip becomes very problematic. Further gain of computing power of supercomputers is connected with significant increase in the number of nodes and the search of more sophisticated technologies.

One of the directions of developement is the widespread use of accelerators such as GPUs that is the transition to heterogeneous systems. Despite the fact that GPU originally were created for mass market, their attractive price-to-performance ratio lead to the adoption of such devices for high performance computations with the development of GP-GPU technologies. Another trend is the increasing attention to energy-efficiency and the rise of ARM architecture on the server market [1].

If you use coprocessors the limitation is usually in the fact that these devices are physically separated from the CPU. Communication could become much easier by placing the CPU and co-processor on a single chip. Such systems-on-chip (SoC) exist, e.g. AMD Accelerated Processing Units, ARM SOCs with

V. Voevodin and S. Sobolev (Eds.): RuSCDays 2016, CCIS 687, pp. 199–211, 2016.
DOI: 10.1007/978-3-319-55669-7_16

Mali GPUs, Intel CPUs with HD Graphics. Recently, Nvidia combined these two trends in one device and released Tegra K1 SoC and then Tegra X1 SoC. These are SoCs that combine multiple ARM Cortex-A cores and Nvidia GPU(s). After nine years of development, the Nvidia CUDA technology has appeared in many scientific and engineering programs. The novel 64-bit architecture ARMv8 demonstrates the potential for high performance computing [2]. With low power consumption, Tegra SoCs could be considered as possible prototypes of the future HPC hardware.

The rapid development of hardware increases the significance of the efficiency tests of new architectures. Benchmarking of (super) computer system has several decades of history (see [3–6]). The spectrum of the computational algorithms became wide enough so the carefully selected set of tests are an important tool in the development of new supercomputers. Measurement and presentation of the results of performance tests of parallel computer systems become more and more often evidence-based [7], including the measurement of energy consumption [8], which is crucial for the development of exascale supercomputers [9].

2   Literature Review

An attempt to compare the MD algorithm efficiency of different architectures (CPU and CPU + GPU) has been made previously [10,11]. This comparison has been done in a rather empirical way without an in-depth consideration of the underlying connection between the software and the hardware. A systematic comparison of hybrid systems with different accelerators has been presented recently in [12] using the LULESH proxy application. The efficiency of the astrophysical code on different accelerators has been compared in [13] with the focus on Intel Xeon Phi. The empirical tuning of the GPU algorithm parameters provides a significant speedup and better utilization of GPU [14].

Some preliminary results on the floating-point performance of ARM cores and their efficiency in classical molecular dynamics has been published in [15]. The recent paper of Laurenzano et al. [16] is dedicated to new ARMv8 processors comparison with Atom and Xeon CPUs. It collects a huge set of test results, which have been analyzed using statistical methods. An original approach is used to identify the bottlenecks of each CPU architecture.

Recently, there have been several studies of Tegra SoCs for HPC applications. Using unified memory (supported since CUDA 6.5), Ukidave et al. showed increasing CPU-GPU data throughput in Tegra K1 in comparison with the data transfer rate over the PCIe bus [17]. Haidar et al. considered the performance and the energy efficiency by using the MAGMA linear algebra library on Jetson TK1 [18]. Stone et al. measured the performance of the basic packages of computational biology on some of the new heterogeneous platforms, Jetson TK1 and Jetson TX1 including [19]. Our preliminary results for Nvidia Jetson TK1 and TX1 have been published in [20].

3   Test Setup

3.1   Hardware

Nvidia Jetson TK1. Nvidia Jetson TK1 is a developer board based on the 32-bit Tegra K1 SoC with LPDDR3 (930 MHz). Tegra K1 CPU complex includes 4 Cortex-A15 cores running at 2.3 GHz, the 5-th low power companion Cortex core designed to replace the basic cores in the low load mode to reduce power consumption and heat generation. The chip includes one GPU Kepler streaming multiprocessor (SM) running at 852 MHz (128 CUDA cores). Each Cortex-A15 core has 32 KB L1 instruction and 32 KB L1 data caches. 4-core cluster has 2 MB of shared L2 cache.

The program environment of the device consists of Linux Ubuntu 14.04.1 LTS (GNU/Linux 3.10.40-gdacac96 armv7l). The toolchain includes GCC ver. 4.8.4 and CUDA Toolkit 6.5.

Nvidia Jetson TX1. Jetson TX1 is based on the 64-bit Tegra X1 SoC with LPDDR4 memory (1600 MHz). Tegra X1 includes 4 Cortex-A57 cores running at 1.9 GHz, 4 slower Cortex-A53 in big.LITTLE configuration and two GPU Maxwell SMs running at 998 MHz (256 CUDA cores). Each Cortex-A57 core has 48 KB L1 instruction cache, 32 KB L1 data cache and 2 MB of shared L2 cache.

The operation system is Linux Ubuntu 14.01.1 LTS with 64-bit core built for aarch64. Nevertheless we use the 32-bit toolchain and software environment (same as for Nvidia Jetson TK1), except for the newer CUDA Toolkit 7.0.

In summer 2016, the 64-bit userspace and toolchain have been released. Preliminary tests show that the new 64-bit software can be noticeably faster in some rare cases only.

Server with Nvidia Tesla K80. The high-performance server based on two Intel Xeon E5-2697 v3 CPUs is used in this paper for the representative comparison. These CPUs are x86_64 Haswell processors with 14 hardware cores that support hyper-threading and run at 2.8 GHz. The server includes one Nvidia Tesla K80 accelerator. In has about 5000 CUDA cores running at 627 MHz (the frequency is increased in the turbomode).

3.2   Software

Empirical Roofline Toolkit. In this work we use Empirical Roofline Toolkit 1.1.0, its source codes are available on Bitbucket [21].

The GCC keys for Jetson TK1 are '-O3 -march=armv7-a -mtune= cortex-a15 -mfpu=neon-vfpv4 -ffast-math'.

For TX1 we use '-O3 -macrh=armv8-a -mfpu=neon-vfpv4 -ffast-math'.

For Haswell CPU we use '-O3 -march=haswell -mtune=haswell -mavx2 -mfma -ffast-math'.

LAMMPS. To compile LAMMPS for Cortex-A15/A57 cores of Tegra K1/X1 we use the same options as for the Empirical Roofline Toolkit. We compile LAMMPS with the USER-OMP package for OpenMP support. For the GPU package we use 'arch=sm_32' for K1 and 'arch=sm_53' for X1. For the USER-CUDA package we use 'arch=21' for both K1 and X1.

There are differences in LAMMPS compilation with GPU or USER-CUDA packages. For the GPU package there is a straightforward way to use any CUDA architecture. However the developers of the currently unsupported USER-CUDA provided scripts and documentation only for few outdated architectures. Despite this fact, we compiled USER-CUDA for sm_32 and sm_53 architectures in order to test Jetsons TX1 and TK1. Measurements revealed that the performance of USER-CUDA depends very little on the architecture type (at the level of a measurement accuracy) that is why we use 'arch=21' for all reported USER-CUDA benchmarks.

4   Roofline Performance Model

4.1   Roofline

The Roofline performance model [6] was developed in the LBNL "Performance and Algorithms Research" group. It can represent on a single plot limitations caused by processors peak performance, memory bandwidth and effects of cache memory hierarchy and vectorization.

The key concept for understanding the Roofline model is the arithmetic intensity of an algorithm. The definition of this quantity is as follows: the arithmetic intensity is the ratio of the total number of arithmetic operations to the total number of transmitted data bytes. For visualization of the model it is necessary to depict the computer's performance in GFlops/sec as a function of the arithmetic intensity in Flops/byte. The resulting curve limits the area where all kinds of algorithms lie for the computing system considered. Moreover, these curves are displayed in straight lines when the double logarithmic scale is applied.

In this work the Empirical Roofline Toolkit is used to build Roofline models. The basic idea is in the calculations of simple arithmetic operations on the elements of an array of a certain length with the corresponding time measurement of the task execution time. The arithmetic intensity and the size of the working set vary on the separate steps of the test. This allows one to plot the dependence of performance on the arithmetic intensity taking into account the effects of the memory hierarchy.

4.2   Theoretic Estimates

This section shows how peak values are estimated in this study.

The theoretical bandwidth of the L1 cache can be estimated as CPU_{freq} [GHz] $* 32$[Bits] $* (1$[Input] $+ 1$[Output]). The theoretical bandwidth of DRAM transfers is much lower $DRAM_{freq}$[GHz] $* 8$[Byte] $* (1$[Input] $+ 1$[Output]).

The aggregate L2 bandwidths is estimated as[Number_of_SMs] * SM_{freq} [GHz] * 32[Banks] * 4[Bytes] * 0.5 (54.4 GB/s for one Kepler SM of Tegra K1 and 128 GB/s for two Maxwell SMs of Tegra X1).

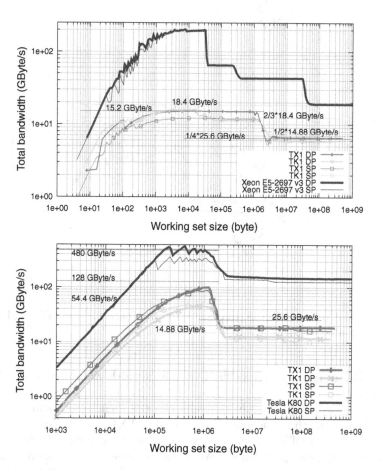

Fig. 1. Dependence of the CPU (top) and GPU (bottom) memory bandwidth against the working set size. For comparison we provide plots built on the Xeon processor with accelerator

4.3   Analysis

Figure 1 shows the dependence of the bandwidth on the amount of data to be processed by SoC.

The bandwidth data for one CPU core of Tegra K1 show that the sustained bandwidth is about 85% of the L1 peak value. The sustained bandwidth for L2 is about 2/3 smaller. The bandwidth data for Tegra X1 show a very close agreement with the theoretical value and no difference between L1 and L2. In

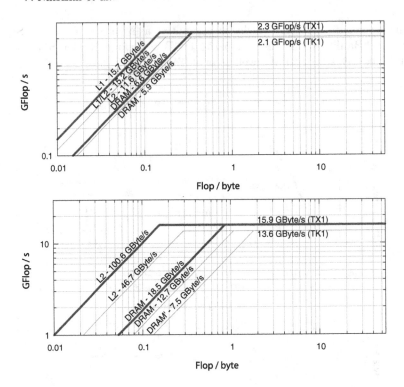

Fig. 2. Roofline model for CPU (top) and GPU (bottom) in Jetson TX1 and Jetson TK1 using double precision

single precision Tegra X1 shows the L1/L2 bandwidth about 5/6 of the double precision value.

The sustained DRAM bandwidth values for the single core Roofline tests are 1/2 and 1/4 of the theoretical peak for TK1 and TX1 respectively.

For the GPU Roofline benchmark, we observe the aggregate L2 bandwidths close to the estimated peak values. We do not observe any significant difference between double and single precision in this case. Figure 1 illustrates the small difference in GPU L2 cache sizes for Kepler (1.5 Mb) and Maxwell (2 Mb) SMs. So far it remains unclear why the memory bandwidth is reduced when the working set size is above 50–70 MB.

Figure 2 shows that the peak performance values of A15 and A57 ARM cores are very close: 0.9 Flops/cycle for A15 (close to the previous estimate [15]) and 1.2 Flops/cycle for A57 (our preliminary benchmark for TX1 with the unofficial 64-bit libraries has shown the same value). For single precision, we get 5.0 GFlops/sec for A15 and 5.1 GFlops/sec for A57.

The disassembly shows that the Roofline binary uses FMA instructions on Cortex-A5/15/57 but at the same time requests an additional data movement operation that leads to lower R_{peak} results. In the ARM instruction set FMA operations are represented only as $a = a + b \cdot c$, while in FMA3 (for example in

Haswell CPUs) there are two other forms $a = a \cdot c + b$ and $a = a \cdot b + c$. The GCC compiler considered does not perform effective vectorization using FMA operations despite the fact that they are available and could provide x2 or x4 performance increase.

For GPUs, the observed single precision peak performance is 209.9 GFlops/sec for one Kepler SM and 485.1 GFlops/sec for two Maxwell SMs. These values are close to the theoretical peak of 218 GFlops/sec and 511 GFlops/sec respectively. The double precision peak performance values are 13.6 GFlops/sec and 15.9 GFlops/sec respectively. They agree with the expected value of 16 GFlops/cycle in double precision that is the same both for Tegra K1 and X1 (the FP64 rates are 1/24 the FP32 rate for K1 Kepler and 1/32 the FP32 rate for TX1 Maxwell).

5   Classic Molecular Dynamics and LAMMPS

5.1   Molecular Dynamics Method

The molecular dynamic method is based on solving the Newtonian equations of motion of individual particles and it is a research instrument of the greatest importance. The computational performance and the efficiency of parallelization are the main factors that limit spacial and temporal scales available for the MD calculations. The currently achieved limits are trillions of atoms [22] and milliseconds of time [23] with a typical MD step about 1 femtosecond. The bottleneck in the MD algorithm is the computational complexity of interatomic potentials. Hybrid architectures are considered as the main opportunity to speed up supercomputer nodes. MD algorithms on graphics accelerators are a particular case of porting MD algorithms on SIMD architectures.

5.2   LAMMPS and Hybrid Architectures

LAMMPS package is used in this paper, it is a flexible tool for building models of classical MD in materials science, chemistry and biology [24]. LAMMPS is not the only MD package that is ported to the hybrid architecture (for example HOOMD [25] was originally designed with the perspective to run it on GPU accelerators). The USER-CUDA [26] and GPU [27,28] packages were the first implementations of the MD algorithm for hybrid architectures in LAMMPS. Another hybrid implementation of the MD algorithm in LAMMPS in based on the KOKKOS C++ library [29].

The USER-CUDA package supports only CUDA-compatible devices. Unlike the GPU-package, this package is designed to allow an entire LAMMPS calculation for many time steps to run entirely on the GPU (except for inter-nodes MPI communication), so that atom-based data do not have to move back-and-forth between the CPU and the GPU. Neighbor lists are also constructed on the GPU, while in the GPU package either CPU or GPU can be deployed for the neighbor list construction. Nevertheless the non-GPU operation calls in the

LAMMPS input script make data move back to the CPU, which may result in performance degradation. The USER-CUDA package is tested in this work in the double, single and mixed precision modes.

The GPU package supports both CUDA and OpenCL. The package was designed to exploit common GPU hardware configurations where one or more GPUs are coupled to one or more multi-core CPUs. The specifics of the algorithm is that data move from the host to the GPU every time step. Unlike USER-CUDA, the GPU package allows to start multiple CPU threads per one GPU module and can perform force computation on CPU and GPU simultaneously (however in our tests any combination other than 1 CPU core per 1 GPU leads to performance degradation). In this work we use the CUDA version of the GPU package, building the library for this package with all three supported precisions in the double, single and mixed modes.

Another hybrid implementation of the MD algorithm in LAMMPS is based on the KOKKOS library that can be deployed with pthreads, OpenMP, and CUDA back-ends. As in the GPU package, MD computations are off-loaded from CPU to GPU. This implementation supports calculations in double precision only.

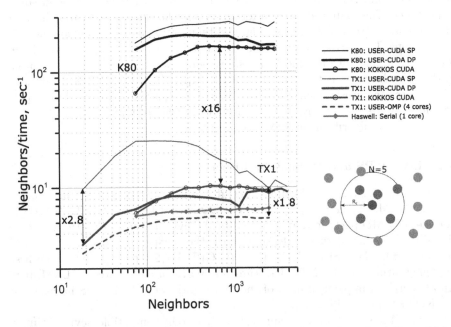

Fig. 3. Performance of MD algorithms as a function of the number of neighbors M in the Lennard-Jones liquid model (the scheme illustrate the number of neighbors $M = 5$ for the given cut-off radius R_{cut}).

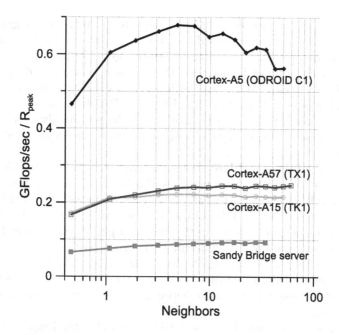

Fig. 4. The normalized sustained performance of LAMMPS for four systems with different CPU types as a function of the number of nearest neighbors in the Lennard-Jones liquid model.

5.3   Test Model

The Lennard-Jones liquid model is used as benchmark with the density $0.8442\sigma^{-3}$, $N = 108000$ particles and 250 timesteps with the NVE integrating scheme. The minimum cut-off radius considered is 1.8σ that corresponds to about $M = 18$ nearest neighbors per particle. In this study we consider different cases with increasing cut-off radius (up to about $M \sim 3000$ nearest neighbors per particle).

The cut-off radius defines the amount of calculations with the same data set of N particles. In terms of the Roofline model, we can say that the arithmetic intensity of the MD algorithm is directly proportional to the number of neighbors M. Thus, the ratio of M over the total calculation time should be proportional to the number of Flops spent for the MD algorithm computations (for the Lennard-Jones model in LAMMPS every pair interaction takes 23 Flops, or a half of this number if the Newton's third law is taken into account). This is an analogue of the R_{max} value of the standard High Performance Linpack benchmark.

These data are displayed on Fig. 3 for the USER-CUDA package in double and single precision, the USER-OMP package (based on OpenMP) in double precision and KOKKOS in the CUDA mode and the OpenMP mode.

The USER-CUDA algorithm is memory bound for small number of neighbors M that is why the ratios of the performance in double precision to the performance in single precision (x2.5 for Tegra K1 and x2.8 for Tegra X1) are determined by the memory bandwidth ratios (in single precision there are twice less data and the GPU L2 bandwidth is slightly higher as well). We see the GPU performance is x1.8 times higher than the 4-cores performance obtained with USER-OMP that is close to the peak performance ratio of the Maxwell GPU and 4 Cortex-A57 cores 15.9/(4*2.3)~1.7. USER-CUDA demonstrates quite poor performance in single precision: the algorithm is always strongly memory-bound since the increase of the number of neighbors results in complicated memory access patterns.

Comparison of TX1 and K80 shows that the performance is still memory bandwidth limited even for large numbers of neighbors. For example, this ratio for the KOKKOS case is ~16 (Fig. 3). At the same time the ratio of the memory bandwidths is ~5–8 and the ratio of the peak performance values is ~350.

On Fig. 4, we show the values of the sustained performance normalized over the peak performance for four different CPU types. We see that the weakest Cortex-A5 core (see [15]) is the most effective in terms of the peak performance utilization. The explanation of this fact is that the Cortex-A5 based ODROID-C1 minicomputer has better balance between the memory subsystem speed and the CPU speed.

The analysis made with the Haswell CPU shows that the Empirical Roofline Test binary uses in the main computational cycle vector FMA instructions from the AVX2 set, which provide an eight-fold acceleration. At the same time, the GCC compiler uses only two FMA instructions per a Lennard-Jones interaction in LAMMPS. This fact shows that the increased peak performance of novel Intel CPU architectures is not transferred easily into the sustained performance of such a software as LAMMPS.

6   Conclusion

In this work we quantify the Jetson TK1 and novel TX1 boards performance using the Roofline model and the Empirical Roofline Toolkit. We consider both single and double precision performance. Different implementations of the MD algorithm in LAMMPS are used for benchmarking. We have collected data about memory bandwidth and cache hierarchy. These results are compared with theoretic estimates.

For benchmarking we use 3 hybrid implementations of MD algorithm from LAMMPS package and the OpenMP parallel version. We have proposed a new method for varying arithmetic intensity of the MD task by changing the cut-off radius of the pair potential in the MD model. Thus, we show the transition of MD calculation from memory-bound to compute-bound mode. The comparison of the hybrid MD algorithm data for Tegra TX1 with the data for the Xeon-based server with Tesla K80 shows that the hybrid MD algorithm becomes partially memory-bound on K80, even for large cut-off radii.

We consider the sustained performance normalized over the peak performance for Cortex-A5, A15 and A57 systems as well as for the Sandy Bridge system. The results show that the slowest system based on Cortex-A5 is the most efficient in terms of the sustained performance ratio over the peak performance. This result illustrates the need for the balance between the performance of the memory subsystem and the CPU.

Acknowledgments. HSE and MIPT provided funds for purchasing the hardware used in this study. The authors are grateful to the Forsite company for the access to the server with Nvidia Tesla K80. The authors acknowledge Joint Supercomputer Centre of RAS for the access to MVS-100K and MVS-10P supercomputers. The work was supported by the grant No. 14-50-00124 of the Russian Science Foundation.

References

1. Mitra, G., Johnston, B., Rendell, A., McCreath, E., Zhou, J.: Use of SIMD vector operations to accelerate application code performance on low-powered ARM and Intel platforms. In: 2013 IEEE 27th International Parallel and Distributed Processing Symposium Workshops PhD Forum (IPDPSW), pp. 1107–1116 (2013). doi:10.1109/IPDPSW.2013.207

2. Keipert, K., Mitra, G., Sunriyal, V., Leang, S.S., Sosonkina, M., Rendell, A.P., Gordon, M.S.: Energy-efficient computational chemistry: comparison of x86 and ARM systems. J. Chem. Theory Comput. **11**(11), 5055–5061 (2015). doi:10.1021/acs.jctc.5b00713

3. Curnow, H.J., Wichmann, B.A.: A synthetic benchmark. Comput. J. **19**(1), 43–49 (1976)

4. Strohmaier, E., Hongzhang, S.: Apex-Map: a global data access benchmark to analyze HPC systems and parallel programming paradigms. In: Proceedings of the ACM/IEEE SC 2005 Conference (2005). doi:10.1109/SC.2005.13

5. Heroux, M.A., Doerfler, D.W., Crozier, P.S., Willenbring, J.M., Edwards, H.C., Williams, A., Rajan, M., Keiter, E.R., Thornquist, H.K., Numrich, R.W.: Improving performance via mini-applications. Technical report, Sandia National Laboratories (2009)

6. Williams, S., Waterman, A., Patterson, D.: Roofline: an insightful visual performance model for multicore architectures. Commun. ACM **52**(4), 65–76 (2009)

7. Hoefler, T., Belli, R.: Scientific benchmarking of parallel computing systems: twelve ways to tell the masses when reporting performance results. In: Proceedings of the International Conference for High Performance Computing, Networking, Storage and Analysis, SC 2015, pp. 73:1–73:12 (2015). http://doi.acm.org/10.1145/2807591.2807644

8. Pruitt, D.D., Freudenthal, E.A.: Preliminary investigation of mobile system features potentially relevant to HPC. In: Proceedings of the 4th International Workshop on Energy Efficient Supercomputing, E2SC 2016, pp. 54–60. IEEE Press, Piscataway, NJ, USA (2016). doi:10.1109/E2SC.2016.13

9. Scogland, T., Azose, J., Rohr, D., Rivoire, S., Bates, N., Hackenberg, D.: Node variability in large-scale power measurements: perspectives from the Green500, Top500 and EEHPCWG. In: Proceedings of the International Conference for High Performance Computing, Networking, Storage and Analysis, SC 2015 (2015). http://doi.acm.org/10.1145/2807591.2807653

10. Stegailov, V.V., Orekhov, N.D., Smirnov, G.S.: HPC hardware efficiency for quantum and classical molecular dynamics. In: Malyshkin, V. (ed.) PaCT 2015. LNCS, vol. 9251, pp. 469–473. Springer, Heidelberg (2015). doi:10.1007/978-3-319-21909-7_45

11. Smirnov, G.S., Stegailov, V.V.: Efficiency of classical molecular dynamics algorithms on supercomputers. Math. Models Comput. Simul. 8(6), 734–743 (2016). doi:10.1134/S2070048216060156

12. Gallardo, E., Teller, P.J., Argueta, A., Jaloma, J.: Cross-accelerator performance profiling. In: Proceedings of the XSEDE16 Conference on Diversity, Big Data, and Science at Scale XSEDE 2016, pp. 19:1–19:8. ACM, NY, USA (2016). doi:10.1145/2949550.2949567

13. Glinsky, B., Kulikov, I., Chernykh, I., Weins, D., Snytnikov, A., Nenashev, V., Andreev, A., Egunov, V., Kharkov, E.: The co-design of astrophysical code for massively parallel supercomputers. In: Carretero, J., et al. (eds.) ICA3PP 2016. LNCS, vol. 10049, pp. 342–353. Springer, Heidelberg (2016). doi:10.1007/978-3-319-49956-7_27

14. Rojek, K., Wyrzykowski, R., Kuczynski, L.: Systematic adaptation of stencil-based 3D MPDATA to GPU architectures. Concurr. Comput.: Pract. Exp. (2016). doi:10.1002/cpe.3970

15. Nikolskiy, V., Stegailov, V.: Floating-point performance of ARM cores and their efficiency in classical molecular dynamics. J. Phys.: Conf. Ser. 681(1) (2016). Article ID 012049. http://stacks.iop.org/1742-6596/681/i=1/a=012049

16. Laurenzano, M.A., Tiwari, A., Cauble-Chantrenne, A., Jundt, A., Ward, W.A., Campbell, R., Carrington, L.: Characterization and bottleneck analysis of a 64-bit ARMv8 platform. In: 2016 IEEE International Symposium on Performance Analysis of Systems and Software (ISPASS), pp. 36–45 (2016). doi:10.1109/ISPASS.2016.7482072

17. Ukidave, Y., Kaeli, D., Gupta, U., Keville., K.: Performance of the NVIDIA Jetson TK1 in HPC. In: 2015 IEEE International Conference on Cluster Computing, pp. 533–534 (2015)

18. Haidar, A., Tomov, S., Luszczek, P., Dongarra, J.: Magma embedded: towards a dense linear algebra library for energy efficient extreme computing. In: High Performance Extreme Computing Conference (HPEC), pp. 1–6. IEEE (2015)

19. Stone, J.E., Hallock, M.J., Phillips, J.C., Peterson, J.R., Luthey-Schulten, Z., Schulten, K.: Evaluation of emerging energy-efficient heterogeneous computing platforms for biomolecular and cellular simulation workloads. In: International Parallel and Distributed Processing Symposium Workshop (IPDPSW). IEEE (2016)

20. Nikolskiy, V.P., Stegailov, V.V., Vecher, V.S.: Efficiency of the Tegra K1 and X1 systems-on-chip for classical molecular dynamics. In: 2016 International Conference on High Performance Computing Simulation (HPCS), pp. 682–689 (2016). doi:10.1109/HPCSim.7568401

21. Lo, Y.J., et al.: Roofline model toolkit: a practical tool for architectural and program analysis. In: Jarvis, S.A., Wright, S.A., Hammond, S.D. (eds.) PMBS 2014. LNCS, vol. 8966, pp. 129–148. Springer, Heidelberg (2015). doi:10.1007/978-3-319-17248-4_7

22. Eckhardt, W., et al.: 591 TFLOPS multi-trillion particles simulation on SuperMUC. In: Kunkel, J.M., Ludwig, T., Meuer, H.W. (eds.) ISC 2013. LNCS, vol. 7905, pp. 1–12. Springer, Heidelberg (2013). doi:10.1007/978-3-642-38750-0_1

23. Piana, S., Klepeis, J.L., Shaw, D.E.: Assessing the accuracy of physical models used in protein-folding simulations: quantitative evidence from long molecular dynamics simulations. Curr. Opin. Struct. Biol. **24**, 98–105 (2014). doi:10.1016/j.sbi.2013.12.006

24. Plimpton, S.: Fast parallel algorithms for short-range molecular dynamics. J. Comput. Phys. **117**(1), 1–19 (1995). doi:10.1006/jcph.1995.1039

25. Glaser, J., Nguyen, T.D., Anderson, J.A., Lui, P., Spiga, F., Millan, J.A., Morse, D.C., Glotzer, S.C.: Strong scaling of general-purpose molecular dynamics simulations on GPUs. Comput. Phys. Commun. **192**, 97–107 (2015). doi:10.1016/j.cpc.2015.02.028

26. Trott, C.R., Winterfeld, L., Crozier, P.S.: General-purpose molecular dynamics simulations on GPU-based clusters. ArXiv e-prints arXiv:1009.4330 (2010)

27. Brown, W.M., Wang, P., Plimpton, S.J., Tharrington, A.N.: Implementing molecular dynamics on hybrid high performance computers – short range forces. Comput. Phys. Commun. **182**(4), 898–911 (2011). doi:10.1016/j.cpc.2010.12.021

28. Brown, W.M., Kohlmeyer, A., Plimpton, S.J., Tharrington, A.N.: Implementing molecular dynamics on hybrid high performance computers – particle–particle particle-mesh. Comput. Phys. Commun. **183**(3), 449–459 (2012). doi:10.1016/j.cpc.2011.10.012

29. Edwards, H.C., Trott, C.R., Sunderland, D.: Kokkos: enabling manycore performance portability through polymorphic memory access patterns. J. Parallel Distrib. Comput. **74**(12), 3202–3216 (2014). doi:10.1016/j.jpdc.2014.07.003

Revised Pursuit Algorithm for Solving Non-stationary Linear Programming Problems on Modern Computing Clusters with Manycore Accelerators

Irina Sokolinskaya and Leonid Sokolinsky[(✉)]

South Ural State University,
76, Lenin prospekt, Chelyabinsk, Russia 454080
{Irina.Sokolinskaya,Leonid.Sokolinsky}@susu.ru

Abstract. This paper is devoted to the new edition of the parallel *Pursuit* algorithm proposed the authors in previous works. The *Pursuit* algorithm uses Fejer's mappings for building pseudo-projection on polyhedron. The algorithm tracks changes in input data and corrects the calculation process. The previous edition of the algorithm assumed using a cube-shaped pursuit region with the number of K cells in one dimension. The total number of cells is K^n, where n is the problem dimension. This resulted in high computational complexity of the algorithm. The new edition uses a cross-shaped pursuit region with one cross-bar per dimension. Such a region consists of only $n(K - 1) + 1$ cells. The new algorithm is intended for cluster computing system with Xeon Phi processors.

Keywords: Non-stationary linear programming problem · Fejer's mappings · Pursuit algorithm · Massive parallelism · Cluster computing systems · MIC architecture · Intel Xeon Phi · Native mode · OpenMP

1 Introduction

In the papers [7,8], the authors proposed the new *Pursuit* algorithm for solving high-dimension, non-stationary, linear programming problem. This algorithm is focused on cluster computing systems. High-dimensional, non-stationary, linear programming problems with quickly-changing input data are often seen in modern economic-mathematical simulations. The non-stationary problem is characterized by the fact that the input data is changing during the process of its solving. One example of such problem is the problem of investment portfolio management by using algorithmic trading methods (see [1,2]). In such problems, the number of variables and inequalities in the constraint system can be in

---

I. Sokolinskaya—The reported study was partially funded by RFBR according to the research project No. 17-07-00352-a, and by Act 211 Government of the Russian Federation according to the contract No. 02.A03.21.0011.

© Springer International Publishing AG 2016
V. Voevodin and S. Sobolev (Eds.): RuSCDays 2016, CCIS 687, pp. 212–223, 2016.
DOI: 10.1007/978-3-319-55669-7_17

the tens and even hundreds of thousands, and the period of input data change is within the range of hundredths of a second. The first version of the algorithm designed by the authors used a cubic-shaped pursuit region with the quantity of K cells in one dimension. In this case, the total number of cells is equal to K^n, where n is the dimension of the problem. This results in the high computational complexity of the algorithm. In this paper, we describe a new edition of the *Pursuit* algorithm, which uses a cross-shaped pursuit region with one cross-bar per dimension and containing only $n(K-1)+1$ cells. The main part of the *Pursuit* algorithm is a subroutine of calculating the pseudoprojection on the polyhedron. Pseudoprojection uses Fejer's mappings to substitute the projection operation on a convex set [4]. The authors implemented this algorithm in C++ language parallel programming technology OpenMP 4.0 [6] and the vector instruction set of Intel C++ Compiler for Xeon Phi [9]. The efficiency of the algorithm implementation for coprocessor Xeon Phi with KNC architecture [10] was investigated using a scalable synthetic linear programming problem. The results of these experiments are presented in this paper. The rest of this paper is organized as follows. In Sect. 2, we give a formal statement of a linear programming problem and define Fejer's process and the projection operation on a polyhedron. Section 3 describes the new version of the algorithm with a cross-shaped pursuit region. Section 4 provides a description of the main subroutine and subroutine for calculating the pseudoprojection of the revised algorithm by using UML activity diagrams. Section 5 is devoted to investigation of the efficiency of Intel Xeon Phi coprocessor usage for computing pseudoprojection. In conclusion, we summarize the results obtained and propose the directions for future research.

2   Problem Statement

Given a linear programming problem

$$\max\left\{\langle c, x\rangle \,|\, Ax \leq b,\ x \geq 0\right\}. \tag{1}$$

Let us define the Fejer's mapping $\varphi : \mathbb{R}^n \to \mathbb{R}^n$ as follows:

$$\varphi(x) = x - \sum_{i=1}^{m} \alpha_i \lambda_i \frac{\max\left\{\langle a_i, x\rangle - b_i, 0\right\}}{\|a_i\|^2} a_i. \tag{2}$$

Let M be a polyhedron defined by the constraints of the linear programming problem (1). This polyhedron is always convex. It's known [3] that φ will be a single-valued continuous M-fejerian mapping for any $\alpha_i > 0$ $(i = 1, \ldots, m)$, $\sum_{i=1}^{m} \alpha_i = 1$, and $0 < \lambda_i < 2$. Putting in formula (2) $\lambda_i = \lambda$ and $\alpha_i = 1/m$ $(i = 1, \ldots, m)$, we get the formula

$$\varphi(x) = x - \frac{\lambda}{m} \sum_{i=1}^{m} \frac{\max\left\{\langle a_i, x\rangle - b_i, 0\right\}}{\|a_i\|^2} a_i, \tag{3}$$

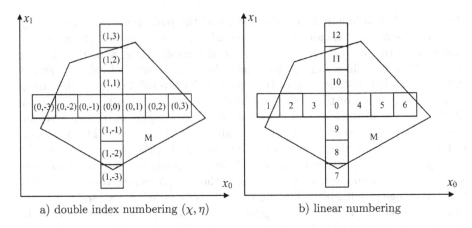

a) double index numbering (χ, η) b) linear numbering

Fig. 1. Cross-shaped pursuit region ($n = 2$, $K = 7$).

which is used in the *Pursuit* algorithm.

Let us set

$$\varphi^s(x) = \underbrace{\varphi \ldots \varphi}_{s}(x). \tag{4}$$

Let the Fejerian process generated by mapping φ from an arbitrary initial approximation $x_0 \in \mathbb{R}^n$ to be a sequence $\{\varphi^s(x_0)\}_{s=0}^{+\infty}$. It is known that this Fejerian process converges to a point belonging to the set M:

$$\{\varphi^s(x_0)\}_{s=0}^{+\infty} \to \bar{x} \in M. \tag{5}$$

Let us denote this concisely as follows: $\lim_{s \to \infty} \varphi^s(x_0) = \bar{x}$.

Let φ-*projection* (*pseudoprojection*) of point $x \in \mathbb{R}^n$ on polyhedron M be understood as the mapping $\pi_M^\varphi(x) = \lim_{s \to \infty} \varphi^s(x)$.

3   Description of the Revised Algorithm

Without losing of generality, we may suppose all the processes are carried out in the region of positive coordinates.

Let n be the dimension of solution space. The new edition of the algorithm uses a cross-shaped pursuit region. This region consists of $n(K-1)+1$ hypercubical cells of equal size. The edges of all cells are codirectional with the coordinate axis. One of these cells designates the center. We will call this cell *central*. The remaining cells form an axisymmetrical cross-shaped figure around the central cell. An example of a cross-shaped pursuit region in a two-dimensional space is presented in Fig. 1. The total number of cells in the cross-shaped pursuit region can be calculated by the following formula:

$$P = n(K - 1) + 1. \tag{6}$$

Each cell in the cross-shaped pursuit region is uniquely identified by a label being a pair of integer numbers (χ, η) such that $0 \leq \chi < n$, $|\eta| \leq (K-1)/2$. From an informal point of view, χ specifies the cell column codirectional to the coordinate axis indexed by χ, and η specifies the cell sequence number in the column in relation to the center cell. The corresponding double index numbering is shown in Fig. 1(a).

We will call the vertex closest to the origin a *zero vertex*. Let $(g_0, \ldots, g_{n-1})$ be the Cartesian coordinates of the central cell zero vertex. Let us denote by s the cell edge length. Then the Cartesian coordinates $(y_0, \ldots, y_{n-1})$ of the zero vertex of the cell (χ, η) are defined by the following formula:

$$
y_j = \begin{cases} g_\chi + \eta s, & \text{if } j = \chi \\ g_j, & \text{if } j \neq \chi \end{cases} \tag{7}
$$

for all $j = 0, \ldots, n - 1$.

Informally, the algorithm with cross-shaped pursuit region can be described by the following sequence of steps.

1. Initially, we choose a cross-shaped pursuit region which has K cells in one dimension, with the cell edge length equal to s, in such a way, that the central cell has nonempty intersection with the polyhedron M.
2. The point $z = g$ is chosen as an initial approximation.
3. Given dynamically changing input data (A, b, c), for all cells of cross-shaped pursuit region, the pseudoprojection from the point z on the intersection of the cell and polyhedron M is calculated. If intersection is empty, then the corresponding cells are discarded.
4. If the obtained set of pseudoprojections is empty then we increase the cell size w times and go to the step 3.
5. If we receive a nonempty set of pseudoprojections then, for each cross bar, we choose the cell for which the cost function takes the maximal value at the point of pseudoprojection if it exist. For the set of cells obtained in such a way, we calculate the centroid and move point z at the position of the centroid.
6. If the distance between centroid and central cell is less than $\frac{1}{4}s$ then we decrease the cell length s 2 times.
7. If the distance between centroid and central cell is greater than $\frac{3}{4}s$ then we increase the cell length s 1.5 times.
8. We translate the cross-shaped pursuit region in such a way that its central point be situated at the centroid point found at the step 5.
9. Go to the step 3.

In the step 3, the pseudoprojections for the different cells can be calculated in parallel without data exchange between MPI-processes. This involves P MPI-processes, where P is determined by the formula (6). We use the linear cell numbering for the distributing the cells on the MPI-processes. Each cell of the cross-shaped pursuit region is assigned an unique number $\alpha \in \{0, \ldots, P-1\}$.

The sequential number α can be uniquely converted to the label (χ, η) by the following formulas[1]:

$$\chi = (\alpha - 1) \div (K - 1) \tag{8}$$

$$\eta = \begin{cases} 0, \text{ if } \alpha = 0 \\ (\alpha - 1) \bmod \frac{K-1}{2} - \frac{K-1}{2}, \text{ if } 0 \le (\alpha - 1) \bmod (K - 1) < \frac{K-1}{2} \\ (\alpha - 1) \bmod \frac{K-1}{2} + 1, \text{ if } (\alpha - 1) \bmod (K - 1) \ge \frac{K-1}{2} \end{cases} \tag{9}$$

The reverse conversion of (χ, η) in α can be performed by the formula

$$\alpha = \begin{cases} 0, \text{ if } \eta = 0 \\ \eta + \frac{K-1}{2} + \chi(K - 1) + 1, \text{ if } \eta < 0 \\ \eta + \frac{K-1}{2} + \chi(K - 1), \text{ if } \eta > 0 \end{cases} \tag{10}$$

Figure 1(b) shows the linear numbering corresponding to the double index numbering shown in Fig. 1(a).

4 Implementation of Revised Algorithm

This section describes the changes in the implementation of the new version of the *Pursuit* algorithm with reference to the description given in the paper [8].

4.1 Diagram of Main Subroutine

The activity diagram of the main subroutine of the *Pursuit* algorithm is shown in Fig. 2. In the loop *until* with label 1, the approximate solution $z = (z_0, \ldots, z_{n-1})$ of the linear programming problem (1) is permanently recalculated according to the algorithm outline presented in the Sect. 3. As an initial approximation, z may be chosen as an arbitrary point.

The main subroutine of the *Pursuit* algorithm is implemented as an independent process, which is performed until the variable *stop* takes the value of *true*.

The initial setting of the variable *stop* to the value *false* is performed by the root process corresponding to the main program. The same root process sets the variable *stop* to the value *true*, when the computations must be stopped.

In the body of the loop *until*, the following steps are performed. In the step 2, the K parallel threads are created. Each of them independently calculates the pseudoprojection from the point z on the intersection of the i-th cell and polyhedron M $(i = 0, \ldots, P - 1)$. Recall that P is equal to the number of MPI-processes that in turn is equal to the total number of cells in the cross-shaped pursuit region calculated by the formula (6). The activity diagram of subroutine π for calculating the pseudoprojection is described in Sect. 4.2.

In the loop *for* with label 3, for each cross bar $\chi = 0, \ldots, n - 1$, we calculate the sequential number α'_χ of the cell in this cross bar, in which the cost function C takes the maximum. It is calculated in the loop with label 5. In order to

---
[1] We use symbol $\div$ to denote the integer division.

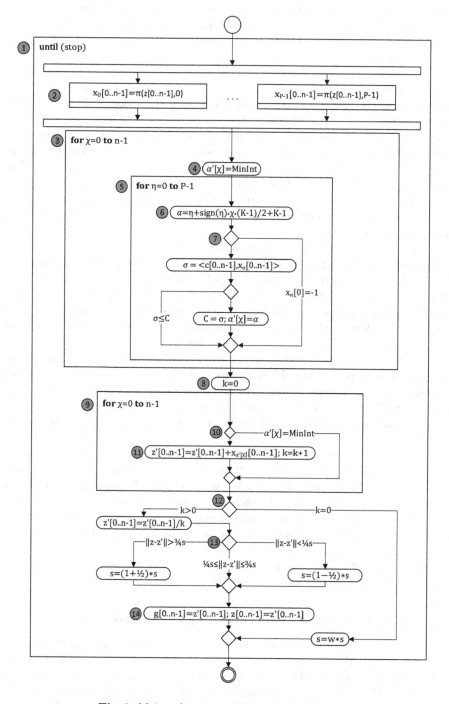

Fig. 2. Main subroutine of *Pursuit* algorithm.

guarantee the correct execution of the cycle 5, we initially assign the value $MinInt$ to variable α'_χ. This value corresponds to the minimal value of the integer type. In the step 6, we calculate the sequential number α for the cell with label (χ, η) by using formula (10).

The subroutine π calculating the point $x_\alpha = (x_0, \ldots, x_{n-1})$ of pseudoprojection from the point z on the intersection of polyhedron M with the cell with number α assigns value -1 to x_0 when the pseudoprojection point x_α does not belong to the polyhedron M. This situation occurs when the intersection of the polyhedron M with the cell with number α is empty. If x_α belongs to the polyhedron then value of x_0 can't be negative because of our assumption that all the processes are carried out in the region of positive coordinates (see Sect. 3). This condition is checked in the step 7. Cases with $x_0 = -1$ are excluded from consideration. If all the cells in the current cross bar of pursuit region have empty intersection with polyhedron M then the variable α'_χ saves the value $MinInt$. This case is fixed in the step 10.

Then, in the loop 9, a new approximate solution z' of the problem (1) is calculated. Variable k takes the value which is equal to the number of cross bars having the nonempty intersections with polyhedron M. For this purpose, in the step 8, it is assigned the zero value. In the step 10, the cross bars having the empty intersections with polyhedron M are excluded from consideration. In the step 11, we calculate the sum of all pseudoprojection points, in which the cost function takes maximum, and assign this value to z'.

If in the step 12 we have $k = 0$, it means that the pursuit region has empty intersection with polyhedron M. In this case, the length s of cell edge is increased w times, and we go back to the step 1. The constant w is a parameter of the algorithm. If in the step 12 we have $k > 0$ then the new approximation z' is assigned the value which is equal to the centroid of all the cell selected in the loop labeled 9.

In the step 13, we investigate how far the new approximation z' is distant from the previous approximation z. If the distance between z' and z is greater than $\frac{3}{4}s$ then the length s of cell edge is increased 1.5 times. If the distance between z' and z is less than $\frac{3}{4}s$ then the length s of cell edge is decreased 2 times. If the distance between z' and z is greater than or equal $\frac{1}{4}s$ and less than or equal $\frac{3}{4}s$ then the length s of cell edge is unchanged. The values $1/4$ and $3/4$ are the parameters of the algorithm.

In the step 14, the pursuit region is translated by vector $(z' - z)$, z is assigned z', and computation is continued.

4.2   Diagram of Subroutine Calculating Pseudoprojection

In Fig. 3, the activity diagram of the subroutine calculating the pseudoprojection $x = \pi(z, \alpha)$ from the point z on the intersection of the polyhedron M and the cell with number α calculated by the formula (10) is presented. The pseudoprojection is calculated by organizing a Fejerian process (5) (see Sect. 2). In the step 1, the initialization of the variables used in iterative process is performed. The initial value of x is assigned to point z; the zero vertex y of the cell with number α is

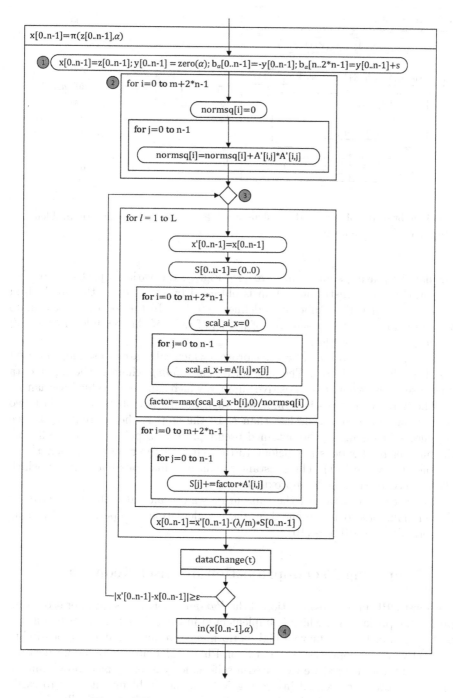

Fig. 3. Subroutine π calculating pseudoprojection.

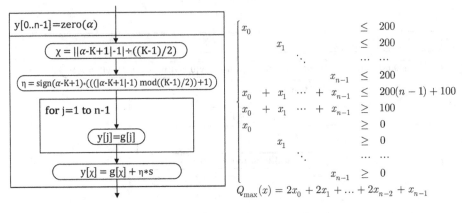

Fig. 4. Subroutine of zero vertex calculation for cell with number α.

Fig. 5. *Model-n* synthetic problem.

calculated by using subroutine *zero* (see Fig. 4); the variable part of extended column b' of the constraint system is obtained by intersecting the polyhedron M and the cell with number α is defined (see [8]). In the loop 2, we calculate *normsq* being a vector of squares of norms of rows of the extended matrix A': $normsq_i = \|a'_i\|^2$ (see [8]).

In step 1, we organize an iterative process which calculates a pseudoprojection based on the formula (3). The subroutine *dataChange* changes the input data every t seconds (where t is a positive number, which can take a value less than 1).

The iterative process is terminated when the distance between the last two approximations x and x' is less than ε. In the step 1, the subroutine *in* (see [8]) checks belonging of the obtained pseudoprojection point x to the cell with the number α. If x does not belong to the cell with the number α, then $x[0]$ is assigned the value (-1). The constant ε defines a small positive number, which allows to correctly handle approximate values.

The activity diagram of the subroutine which calculates the zero vertex of the cell numbered α is presented in Fig. 4. Calculations are performed by using the formulas (8), (9) and (10).

5   Computing Pseudoprojection on Intel Xeon Phi

The most CPU intensive operation of the *Pursuit* algorithm is the operation computing the projections, which is implemented in the subroutine described in the Sect. 4.2. In order to achieve a high performance we investigated the possibility of effective use of coprocessors Intel Xeon Phi to calculate the pseudoprojection.

In our experiments, we exploited a self-made synthetic linear programming problem *Model-n* presented in the Fig. 5. Such the problems allow us to easily calculate the precise solution analytically. Therefore, they are well suited for algorithm validation and scalability evaluation.

We implemented the algorithm in C++ language using OpenMP. The task run was performed on Xeon Phi in native mode [10]. For computational experiments, we used the computer system "Tornado-SUSU" [5] with a cluster architecture. It includes 384 processor units connected by the InfiniBand QDR and Gigabit Ethernet. One processor unit includes two six-core CPU Intel Xeon X5680, 24 GB RAM and coprocessor Intel Xeon Phi SE10X (61 cores, 1.1 GHz) connected by PCI Express bus.

In the first series of experiments, we investigated the efficiency of parallelization of calculating pseudoprojection for different numbers of threads. The results are presented in Fig. 6. The pseudoprojections were calculated on the intersection of the polyhedron defined by constrains of linear programming problem *Model-n* and the cell with edge length $s = 20$ having coordinates of zero vertex equaling to $(100, \dots, 100)$. The calculations were conducted for the dimensions $n = 1200, 9600, 12000$. The graphs show that the parallelization efficiency is strongly depends on the dimension of the problem. So, for the dimension $n = 1200$, the speedup curve actually stops growing after 15 threads. This means that a problem of such dimension cannot fully load all cores of Xeon Phi. The situation is changed when $n = 9600$ and more. Speedup becomes near-linear up to 60 threads, which is equal to the number of cores in Xeon Phi. Then parallelization efficiency is decreased, and for the dimension $n = 9600$, we even observe a performance degradation. The degradation is most evident at the point corresponding to the use of 180 threads. This dip is due to the fact that is not divisible by the dimension of 9600 divisible by 180, hence the compiler cannot uniformly distribute the iterations of the *parallel for* cycle between threads. The same situation takes place at the point "45 threads" for the dimensions 1200 and 9600. However, if the dimension is increased up to 12000, this effect is weakened.

In the second series of experiments, we compared the performance of two CPUs Intel Xeon and coprocessor Intel Xeon Phi. The results are presented in Fig. 7. The calculations were made for the dimensions $n = 9600, 12000, 19200$. For the Intel Xeon Phi, we made two builds: without the vectorization (MIC) and with the vectorization including data alignment (MIC+VECTOR). In all the cases, for the 2×CPU runs we used 12 threads, and for Xeon Phi runs we used 240 threads. The experiments show that for the dimension $n = 9600$ the 2×CPU outperform the coprocessor Xeon Phi, for the dimension $n = 12000$ the 2×CPU demonstrate the same performance as the coprocessor Xeon Phi does, and for the dimension $n = 19200$ the coprocessor Xeon Phi noticeably outperforms 2×CPU. At the same time, for the dimensions 9600, 12000 and 19200, the vectorization and data alignment provides a performance boost of 12%, 12.3% 20% correspondingly. Thus, we can conclude that the efficiency of the Xeon Phi coprocessor usage increases with the growth of the problem dimension. Simultaneously the significance of the vectorization and data alignment is increased.

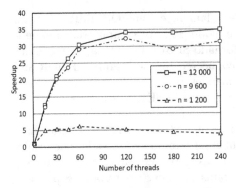

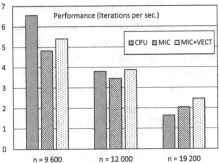

Fig. 6. Speedup of computing pseudo-projection on Xeon Phi.

Fig. 7. Performance comparison of CPU and Xeon Phi (MIC).

6   Conclusion

The paper describes a new version of the *Pursuit* algorithm for solving high-dimension, non-stationary linear programming problem on the modern cluster computing systems. The distinctive feature of the new version is that it uses a cross-shaped pursuit region consisting of $n(K-1)+1$ cells, where n – dimension of the problem, K – number of cells in one cross-bar. The previous version of the algorithm uses a cube-shaped pursuit region consisting of K^n cells that results in high computational complexity of the algorithm. The results of the computational experiments investigating the efficiency of the coprocessor Xeon Phi use for pseudoprojection computation were presented. In these experiments, a synthetic linear programming problem was used. Studies have shown that the use of Intel Xeon Phi coprocessors is effective for high-dimension problems (over 10000). Our future goal is to investigate the efficiency of the proposed algorithm on the cluster computing systems using MPI technology.

References

1. Ananchenko, I.V., Musaev, A.A.: Torgovye roboty i upravlenie v khaoticheskikh sredakh: obzor i kriticheskiy analiz [Trading robots and management in chaotic environments: an overview and critical analysis]. Trudy SPIIRAN [SPIIRAS Proc.] **34**(3), 178–203 (2014)
2. Dyshaev, M.M., Sokolinskaya, I.M.: Predstavlenie torgovykh signalov na osnove adaptivnoy skol'zyashchey sredney Kaufmana v vide sistemy lineynykh neravenstv [Representation of trading signals based on Kaufman's adaptive moving average as a system of linear inequalities]. Vestnik Yuzhno-Ural'skogo gosudarstvennogo universiteta **2**(4), 103–108 (2013). Seriya: Vychislitel'naya matematika i informatika [Bulletin of South Ural State University. Series: Computational Mathematics and Software Engineering]
3. Eremin, I.I.: Fejerovskie metody dlya zadach linejnoj i vypukloj optimizatsii [Fejer's Methods for Problems of Convex and Linear Optimization], 200 p. Publishing of the South Ural State University, Chelyabinsk (2009)

4. Ershova, A.V., Sokolinskaya, I.M.: O skhodimosti masshtabiruemogo algoritma postroeniya psevdoproektsii na vypukloe zamknutoe mnozhestvo [About convergence of scalable algorithm of constructing pseudo-projection on convex closed set]. Vestnik YuUrGU. Seriya "Matematicheskoe modelirovanie i programmirovanie" [Bulletin of South Ural State University. Series: Mathematical simulation and programming], vol. 10, no. 37(254), pp. 12–21 (2011)

5. Kostenetskiy, P.S., Safonov, A.Y.: SUSU supercomputer resources. In: Proceedings of the 10th Annual International Scientific Conference on Parallel Computing Technologies (PCT 2016), CEUR Workshop Proceedings, vol. 1576, pp. 561–573. CEURWS (2016)

6. OpenMP Application Program Interface. Version 4.0, July 2013. http://www.openmp.org/mp-documents/OpenMP4.0.0.pdf

7. Sokolinskaya, I., Sokolinsky, L.: Solving unstable linear programming problems of high dimension on cluster computing systems. In: Proceedings of the 1st Russian Conference on Supercomputing – Supercomputing Days (RuSCDays 2015), Moscow, Russian Federation, 28–29 September 2015, CEUR Workshop Proceedings, vol. 1482, pp. 420–427. CEURWS (2015)

8. Sokolinskaya, I., Sokolinsky, L.: Implementation of parallel pursuit algorithm for solving unstable linear programming problems. In: Proceedings of the 10th Annual International Scientific Conference on Parallel Computing Technologies (PCT 2016), Arkhangelsk, Russia, 29–31 March 2016, CEUR Workshop Proceedings, vol. 1576, pp. 685–698. CEURWS (2016)

9. Supalov, A., Semin, A., Klemm, M., Dahnken, C.: Optimizing HPC Applications with Intel Cluster Tools. 269 p. Apress (2014). doi:10.1007/978-1-4302-6497-2

10. Thiagarajan, S.U., Congdon, C., Naik, S., Nguyen, L.Q.: Intel Xeon Phi coprocessor developer's quick start guide. White Paper, Intel (2013) https://software.intel.com/sites/default/files/managed/ee/4e/intel-xeon-phi-coprocessor-quick-start-developers-guide.pdf

Solving Multidimensional Global Optimization Problems Using Graphics Accelerators

Konstantin Barkalov[✉] and Ilya Lebedev

Lobachevsky State University of Nizhny Novgorod, Nizhny Novgorod, Russia
{konstantin.barkalov,ilya.lebedev}@itmm.unn.ru

Abstract. In the present paper an approach to solving the global optimization problems using a nested optimization scheme is developed. The use of different algorithms at different nesting levels is the novel element. A complex serial algorithm (on CPU) is used at the upper level, and a simple parallel algorithm (on GPU) is used at the lower level. This computational scheme has been implemented in ExaMin parallel solver. The results of computational experiments demonstrating the speedup when solving a series of test problems are presented.

Keywords: Global optimization · Multiextremal functions · Dimension reduction · Parallel algorithms · Speedup · Graphics accelerators

1 Introduction

Let us consider the problem of search for a global minimum of an N-dimensional function $\varphi(y)$ within a hyperinterval D

$$\varphi^* = \varphi(y^*) = \min \{\varphi(y) : y \in D\}, \tag{1}$$

$$D = \{y \in R^N : a_i \leq y_i \leq b_i, 1 \leq i \leq N\}, \tag{2}$$

where $a, b \in R^N$ are given vectors.

The numerical solving of problem (1) is reduced to building an estimate $y_k^* \in D$ corresponding to some measure of proximity to point y^* (for example, $\|y^* - y_k^*\| \leq \epsilon$, where $\epsilon > 0$ is given accuracy) based on a finite number k of computations of the objective function values. With respect to the class of considered problems, the fulfillment of two important conditions is suggested.

First, it is suggested that the objective function $\varphi(y)$ may be defined not analytically (as a formula) but algorithmically as a result of execution of some subroutine or library.

Second, it is suggested that the function $\varphi(y)$ satisfies a Lipschitz condition

$$|\varphi(y_1) - \varphi(y_2)| \leq L \|y_1 - y_2\|, \; y_1, y_2 \in D, \; 0 < L < \infty, \tag{3}$$

with an a priori unknown constant L. This suggestion is typical for many approaches to solving the global optimization problems (see, for example [1–5]). It can be interpreted (with respect to the applied problems) as the reflection of the limited power causing the changes in the modeled system.

© Springer International Publishing AG 2016
V. Voevodin and S. Sobolev (Eds.): RuSCDays 2016, CCIS 687, pp. 224–235, 2016.
DOI: 10.1007/978-3-319-55669-7_18

The multiextremal optimization problems are more computational costly essentially as compared to other types of the optimization problems since the global optimum is an integral characteristic of the problem being solved and requires the investigation of the whole search domain. As a result, the search for the global optimum is reduced to the generation of a grid in the parameter domain, and to the choice of the best function value on the grid. At that, the computational costs of solving the problem increase exponentially with increasing dimension.

A novel approach to solving the global optimization problems has been developed under the supervision by prof. R.G. Strongin at Lobachevsky State University of Nizhny Novgorod (see [6–12]). Within the framework of this approach, solving of the multidimensional problems is reduced to solving a series of nested problems with a lower dimension. For the efficient solving of the multidimensional problems with the computationally inexpensive objective function, it is proposed to transfer solving of the nested subproblems to the graphics accelerator completely.

2   Core Global Search Algorithm with Parallel Trials

As a base problem, we will consider a one-dimensional multiextremal optimization problem

$$\varphi^* = \varphi(x^*) = \min\left\{\varphi(x) : x \in [0,1]\right\}, \tag{4}$$

with objective function satisfying the Lipschitz condition.

Let us give the description of Parallel Global Search Algorithm (PGSA) applied to solving above problem (let us formulate it here according to [6]). Let us assume the problem to be solved in a parallel computational system with p processors.

The algorithm for solving problem (4) involves constructing a sequence of points x^i, where the values of the objective function $z^i = \varphi(x^i)$ are calculated. Let us call the function value calculation process the *trial*, and pair (x^i, z^i) the *trial result*. At each iteration of the method p of trials is carried out in parallel, and the set of pairs $(x^i, z^i), 1 \leq i \leq k = np$, makes up the search data collected using the method after carrying out n steps.

At the first iteration of the method the trials are carried out in parallel at arbitrary internal points $x^1, \ldots, x^p$ of the interval $[0,1]$. For example, these points can be uniformly distributed over the interval. The results of the trials $(x^i, z^i), 1 \leq i \leq p$, are saved in the database of the algorithm.

Suppose, now, that $n \geq 1$ iterations of the method have already been executed. The trial points $x^{k+1}, \ldots, x^{k+p}$ of the next $(n+1)$-th iteration are then chosen by using the following rules.

Rule 1. Renumber points of the set

$$X_k = \{x^1, \ldots, x^k\} \cup \{0\} \cup \{1\}, \tag{5}$$

which includes boundary points of the interval $[0, 1]$ and the points $\{x^1, \ldots, x^k\}$ of the previous $k = k(n) = np$ trials, with subscripts in increasing order of coordinate values, i.e.,

$$0 = x_0 < x_1 < \cdots < x_k < x_{k+1} = 1. \tag{6}$$

Rule 2. Supposing that $z_i = \varphi(x_i)$, $1 \leq i \leq k$, calculate values

$$\mu = \max_{2 \leq i \leq k} \frac{|z_i - z_{i-1}|}{\Delta_i}, \quad M = \begin{cases} r\mu, & \mu > 0, \\ 1, & \mu = 0, \end{cases} \tag{7}$$

where $r > 1$ is a preset reliability parameter of the method, and $\Delta_i = x_i - x_{i-1}$.

Rule 3. Calculate a *characteristic* for every interval (x_{i-1}, x_i), $1 \leq i \leq k+1$, according to the following formulae

$$R(1) = 2\Delta_1 - 4\frac{z_1}{M}, \tag{8}$$

$$R(i) = \Delta_i + \frac{(z_i - z_{i-1})^2}{M^2\Delta_i} - 2\frac{z_i + z_{i-1}}{M}, 1 < i < k+1, \tag{9}$$

$$R(k+1) = 2\Delta_{k+1} - 4\frac{z_k}{M}. \tag{10}$$

Step 4. Arrange characteristics $R(i), 1 \leq i \leq k+1$, in decreasing order

$$R(t_1) \geq R(t_2) \geq \ldots \geq R(t_k) \geq R(t_{k+1}) \tag{11}$$

and select p maximum characteristics with interval numbers $t_j, 1 \leq j \leq p$.

Rule 5. Carry out new trials at the points $x^{k+j}, 1 \leq j \leq p$, calculated using the following formulae

$$x^{k+j} = \frac{x_{t_j} + x_{t_j-1}}{2}, \text{ if } t_j = 1 \text{ or } t_j = k+1, \tag{12}$$

$$x^{k+j} = \frac{x_{t_j} + x_{t_j-1}}{2} - \frac{z_{t_j} - z_{t_j-1}}{2M}, \text{ if } 1 < t_j < k+1. \tag{13}$$

The algorithm terminates if the condition $\Delta_{t_j} < \epsilon$ is satisfied at least for one number $t_j, 1 \leq j \leq p$; here $\epsilon > 0$ is the preset accuracy. As *current estimate* of the optimum at the step n we accept the values

$$\varphi_k^* = \min_{1 \leq i \leq k} \varphi(x^i), \tag{14}$$

$$x_k^* = \arg \min_{1 \leq i \leq k} \varphi(x^i). \tag{15}$$

This method of the parallel computations organization has the following substantiation. The characteristics of intervals (8)–(10) used in the algorithm can be considered as some measures of probability of localization of the global minimum point within these intervals. Inequalities (11) arrange the intervals according to the characteristics of these ones, and the trials are executed in parallel in the first p intervals with the highest probabilities to find the global optimum point in. Various modifications of this algorithm and the corresponding theory of convergence are presented in [6].

3   Dimension Reduction

3.1   Dimension Reduction Using Peano Curves

For decreasing the complexity of the global optimization algorithms generating nonuniform coverages of the multidimensional search domain, various dimension reduction schemes are widely used. These schemes allow reducing the solving of the multidimensional problem to solving a family of connected subproblems of lower dimension (in particular, to the one-dimensional problems).

The use of Peano curve $y(x)$

$$\{y \in R^N : -2^{-1} \le y_i \le 2^{-1}, 1 \le i \le N\} = \{y(x) : 0 \le x \le 1\} \qquad (16)$$

unambiguously mapping the interval of real axis $[0,1]$ onto a N-dimensional cube is the first of the dimension reduction methods considered. Problems of numerical construction of Peano-type space filling curves and the corresponding theory are considered in detail in $[6,13]$. Here we will note that a numerically constructed curve (*evolvent*) is 2^{-m} accurate approximation of the theoretical Peano curve, where m is an evolvent construction parameter.

By using this kind of mapping it is possible to reduce the multidimensional problem (1) to a univariate problem

$$\varphi(y^*) = \varphi(y(x^*)) = \min\{\varphi(y(x)) : x \in [0,1]\}. \qquad (17)$$

An important property of such mapping is preservation of boundedness of function relative differences (see $[6,13]$). If the function $\varphi(y)$ in the domain D satisfies the Lipschitz condition, then the function $\varphi(y(x))$ on the interval $[0,1]$ will satisfy a uniform Hölder condition

$$|\varphi(y(x_1)) - \varphi(y(x_2))| \le H |x_1 - x_2|^{1/N}, \qquad (18)$$

where the Hölder constant H is linked to the Lipschitz constant L by the relation

$$H = 2L\sqrt{N+3}. \qquad (19)$$

Relation (18) allows adopting the algorithm for solving the one-dimensional problems presented in Sect. 2 for solving the multidimensional problems reduced to the one-dimensional ones. For this, the lengths of intervals Δ_i involved into rules (3)–(5) of the algorithm are substituted by the lengths in a new metrics

$$\Delta_i = (x_i - x_{i-1})^{1/N} \qquad (20)$$

and the following expression is introduced instead of formula (13):

$$x^{k+j} = \frac{x_{t_j} + x_{t_j-1}}{2} - \text{sign}(z_{t_j} - z_{t_j-1})\frac{1}{2r}\left[\frac{|z_{t_j} - z_{t_j-1}|}{\mu}\right]^N, \text{if } 1 < t_j < k+1.$$

$$\qquad (21)$$

3.2   Nested Optimization Scheme

Nested optimization scheme is based on a well known relation (see [15])

$$\min_{y \in D} \varphi(y) = \min_{a_1 \leq y_1 \leq b_1} \min_{a_2 \leq y_2 \leq b_2} \cdots \min_{a_N \leq y_N \leq b_N} \varphi(y), \qquad (22)$$

which allows replacing the solving of multidimensional problem (1) by solving a family of one-dimensional subproblems related to each other recursively.

Let us introduce a set of functions

$$\varphi_N(y_1, \ldots, y_N) = \varphi(y_1, \ldots, y_N), \qquad (23)$$

$$\varphi_i(y_1, \ldots, y_i) = \min_{a_{i+1} \leq y_{i+1} \leq b_{i+1}} \varphi_{i+1}(y_1, \ldots, y_i, y_{i+1}), 1 \leq i \leq N - 1. \qquad (24)$$

into consideration. Then, according to relation (22), the solving of initial problem (1) is reduced to solving a one-dimensional problem

$$\varphi_1(y_1^*) = \min_{a_1 \leq y_1 \leq b_1} \varphi_1(y_1). \qquad (25)$$

However, at that, every calculation of a value of one-dimensional function $\varphi_1(y_1)$ in a certain point implies solving a one-dimensional minimization problem

$$\varphi_2(y_1, y_2^*) = \min_{a_2 \leq y_2 \leq b_2} \varphi_2(y_1, y_2), \qquad (26)$$

etc. up to calculation of φ_N.

For the nested scheme presented above, a generalization (*block nested optimization scheme*), which combines the use of evolvents and the nested scheme has been proposed in [14] with the purpose of efficient parallelization of the computations.

Let us consider vector y as a vector of block variables

$$y = (y_1, y_2, \ldots, y_N) = (u_1, u_2, \ldots, u_M), \qquad (27)$$

where the i-th block variable u_i is a vector of dimension N_i of components of vector y, taken serially i.e. $u_1 = (y_1, y_2, \ldots, y_{N_1})$, $u_2 = (y_{N_1+1}, y_{N_1+2}, \ldots, y_{N_1+N_2}), \ldots, u_M = (y_{N-N_M+1}, y_{N-N_M+2}, \ldots, y_N)$, at that $N_1 + N_2 + \ldots + N_M = N$.

Using the new variables, main relation of the nested scheme (22) can be rewritten in the form

$$\min_{y \in D} \varphi(y) = \min_{u_1 \in D_1} \min_{u_2 \in D_2} \cdots \min_{u_M \in D_M} \varphi(y), \qquad (28)$$

where the subdomains $D_i, 1 \leq i \leq M$, are projections of initial search domain D onto the subspaces corresponding to the variables $u_i, 1 \leq i \leq M$.

The formulae defining the method of solving of problem (1) based on relation (28), in general, are the same to the ones of nested scheme (23)–(25). It is only necessary to substitute the original variables $y_i, 1 \leq i \leq N$, by the block variables $u_i, 1 \leq i \leq M$.

At that, the nested subproblems

$$\varphi_i(u_1,\ldots,u_i) = \min_{u_{i+1}\in D_{i+1}} \varphi_{i+1}(u_1,\ldots,u_i,u_{i+1}), 1 \le i \le M-1. \qquad (29)$$

in the block scheme are the multidimensional ones. The dimension reduction method based on Peano curves can be applied to solving these ones. It is a principal difference from the initial scheme.

The number of vectors M and the quantity of components in each vector $N_1, N_2, \ldots, N_M$ are the parameters of block nested scheme and can be used for the forming of the subproblems with necessary properties. For example, if $M = N$ i.e. $u_i = y_i, 1 \le i \le N$, the block scheme is identical to the initial one; each nested subproblem is a one-dimensional one. And if $M = 1$, i.e. $u = u_1 = y$, the solving of the problem is equivalent to solving this one using a single evolvent mapping $[0,1]$ into D; the nested subproblems are absent.

4   Organization of Parallel Computing

For organization of parallel computing, we will use a small (2–3) number of nesting levels. Correspondingly, the initial problem of large dimension is subdivided into 2–3 nested subproblems of lower dimension. Then, applying the parallel methods of global optimization in block nested scheme (28) to the solving of nested problems (29), we obtain a parallel algorithm with a high degree of variability. For example, it is possible to vary the number of processors at different nesting levels, to apply various parallel search methods at different levels, and also to use various types of computing devices.

For instance, for solving the problems with computationally inexpensive functions, one can use Parallel Global Search Algorithm on CPU at the upper nesting level and employ the scanning method on a uniform grid on GPU at the lower one. Note that the implementation of PGSA with the use of computation accelerators in the problems with time-consuming objective functions has been considered in details in [17,18].

The scanning method is featured by the fact that all computations can be executed independently. The parallelization of the scanning method can be organized easily by means of subdivision of the grid into several subdomains of equal size and simultaneous search of the solution in these subdomain in different thread blocks.

Within this approach to the parallelization of the scanning method, the objective function values would be computed many times in every thread. The computations within a thread block are executed in the following way. All threads in a block compute the objective function values in parallel. Then, the synchronization is performed inside the block. After that, the zeroth thread chooses the best value in the block and saves it in the global memory. The process is repeated until all necessary trials are completed. Upon completing the computations, the lowest value among the blocks is selected, and the point of the best value is returned. Shared memory is allocated for the blocks in amount necessary for storing the coordinates of the points computed in parallel and the function values obtained. The data transfer from CPU to GPU will be minimal: it is required to send the fixed coordinates of the trial point to GPU and to receive the coordinates and values of current global minimum point found back.

The general scheme of the computations using several cluster nodes and several GPUs is presented in Fig. 1. The processes of a parallel program will make a tree corresponding to the nesting levels of the subproblems. According to this scheme, the nested subproblems

$$\varphi_i(u_1, \ldots, u_i) = \min_{u_{i+1} \in D_{i+1}} \varphi_{i+1}(u_1, \ldots, u_i, u_{i+1}), i = 1, \ldots, M - 2, \qquad (30)$$

are solved using CPU only. In these subproblems, the values of the function $\varphi(y)$ are not computed directly since the calculation of the values of functions $\varphi_i(u_1, \ldots, u_i)$ is defined when solving the minimization problems of the next level. Each subproblem is solved in a separate process; the exchange of the results is organized by means of MPI.

The subproblem of the lowest $(M - 1)$-th level

$$\varphi_i(u_1, \ldots, u_{M-1}) = \min_{u_M \in D_M} \varphi_M(u_1, \ldots, u_M) \qquad (31)$$

differs from all previous subproblems because the values of the objective function are computed within this one since $\varphi_M(u_1, \ldots, u_M) = \varphi(y_1, \ldots, y_N)$. This subproblem is executed in a separate process also and can be solved on CPU (using Global Search Algorithm) as well as on GPU (using the scanning method over a uniform grid).

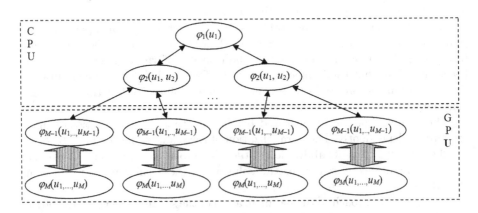

Fig. 1. The scheme of the parallel computing on a cluster

5   Numerical Experiments

Computational experiments were carried out on a high-performance cluster of Lobachevsky State University of Nizhny Novgorod. The cluster node included two Intel Xeon L5630 CPUs, 24 Gb RAM, and two NVIDIA Tesla M2070 GPUs. The CPU had 4 cores, i.e. 8 cores were available in each node. For the implementation of the GPU algorithm, CUDA Toolkit 6.0 was used. To carry out the computational experiments ExaMin parallel solver developed in Lobachevsky State University of Nizhny Novgorod was used.

The efficiency of the parallel algorithm was estimated by solving a set of test problems, selected from some class randomly. At that, each test problem can be considered as particular realization of a random function defined with a special generator. We have used well known GKLS generator of test problems for the multiextremal optimization [16]. This generator allows generating multiextremal optimization problems with known properties (the number of local minima, the size of their domains of attraction, the global minimum point, etc.). The comparison of several serial optimization algorithms carried out using GKLS generator in [17,18] has demonstrated the advantage of serial Global Search Algorithm over the other similar purpose algorithms.

The numerical experiments were carried out using Simple and Hard function classes from [19]. The global minimum y^* was considered to be found if the algorithm generates a trial point y^k in δ-vicinity of the global minimum, i.e. $\|y^k - y^*\| \leq \delta$. The size of the vicinity was selected (according to [19]) as $\delta = \|b - a\| \sqrt[N]{\Delta}$, where N is the problem dimension, a and b are the borders of search domain D, the parameter $\Delta = 10^{-4}$ at $N = 2$, $\Delta = 10^{-6}$ at $N = 3, 4$ and $\Delta = 10^{-7}$ at $N = 5$. When using GSA, the parameter r was selected to be 4.5 for Simple class and 5.6 for Hard class. The evolvent construction parameter was fixed as $m = 10$. The maximum allowable number of iterations was $K_{max} = 10^6$.

In the first series of experiments, 800 problems of various dimensions have been solved in order to estimate the efficiency of the parallel algorithm using the resources of CPU only. At that, block nested optimization scheme has not been used, i.e. core Parallel Global Search Algorithm was used only.

In Table 1, the run time (in seconds) and the averaged number of iterations when running the serial algorithm are presented. The speedup of the parallel algorithm run on CPU using p threads is presented in Table 2.

Table 1. Average time and number of iterations on CPU

	$N = 2$		$N = 3$		$N = 4$		$N = 5$	
	Simple	Hard	Simple	Hard	Simple	Hard	Simple	Hard
t_{av}	0.02	0.05	0.03	0.08	0.19	0.57	0.24	3.84
k_{av}	2349	4731	2129	5382	12558	37410	15538	247784

The results demonstrate an insignificant speedup of the parallel algorithm. This agrees with our assumptions completely since the values of the objective functions generated by GKLS generator are computed fast, and the effect of extra costs for the organization of the parallel computations become essential.

Table 2. Average speedup on CPU

p	$N = 2$		$N = 3$		$N = 4$		$N = 5$	
	Simple	Hard	Simple	Hard	Simple	Hard	Simple	Hard
2	7.38	6.5	1.1	1.01	1.35	1.32	0.68	1.2
4	1.16	0.63	1.33	1.12	1.53	1.37	0.99	1.29
8	6.87	0.9	1.36	1.33	1.49	1.61	0.62	2.22
16	0.78	1.64	1.43	1.29	1.61	1.89	1.71	1.75

The analogous behavior of the algorithm has been observed when running PGSA using GPU. In Table 3, the speedup in time of the algorithm with the use of GPU relative to the serial algorithm is presented. The values are given subject to the number of threads p; both accelerators available on the node have been employed.

Table 3. Average speedup on GPU

p	$N = 2$		$N = 3$		$N = 4$		$N = 5$	
	Simple	Hard	Simple	Hard	Simple	Hard	Simple	Hard
128	4.6	4.97	1.06	1.19	1.12	1.38	0.66	1.82
256	3.37	6.11	1.02	1.34	1.6	1.51	1.02	2.35
512	2.27	4.36	0.76	1.36	1.59	1.53	0.83	1.29

Further, let us consider solving the problems using the parallel scanning method implemented on GPU. The solution found by the scanning method was adjusted by Hooke-Jeeves method (see, for example, [20]). In Table 4, the speedup of scanning method relative to serial GSA on CPU is presented. In the computations, two accelerators have been used as before.

It is seen from the table, that when using the scanning method on GPU, a greater speedup has been achieved than when using GPU for the parallel computation of the function value within the framework of PGSA. Also, the experiments have demonstrated the speedup to decrease with increasing dimension since the increasing of the number of nodes in the uniform grid in the search domain takes place. Consequently, the use of scanning method on GPU is advisable for solving the problems of low dimension.

The main computational experiments has been carried out on a series of 200 multiextremal problems of the dimensions $N = 6$ and $N = 8$ from Simple class using the block nested optimization scheme. When solving the problems from these classes the parameters $r = 4.5$, $\Delta = 10^{-8}$ for $N = 6$ and $\Delta = 10^{-9}$ for $N = 8$ were used. In accordance with block nested scheme (28), two levels of subproblems with the dimensions $N_1 = N_2 = 3$ for the six-dimensional problems and $N_1 = N_2 = 4$ for the eight-dimensional ones were used. The maximum

Table 4. Speedup of GPU scanning method relative to CPU serial GSA

N = 2		N = 3		N = 4	
Simple	Hard	Simple	Hard	Simple	Hard
39.31	15.49	31.28	11.73	7.6	2.6

allowable number of iterations of the algorithm was $K_{max} = 10^6$ at each level. When using the scanning method on GPU, the step of the uniform grid in the search domain was selected to be 0.1; upon completing the computations, the found solution was adjusted by the local method.

Table 5 presents the averaged time (in seconds) of solving the problem in the following modes:

- on CPU using serial Global Search Algorithm (GSA column);
- in the parallel mode using the block nested optimization scheme (B-GSA column). At each nesting level, Parallel Global Search Algorithm was used; two CPU available on the cluster node were employed in the computations;
- in a hybrid mode using the block nested optimization scheme (H-GSA column). At the first nesting level, PGSA on CPU of a cluster node was used; at the second nesting level 2 graphics accelerators available on the same node were employed;
- in the hybrid mode using the block nested optimization scheme (M-GSA column). At the first nesting level, PGSA on CPU of a cluster node was used, at the second level, four cluster nodes and 8 graphics accelerators were employed, which the parallel scanning method was running on.

In Table 6, the speedup of the method in the same modes relative to the serial running is presented. The number of the unsolved problems is given in the braces; the time spent for trying to solve these ones was not taken into account when calculating the averaged run time of the algorithm.

Table 5. Averaged time of solving the problems of dimensions 6 and 8

N	GSA	B-GSA	H-GSA	M-GSA
6	53.5(20)	4.4	1.1	0.4
8	72.6(19)	78.3	10.7	3.1

The results of experiments confirm the use of the block nested optimization scheme in solving the problems of large dimension to give a considerable speedup as compared to the initial serial algorithm. At the same time, the use of the scanning method implemented on GPU at the lower nesting levels allows obtaining an additional speedup for the problems with computationally inexpensive objective function.

Table 6. Speedup for solving the problems of dimensions 6 and 8

N	B-GSA	H-GSA	M-GSA
6	12.1	48.4	133.7
8	0.9	6.8	23.4

6 Conclusion

In the present work, a multilevel scheme of dimension reduction in the global optimization problems combining the use of Peano curves and the nested scheme is considered. For solving the reduced subproblems, Parallel Global Search Algorithm is used. The issues of the efficiency of using the proposed multilevel scheme for the problems with a small time of computing the objective function values are discussed. In order to estimate the speedup of the parallel algorithm experimentally, the computational experiments have been carried out on a series of test problems of various dimension. The results of experiments demonstrate the proposed multilevel scheme to allow employing the heterogeneous resources of modern computer systems (CPU, GPU) efficiently and achieving a considerable speedup.

Acknowledgements. This study was supported by the Russian Science Foundation, project No. 15-11-30022 "Global optimization, supercomputing computations, and applications".

References

1. Pinter, J.D.: Global Optimization in Action (Continuous and Lipschitz Optimization: Algorithms, Implementations and Applications). Kluwer Academic Publishers, Dordrecht (1996)
2. Jones, D.R.: The direct global optimization algorithm. In: Floudas, C.A., Pardalos, P.M. (eds.) The Encyclopedia of Optimization, pp. 431–440. Kluwer Academic Publishers, Dordrect (2001)
3. Žilinskas, J.: Branch and bound with simplicial partitions for global optimization. Math. Model. Anal. **13**(1), 145–159 (2008)
4. Evtushenko, Y.G., Posypkin, M.A.: A deterministic approach to global box-constrained optimization. Optim. Lett. **7**(4), 819–829 (2013)
5. Paulavičius, R., Žilinskas, J., Grothey, A.: Investigation of selection strategies in branch and bound algorithm with simplicial partitions and combination of Lipschitz bounds. Optim. Lett. **4**(2), 173–183 (2010)
6. Strongin, R.G., Sergeyev, Y.D.: Global Optimization with Non-convex Constraints: Sequential and Parallel Algorithms. Kluwer Academic Publishers, Dordrecht (2000)
7. Gergel, V.P.: A method of using derivatives in the minimization of multiextremum functions. Comput. Math. Math. Phys. **36**(6), 729–742 (1996)
8. Gergel, V.P., Sergeyev, Y.D.: Sequential and parallel algorithms for global minimizing functions with Lipschitzian derivatives. Comput. Math. Appl. **37**(4–5), 163–179 (1999)

9. Gergel, V.P., Strongin, R.G.: Parallel computing for globally optimal decision making on cluster systems. Future Gener. Comput. Syst. **21**(5), 673–678 (2005)
10. Gergel, V., Grishagin, V., Israfilov, R.: Local tuning in nested scheme of global optimization. Procedia Comput. Sci. **51**(1), 865–874 (2015)
11. Gergel, V., Grishagin, V., Gergel, A.: Adaptive nested optimization scheme for multidimensional global search. J. Global Optim. **66**(1), 35–51 (2016)
12. Lebedev, I., Gergel, V.: Heterogeneous parallel computations for solving global optimization problems. Procedia Comput. Sci. **66**, 53–62 (2015)
13. Sergeyev, Y.D., Strongin, R.G., Lera, D.: Introduction to Global Optimization Exploiting Space-Filling Curves. Springer, New York (2013)
14. Barkalov, K.A., Gergel, V.P.: Multilevel scheme of dimensionality reduction for parallel global search algorithms. In: Proceedings of the 1st International Conference on Engineering and Applied Sciences Optimization - OPT-i 2014, pp. 2111–2124 (2014)
15. Sergeyev, Y.D., Grishagin, V.A.: Parallel asynchronous global search and the nested optimization scheme. J. Comput. Anal. Appl. **3**(2), 123–145 (2001)
16. Gaviano, M., Kvasov, D.E., Lera, D., Sergeyev, Y.D.: Software for generation of classes of test functions with known local and global minima for global optimization. ACM Trans. Math. Softw. **29**(4), 469–480 (2003)
17. Barkalov, K., Gergel, V., Lebedev, I.: Use of xeon phi coprocessor for solving global optimization problems. In: Malyshkin, V. (ed.) PaCT 2015. LNCS, vol. 9251, pp. 307–318. Springer, Heidelberg (2015). doi:10.1007/978-3-319-21909-7_31
18. Barkalov, K., Gergel, V.: Parallel global optimization on GPU. J. Global Optim. **66**(1), 2–20 (2016)
19. Sergeyev, Y.D., Kvasov, D.E.: Global search based on efficient diagonal partitions and a set of Lipschitz constants. SIAM J. Optim. **16**(3), 910–937 (2006)
20. Himmelblau, D.M.: Applied Nonlinear Programming. McGraw-Hill, New York (1972)

Supercomputer Simulation of Physicochemical Processes in Solid Fuel Ramjet Design Components for Hypersonic Flying Vehicle

Vadim Volokhov, Pavel Toktaliev, Sergei Martynenko$^{(\boxtimes)}$,
Leonid Yanovskiy, Aleksandr Volokhov, and Dmitry Varlamov

Institute of Problems of Chemical Physics, Chernogolovka, Russia
{vvm,vav,dima}@icp.ac.ru,
{pavel_d_m,martyn_s}@mail.ru, Yanovskiy@ciam.ru

Abstract. A step-by-step computer simulation variant for making scramjet mathematical model is offered. The report considers an approach related to 3D mathematical models development of scramjet components further reduced to 1D models. Mathematical models of physicochemical processes in combustor cooling system are discussed with the aim of subsequent engine performance optimization depending on fuels used. Then 1D separate component models are used to make up a full-scale scramjet model. The one-dimensional models allow calculation times significantly reduce, and the simulation accuracy is conditioned by precision of 3D models to 1D models reduction.

Keywords: Supercomputer simulation · Scramjet · 3D model · 1D model · Models reduction

1 Introduction

Fast development of computer technology has resulted in a new research method appearance - a computer experiment based on the triad "model-algorithm-program" [1]. As a rule, a mathematical model consists of systems of nonlinear differential equations in partial derivatives, integral or integral-differential equations together with boundary and initial conditions. These equations usually express the fundamental conservation laws of main physical values (energy, momentum, mass, etc.). The computing algorithm implies operational procedures by means of which numerical solutions of mathematical model equations are found.

Originally, the mathematical models contained many admissions, which allowed finding solutions of background equations analytically or with minimum calculations. However, as far as computation capability was growing the mathematical models got more complex and permitted explication of physicochemical process particularities running in hardware design elements. Full-scale simulation based on 3D (non) steady-state mathematical model of entire technical unit often with inclusion of its adjacent space is most valuable. Such simulation allows realizing technical unit optimization, revealing its physicochemical process features at different working state, studying various factors impact on efficiency, and getting other useful information on

© Springer International Publishing AG 2016
V. Voevodin and S. Sobolev (Eds.): RuSCDays 2016, CCIS 687, pp. 236–248, 2016.
DOI: 10.1007/978-3-319-55669-7_19

technical unit under investigation. Besides, full-scale modeling greatly reduces experimental study volumes and cuts development and design time for technical device with optimum performance.

Full-scale simulation shortages should include mathematical models complexity tied with computing domain features and/or running physicochemical processes, as well as large computation work volume necessary for conducting numerical solutions of basic equations. As a rule, mathematical models comprise boundary (stationary processes) or initial-boundary (unsteady processes) tasks, which, due to approximation and linearization, reduce to systems of linear algebraic equations (SLAE) with sparse and ill-conditioned matrix coefficients. Modern models lead to SLAE consisting of 10^8 equations and more. At present, the computing algorithms efficiency designed for such SLAE solutions completely depends on mesh performance. Quite recently a multigrid technology is designed for solving a large class of boundary and initial-boundary problems on structured meshes both in sequential and parallel performances [2]. The computation scope for solution of large class of (non)linear boundary value problems is shown close to optimum and makes up $O(N \cdot \lg N)$ arithmetical operations, where N is the number of unknowns. We emphasize that multigrid technology contains the minimum problem-dependent components, and a close-to-optimal computing efficiency is achieved without algorithm adapting to boundary tasks under solving. In other words, the present multigrid technology is specially designed for autonomous software.

Meanwhile a computing mesh with the specified characteristics can be built in rather simple geometry domains. If the domain geometry is complex, unstructured meshes are usually used. At present an efficient algorithm for solving boundary and initial-boundary problems on unstructured meshes still is not designed, thus a forced step is transition from full-scale to bit-by-bit simulation.

The step-by-step simulation is based on breaking the original unit in separate nodes (elements). A mathematical model is built per any element; separate models interaction is taken into account with the help of boundary conditions. The present approach capability is limited by inaccuracy of boundary conditions formulation, so step-by-step simulation fails to realize technical unit optimization, especially when elements number is great enough.

The project studies possible creation of high-speed ramjet on solid fuel with system of active cooling combustor (Fig. 1). Gases generated during solid propellant combustion in autonomous gas-generator (1) are used for solid fuel gasification (2) for ramjet. Gas temperature reduces, since heat is partly absorbed in solid fuel gasification process. Further gas mixture enriched by solid fuel gasification products enters the combustor cooling system. Gas mixture in cooling system ducts is heated and thermodestructed generating lighter hydrocarbon compounds and radicals. The thermodestruction process runs with significant heat absorption providing ramjet combustor efficiency. Besides, thermodestruction products possess increased reactivity that makes much easier their ignition and combustion in air flow. The heat absorbed during fuel destruction process in cooling system ducts returns in combustor providing high propellant combustion completeness. Then thermodestruction products enter combustor burning and creating ramjet thrust. The proposed ramjet scheme is featured by using solid fuel, which allows creating unique flying vehicles, having no analogues in the world, with increased shelf time and high operative readiness. At present a range

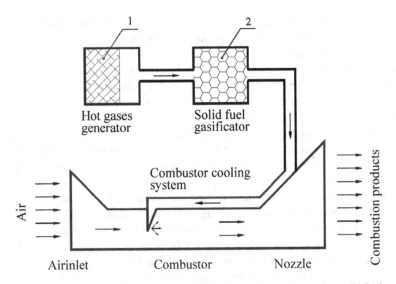

Fig. 1. Ramjet basic diagram: 1 – solid fuel propellant, 2 – ramjet solid fuel

of scramjet designs is considered using cryogenic fuels. Cryogenic hydrogen most often is in the first place. Practically, they speak about hypersonic flying vehicles having sufficient "cold reserve" on board for cooling heat stressed elements of engine and airframe design. However, cryogenic fuel scramjet usage is complicated by significant overhead expenses and technical difficulties connected with fuel low temperature.

When simulating a contradictory situation often appears: on the one hand, full-scale simulation can't be realized because of limited computer capability and absences of efficient numerical solution methods of basic equations, but, from the other side, step-by-step modeling fails to get answers to interesting issues due to inaccurate boundary conditions statement at element junctions. The only contradictory situation outcome is: at first to develop 3D mathematical models of separate technical unit elements, which hereinafter are reduced to 1D models. Further separate element 1D models form full-scale 1D model of the entire technical unit. Usage of 1D models greatly shorten computation time, and simulation accuracy is conditioned by reducing 3D models to 1D ones.

The present work aims at development and testing reduction methodology of 3D mathematical models to 1D models using as an example one of main ramjet component - combustor cooling system.

2 Processes Simulating in Cooling System

2.1 Cooling System Panel

The ramjet combustor is the most heat stressed element of high speed flying vehicle with atmospheric operation zone. The ramjet combustor walls are cooled by fuel further entering through pylons system combustor, where it mixes with atmospheric bleed air and burns creating engine thrust. The ramjet cooling system should provide the required temperature mode of combustor walls, i.e. to save it from destruction due to overheat.

The combustor cooling system appearance is singular defined by engine performance depending on flying vehicle designation. To avoid specific engine linkage works [3, 4] offer for cooling system a panel design way: combustor walls are cooled by special panels, which number depends on fuel consumption (i.e. on flying vehicle designation). Thus, the same panels can be used in various ramjet cooling systems.

The panel used is built on sectional principle (Fig. 2), which has indisputable advantages at combustor cooling system development. Strictly speaking, cooling system design depends on hypersonic flying vehicle purpose. Exactly, the sectional principle allows physicochemical processes in scramjet combustor cooling system to research without linkage to specific flying vehicle. In particular, a three-sectional panel scheme is shown in Fig. 3.

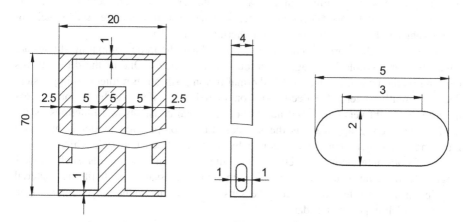

Fig. 2. Heat exchange panel section of cooling system

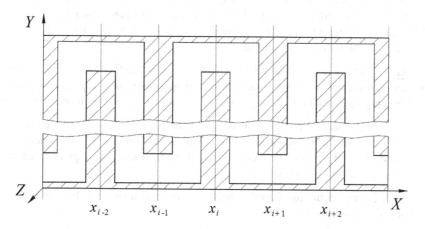

Fig. 3. Cooling system three-sectional panel

Hereinafter we use for computation 20-section panel. Hydrocarbon fuel enters cooling system duct, where it is heated and possibly subjected to thermal decomposition (Fig. 3). Duct geometry is such chosen to provide intensive heat exchange and sufficient heat-transfer agent dwell time under relatively small hydraulic resistance. The washed wall roughness of cooling system panel internal duct is accepted $4 \cdot 10^{-5}$ m. In more details the panel used and simulation results of heat exchange associated are described in [3, 4].

2.2 Simplified 3D Model of Associated Heat Exchange

Individual natural hydrocarbons such as ethane (C_2H_6), propane (C_3H_8), butane (C_4H_{10}), and pentane (C_5H_{12}) when heating decompose on more simple hydrocarbons, and moreover this decomposition process is endothermic, i.e. it runs with heat absorption. Under raised pressure these hydrocarbons easy condensate in liquids, so they might be kept as fluids in a sealed container at ambient temperature. The hydrocarbons, which absorb heat under thermal decomposition and are used as fuels for flying vehicles, have got the name "endothermic fuels" (EF).

At present in the world extensive experimental and theoretic-computing studies are conducted for possible EF usage in various flying vehicles. One of the difficulties is building mathematical models of EF thermal decomposition in ramjet cooling system ducts. Presently several EF decomposition models are offered differing on process description depth, versatility, and number of empirical constants and functions used.

For engineering applications the simplest EF decomposition model is based on replacing actual hydrocarbon compound by some fictitious material, which decomposes without intermediate reactions. The literature sometimes calls such models single-staged. Empirical constants and functions necessary for fictitious material description are so selected that in some sense get the coincidence of computed data with actual EF experimental data.

Works [3, 4] show that different mathematical models of physicochemical processes at EF decomposition in forced flow conditions of heated rough ducts can be reduced to single-stage models with a single additional convective-diffusive equation. The mathematical model presented of associated heat exchange under turbulent flows of EF decaying in heated rough curvilinear ducts of scramjet combustor cooling system is founded on the following positions:

Position 1. Initial EF is substituted by some fictitious media (FM), which thermal characteristics (density, viscosity, thermal conductivity, and heat capacity) depend on pressure, temperature, and function Ψ, hereinafter named local decomposition degree.

Position 2. Mathematical model of associated heat exchange under turbulent flow of decaying EF, apart from equations of continuity, motion, turbulence, energy (for flows), and thermal conductivity (for duct), should contain the equation for local decomposition degree computation.

Position 3. Energy equation should contain a source member providing endothermic effect at FM decomposition.

The equation for function Ψ calculation has the form

$$\frac{\partial(\rho\hat{\Psi})}{\partial t} + \nabla(\rho\vec{V}\hat{\Psi}) = \nabla(D_{\hat{\psi}}\nabla\hat{\Psi}) - \rho f(p,T,\hat{\Psi}),$$

where $D_{\hat{\psi}}$ is "diffusion" factor, and function $\hat{\Psi}$ is connected with decomposition degree by relation $\hat{\Psi} = ln(1 - \Psi)$.

The energy equation at presence of endothermic reactions has the form

$$\frac{\partial(\rho i)}{\partial t} + \nabla(\rho\vec{V}i) = \nabla(\lambda\nabla T) - S,$$

where i is EF enthalpy. Source member $S \geq 0$ in this case depends on fuel decomposition degree and temperature as follows

$$S = AB\rho exp\left(\hat{\Psi} - \frac{E}{RT}\right),$$

where B is empirical factor dependent on EF type.

Thereby, any spatial model of associated heat exchange under turbulent flow of decaying EF in heated curvilinear ducts can be reduced to a simplified model, which contains only three empirical parameters A, B and E [3, 4].

2.3   Reduction of 3D (Three-Dimensional) Model to 1D (One-Dimensional) Model

The present work plots 1D mathematical model of hydrodynamics and associated heat exchange in 20-section panel based on 3D simulation results without and providing hydrocarbon fuel thermal decomposition.

We use the following 3D model equations:

(a)   thermal conductivity equation (in panels metal)

$$0 = \frac{\partial}{\partial x}\left(\lambda\frac{\partial T}{\partial x}\right) + \frac{\partial}{\partial y}\left(\lambda\frac{\partial T}{\partial y}\right) + \frac{\partial}{\partial z}\left(\lambda\frac{\partial T}{\partial z}\right),$$

(b)   energy equation (in panel duct flowpath)

$$\frac{\partial(\rho uI)}{\partial x} + \frac{\partial(\rho vI)}{\partial y} + \frac{\partial(\rho\omega I)}{\partial z} = \frac{\partial}{\partial x}\left(\lambda\frac{\partial T}{\partial x}\right) + \frac{\partial}{\partial y}\left(\lambda\frac{\partial T}{\partial y}\right) + \frac{\partial}{\partial z}\left(\lambda\frac{\partial T}{\partial z}\right) + S,$$

The *assumptions* used at plotting 1D mathematical model of associated heat exchange in cooling system panel:

Assumption 1. Only one panel side is heated, the rest sides are heat isolated. Specific flow heat density q_w used for panel heating can change only along axis X (Fig. 4).

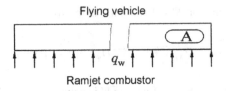

Fig. 4. Ramjet combustor wall cooling system by panel

Assumption 2. Along axis X convective heat transfer is dominant in contrast with thermal conductivity.

Assumption 3. Pressure drop between inlet and exit panel sections of cooling system is relatively small, so we expect that media thermal characteristics (density, viscosity, thermal conductivity, heat capacity) depend only on temperature. As a rule, media thermal characteristics are approximated by polynomials of the type

$$c_p(T) = \sum_{k=0}^{K} \theta_k T^k \tag{1}$$

where c_p is specific media heat capacity under constant pressure, T is absolute temperature, θ_k, $k = 0,1,K$, K are known polynomial factors.

Providing Assumption 2 the energy/thermal conductivity equations are

$$\frac{\partial(\rho u I)}{\partial x} + \frac{\partial(\rho v I)}{\partial y} + \frac{\partial(\rho \omega I)}{\partial z} = \frac{\partial}{\partial y}\left(\lambda \frac{\partial T}{\partial y}\right) + \frac{\partial}{\partial z}\left(\lambda \frac{\partial T}{\partial z}\right) + S \tag{2}$$

where u,v and ω are velocity vector components in directions x, y and z, correspondingly, I is enthalpy, λ is thermal conductivity factor, ρ is density, S is source member conditioned by possible flow physicochemical conversions. Since panel domain Ω consists of metallic case Ω_M and flow path part Ω_F i.e. $\Omega = \Omega_M \cup \Omega_F$, than functions (2) are:

$$u = \begin{cases} u, (x,y,z) \in \Omega_F \\ 0, (x,y,z) \notin \Omega_F \end{cases}, \quad \omega = \begin{cases} \omega, (x,y,z) \in \Omega_F \\ 0, (x,y,z) \notin \Omega_F \end{cases}$$

$$v = \begin{cases} v, (x,y,z) \in \Omega_F \\ 0, (x,y,z) \notin \Omega_F \end{cases}, \quad S = \begin{cases} S, (x,y,z) \in \Omega_F \\ 0, (x,y,z) \notin \Omega_F \end{cases}$$

$$\rho = \begin{cases} \rho_F, (x,y,z) \in \Omega_F \\ \rho_M, (x,y,z) \notin \Omega_M \end{cases}, \quad \lambda = \begin{cases} \lambda_F, (x,y,z) \in \Omega_F \\ \lambda_M, (x,y,z) \notin \Omega_M \end{cases}$$

Here lower indexes M and F attribute to panel metal and heat-transfer agent accordingly. Thereby, defining separate functions we can combine the energy equation for heat-transfer agent and thermal conductivity equation of panel material in a united energy/thermal conductivity equation.

We divide the panel in control volumes

$$\Omega_i = \{(x, y, z) : x \in [x_i, x_{i+1}], y \in [0, L_Y], z \in [0, L_Z]\}$$

where $L_Y = 0.07$ m and $L_Z = 0.04$ m are maximum panel dimensions in directions Y and Z accordingly, x_i is typical sections shown in Fig. 3, moreover $i = 1,2,K, N$, where $N = 41$ for 20-section panel. Integrating energy/thermal conductivity Equation (2) on volumes Ω_i we obtain the following form of energy/thermal conductivity equation

$$\int_0^{L_Y} \int_0^{L_Z} (\rho u I)|_{x_i}^{x_{i+1}} dz dy + \int_{x_i}^{x_{i+1}} \int_0^{L_Z} (\rho v I)|_0^{L_Y} dz dx$$

$$+ \int_{x_i}^{x_{i+1}} \int_0^{L_Y} (\rho \omega I)|_0^{L_z} dy dx$$

$$= \int_{x_i}^{x_{i+1}} \int_0^{L_Z} \left(\lambda \frac{\partial T}{\partial y}\right)\Big|_0^{L_Y} dz dx + \int_{x_i}^{x_{i+1}} \int_0^{L_Y} \left(\lambda \frac{\partial T}{\partial z}\right)\Big|_0^{L_Z} dy dx$$

$$+ \int_{x_i}^{x_{i+1}} \int_0^{L_Y} \int_0^{L_Z} S(x, y, z) dz dy dx$$

After transformations the energy equation takes form

$$\langle I \rangle_{i+1} - \langle I \rangle_i = \frac{L_Y(x_{i+1} - x_i)}{2G} (q_w(x_i) + q_w(x_{i+1})) + \frac{V}{G} S_*$$

where G is mass heat-transfer agent consumption, $\langle I \rangle_i$ is mean mass enthalpy in i-th section

$$\langle I \rangle_i = \frac{1}{G} \int_A (\rho u I)|_{x_i} da$$

The last member type follows from the mean value theorem

$$\int_{x_i}^{x_{i+1}} \int_0^{L_Y} \int_0^{L_Z} S(x, y, z) dz dy dx = V S_*$$

where V is duct flowpath volume between sections x_i and x_{i+1}, S_* is mean source member value in volume V.

First we consider the simplest case, when there is no heat release or absorption in heat-transfer agent flow: $S = 0$. Knowing heat load on cooling panel we easily define mean mass heat-transfer agent enthalpy. However for calculation of mean mass heat-transfer agent temperature it is necessary to use the relation known from thermodynamics for isobaric process

$$dI = c_p(T)dT \Rightarrow I(T) - I(T_0) = \int_{T_0}^{T} c_p(T)dT$$

where T_0 is heat transfer agent temperature at cooling panel duct inlet. Since enthalpy is accurately determined to additive constant, $I(T_0) = 0$ can be taken, i.e.

$$I(T) = \int_{T_0}^{T} c_p(T)dT = \sum_{k=0}^{K} \theta_k \int_{T_0}^{T} T^k dT = \sum_{k=0}^{K} \frac{\theta_k}{k+1}\left(T^{k+1} - T_0^{k+1}\right) \qquad (3)$$

Formally, knowing mean mass heat-transfer agent enthalpy I its mean mass temperature can be found as transcendental Equation (3) solution. Much handy, exact, and quicker to tabulate $I(T)$ function and approximate it by a polynomial of the type

$$T(I) = \sum_{m=0}^{M} \gamma_m I^m.$$

So hereinafter consider known the dependency

$$\langle T \rangle_i = \sum_{m=0}^{M} \gamma_m \langle I \rangle_i^m.$$

To find the panel heated wall temperature use the Newton law:

$$q_w(x_i) = \alpha(x_i)\left(T_w(x_i) - \langle T \rangle_i\right).$$

First believe that 3D calculation is executed and wall temperature $T(x_i)$ and mean mass heat-transfer agent temperature $\langle T \rangle_i$ known. Specific flow heat density $q_w(x_i)$ is also known (boundary conditions). It is easy to find on mean mass temperature values of dynamic viscosity $\mu(x_i)$, thermal conductivity $\lambda(x_i)$, and specific heat capacity under constant pressure $c_p(x_i)$ in section considered x_i (Fig. 3). Then for each section we can compute numbers of Nusselt (Nu), Reynolds (Re), and Prandtl (Pr):

$$Nu = \frac{\alpha d_e}{\lambda}, Re = \frac{4\,G}{P\,\mu}, Pr = \frac{\mu c_p}{\lambda} \qquad (4)$$

where d_e and P are equivalent hydraulic diameter and washed perimeter, accordingly. The results of 3D simulation approximate by the following functional dependency

$$Nu = \chi_1 Re^{\chi_2} Pr^{\chi_3} \qquad (5)$$

where χ_1, χ_2, и χ_3 are empirical constants depending on panel geometry, heat-transfer agent, and boundary conditions.

2.4   One-Dimensional Simulation of Associated Heat Exchange in Ramjet Cooling System Panel Without Providing Hydrocarbon Fuel Thermal Decomposition

Initial data for heat exchange computation in 20-section panel (Fig. 2) are:

- hydrocarbon fuel type: pentane C_2H_5;
- heat-transfer agent mass flow: G = 0.0067 kg/s;
- pentane temperature at cooling system panel duct inlet: T_0 = 300 K;
- pentane pressure at cooling system panel duct inlet: P_0 = 5 MPa;
- mean heat load: $q_w \leq 1$ MWt/m^2.

Three-dimensional computation series is firstly executed without providing pentane thermal decomposition ($S_* = 0$); on the grounds of calculation results the following values of criteria Eq. (5) empirical factors are obtained:

$$Nu = 0.00011 Re^{1.23} Pr^{1.29}.$$

The procedure at 1D simulation is (i = 2, 3, K, N):

1. Calculation of mean mass enthalpy$\langle I \rangle_{i+1}$

$$\langle I \rangle_{i+1} = \langle I \rangle_i + \frac{L_Y(x_{i+1} - x_i)}{2G} (q_w(x_i) + q_w(x_{i+1}));$$

2. Calculation of pentane mean mass temperature $\langle T \rangle_{i+1}$;
3. Calculation of dynamic viscosity μ, thermal conductivity λ, and specific heat capacity under constant pressure c_p on pentane mean mass temperature $\langle T \rangle_{i+1}$ known;
4. Calculation of numbers Reynolds (Re) and Prandtl (Pr) on (4);
5. Calculation of Nusselt number on (5), determination of heat exchange factor α

$$\alpha = Nu \frac{\lambda}{d_e};$$

6. Calculation of ramjet combustor wall temperature under the Newton law

$$T_w(x_i) = \langle T \rangle_i + \frac{q_w(x_i)}{\alpha(x_i)}.$$

Note that T_w is maximum wall temperature, i.e. the temperature of cooling system panel side closest to combustor. Exactly this temperature is a factor limiting ramjet combustor capacity. Temperature of the opposite side can be taken with sufficient accuracy equal to heat-transfer agent mean mass temperature.

Figure 5 presents pentane temperature and maximum temperature of cooling system panel wall (on combustor side) obtained by results of 1D and 3D modeling.

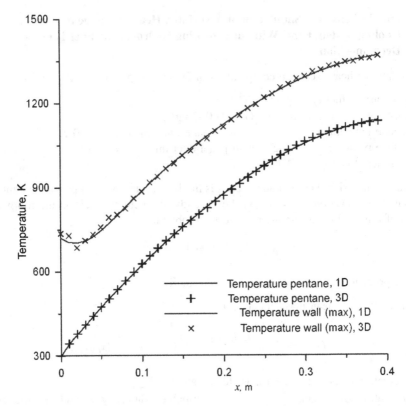

Fig. 5. Results comparison of 1D and 3D modeling

2.5 Computing Experiments

For 3D modeling a computing mesh is used consisting of 3837240 control volumes in flowpath and 12660951 control volumes in cooling panel build (Fig. 6). Computing time for one variant makes up from 10 to 14 days on computing system consisting of 24-48 cores with RAM memory up to 2 Gb/core. Many computing experiments are conducted on clusters of Institute of Problems of Chemical Physics of RAS and high capacity multicore graphic station (GPU acceleration is not applied in the experiments). More than 20 experiments of various computing complexity (defined by mesh nodes calculation accuracy) has been performed. Satisfactory convergence of 3D and 1D reduced computing models is shown.

Similar methodology can be applied for 3D modeling associated heat-exchange in different geometry panels and reducing the results obtained to 1D models. Obviously, a strong variation of heat-transfer agent thermal characteristics prevents usage of alike geometry panels. A typical inverse task is seeking for optimum geometry of panel inner ducts flowpath, which most of all are solved by reducing to direct tasks combination. Application for direct tasks solution of 1D models allows computing volume greatly reduce, and 3D models mainly use for test calculations.

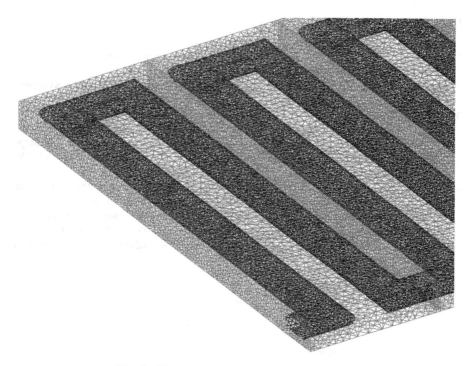

Fig. 6. Unstructured mesh in cooling panel build

3 Conclusion

Thus, the reduction of 3D models to 1D models allows parametric study of complex technical units to conduct at small computation scope. The key reduction moment to mathematical models is getting type (5) relations.

More difficult problem represents provision of heat-transfer agent decomposition since the duct geometry renders greater influence on chemical processes, than on heat exchange. Meanwhile, the mean mass decomposition degree in 1D approach can be computed as follows:

$$\left\langle \breve{\Psi} \right\rangle_{i+1} - \left\langle \breve{\Psi} \right\rangle_i = \frac{AF}{2G} \left[\rho_i \exp\left(-\frac{E}{R\langle T \rangle_i} \right) + \rho_{i+1} \exp\left(-\frac{E}{R\langle T \rangle_{i+1}} \right) \right] \cdot \Upsilon_i,$$

where empirical function Υ_i takes into account spatial effects at decomposition EF.

Mathematical models of complex technical systems such as scramjets can be built as 3D models of separate units with their further reduction to 1D models. Results of supercomputer modeling separate units can be brought to one-dimensional models and subsequent engine optimization under minimum computing efforts.

The activity is a part of the work "Supercomputer simulation of physical and chemical processes in the high-speed direct-flow propulsion jet engine of the hypersonic aircraft on solid fuels" supported by Russian Science Foundation (project no. 15-11-30012).

References

1. Samarskii, A.A.: Theory of Difference Schemes [Теория разностных схем]. Nauka. Fizmatlit, Moscow (1983). (in Russian)
2. Martynenko, S.I.: Multigrid Technology: Theory and Applications [Многосеточная технология: теория и приложения]. Fizmatlit, Moscow (2015). (in Russian)
3. Toktaliev, P.D., Martynenko, S.I.: Mathematical model of cooling system of the combustion chambers of aircraft ramjet engines on endothermic fuels [Математическая модель системы охлаждения камер сгорания авиационных прямоточных двигателей на эндотермических топливах]. Bulletin MSTU, ser. Natural Sciences [Вестник МГТУ. Сер. Естественные науки], vol. 1, pp. 83–97 (2015). (in Russian)
4. Toktaliev, P.D., Babkin, V.I., Martynenko, S.I.: Modeling of the interfaced heat exchange in elements of a design of the cooling system of aviation engines on endothermic fuels [Моделирование сопряжённого теплообмена в элементах конструкции системы охлаждения авиационных двигателей на эндотермических топливах]. Thermal processes in equipment [Тепловые процессы в технике]. 4, vol. 7, pp. 162–165 (2015). (in Russian)

The Future of Supercomputing:
New Technologies

Addition for Supercomputer Functionality

Gennady Stetsyura[✉]

Institute of Control Sciences of Russian Academy of Sciences, Moscow, Russia
gstetsura@mail.ru

Abstract. The addition of an optical wireless switching network with advanced functionalities to a supercomputer system is proposed. The structure of links of nodes (computer devices) a complete graph in which only links are realized is necessary. The switching units are located only in the sources and receivers. The structure of the network links can be changed quickly during execution of the single program instruction. The calculations may be executed for the data in the message without requiring additional time.

Keywords: Wireless optical network · Retroreflector · Dynamical reconfiguration · Distributed synchronization · Barrier synchronization · Distributed computing · Fault tolerance

1 Introduction

The features of supercomputers (SCs) that contain numerous interacting devices are largely determined by a switching system. These switching systems have a fixed topology of network connections, transport data using a switch of messages, eliminate conflicts, and perform intermediate storage of transmission data. The data are processed beyond the switching system. Thus, the functions of SC devices are clearly separated.

This paper, in order to facilitate and accelerate the interaction of SC components, proposes the incorporation of a wireless optical network in the SC switching system. This network integrates the SC facilities for switching and computation. Optoelectronics are used because it is not possible to achieve new networking capabilities using only electronic means.

1. The wireless optical network brings together many nodes (devices of the system) to form a fully connected structure (complete graph).
2. It implements only those connections that are required at the current time. Switching means are located directly at the data source and at the data receiver. Changing the structure of the connections is feasible in the nanosecond range. Thus, the structure of the links can vary not only from program to program but also for the execution of a single program command.
3. Direct interconnection of the nodes allows circuit switching to be used instead of message switching, thus securing the continuous connection of the nodes. Simplification of the process connection allows the exchange of short messages.
4. The opportunities described in points 1–3 make it possible to adapt the structure of the physical connections in the system to the structure of the connections in the

V. Voevodin and S. Sobolev (Eds.): RuSCDays 2016, CCIS 687, pp. 251–263, 2016.
DOI: 10.1007/978-3-319-55669-7_20

program, thereby eliminating the appearance of long chains of connections through the switching system.

5. Fast synchronization of the sources in the network allows the receiver to receive messages from sources at the same time or one after the other without pauses between messages.
6. Sending a message simultaneously to a group of receivers is slightly different in complexity and execution time from sending a message to one receiver.
7. Messages may clash only at the entrance to the receiver, but such conflicts are eliminated quickly.
8. The tools of the network can perform data processing in a message without additional time during this processing. Thus, in the SC that uses the proposed network, there will not be a complete separation of all systems into switching and computing means.
9. There are rapid means of simultaneous notification of all network nodes regarding the current state of the network.

The combination of these features not only enables the network to more flexibly and quickly carry out the exchange of messages but also affects the other SC functions. For example, network tools perform distributed computing, simplify the decentralization of control system performance, and increase the speed of the diagnostic status of the network and of a system.

Thus, the network structure adapts under the program requirements during the execution command of the program, supports fast mass interactions in the SC, implements distributed computing jointly with the transfer of messages, and extends the capabilities of designing algorithms and application software. Thus, this paper examines the network, which combines transportation and data processing (TDP network).

2 Optical Components of TDP Network and Network Connections

2.1 Network Nodes

In the network, there are three types of nodes: the object, the repeater, and the informer of the system [1]. Here and below, we denote the object i as O_i, the repeater j as R_j, and the systems informer as SI. If we do not need to distinguish between the R and SI, they will be denoted as the communication module MS.

The object performs domestic actions (computing, storage). It also performs the organization of interaction between network nodes. The object sends optical signals to the communication modules and receives signals from them. Signals may be of several types that differ qualitatively, for example, by frequency. The node does not distinguish signals of the same type that are received simultaneously.

The object sends a message a specially organized sequence of optical signals—to the nodes. There are two types of signals: pulse signals, the duration of which is known for all components of the network, and continuous signals, the duration of which is variable and is determined by the signal source.

The repeater receives the signals from the object and uses the retroreflector, which reflects without delay each incoming signal to its source. The repeater uses pulsed or continuous signals of one type to modulate the signals to another type. Thus, let a group of objects send continuous signals of f_1-type to a specific repeater and one of the objects send an additional message A of signal-type f_2. Let the repeater modulate signals f_1 by signals f_2, thereby copying the incoming message A. Then, this message will get all objects of the group. Therefore, the repeater does not create new signals for communication between objects. It uses only signals of the objects. The object uses the demultiplexer, which sends signals to the selected repeater or to the group of repeaters or to the system's informer. The object sends to module MS an optical signal $*f$, which prohibits the return to objects signal f_1. The repeater has a memory element. The object sends to the repeater the optical signals $*f_1$ and $*f_2$, which switch the memory element to an "on/off" state, respectively. In the "on" state, the repeater does not return signals f_1 to objects.

The SI is different from the repeater: while receiving the signals from objects, it creates a characteristic only for the non-directional signal f_{si} and sends it to all network objects.

2.2   Communication in the Network (Interaction of Nodes)

The organization of the connections between objects is specific. The object-source signal does not send signals directly to the receiver. Instead, the following procedure is performed. The object-receiver signal selects the repeater through which the receiver will receive signals intended for it, and the receiver sends a continuous signal f_1 into a repeater. The object-source sends a continuous signal f1 and the signal f_2 of a message into a repeater selected by the receiver. The repeater forwards the signals of the source to the receiver by modulating the signals f_1 of the receiver by means of the signal f_2 from the source. An object, as described above, sends signals to a specific repeater, simultaneously to a group of repeaters, or to all repeaters simultaneously. The receiver that transmits the message to the source can act like the source, but it can send a message only into its own MS, which sees the source. Figure 1 shows the types of communication network facilities. For simplicity, only repeaters of the receivers (black circles), which are placed between the source and the receiver, are shown.

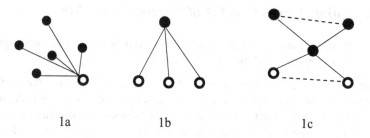

| 1a | 1b | 1c |

Fig. 1. Types of connections in a network

In Fig. 1a, the source (not painted over) sends a message to a randomly selected group of receivers. In Fig. 1b, an arbitrary group of sources sends messages to a single receiver. In Fig. 1c, a group of sources, using a repeater of an object-broker (or an independent repeater), sends a message to a group of receivers.

In Fig. 1a, a single source can simultaneously access a group of receivers. In Fig. 1b, a group of sources has access to a receiver; there is a way to resolve the conflict access of the sources to the receiver. In Fig. 1c, the access conflict of sources to the mediator is eliminated, and the messages are then synchronized. Receivers watch for the mediator and receive from it the messages of the sources.

The topology of connections is changed by sending the appropriate command for a particular object or simultaneously to group of objects or to all objects using the *SI*. Section 7 shows that the existing technical facilities allow you to change the topology quickly. Establishing a direct link between objects allows them to have a connection for a long time (channel switching).

3   Synchronization of the Sources

Synchronous signals from different sources must come simultaneously in an arbitrary *MS* in response to the synchronization signal coming from the *MS* (Sect. 4) [1, 2]. The travel time from object O_i to MS_j and back is denoted as T_{ij}. Methods for determining the time of delivery from the source to the receiver are known. We do not dwell on them and assume that sources know the time of delivery of the signal to each *MS*.

For synchronization, an arbitrary object O_i sends a signal to MS_j with the delay (1).

$$^*T_i = T_{\max} - T_{ij} \tag{1}$$

Here, $T_{max} \geq max\ T_{ij}$. Then, the signals of all identically acting objects arrive simultaneously to MS_j with the same delay T_{max}. As a result, the identical bits of messages from different objects are merged into a single message.

If messages must arrive to the *MS* one after another as a single message without time spaces, each object Oi has to transmit its message with the delay $T_{max} - T_{ij} + Q$, where Q is the total length of messages transmitted by the objects prior to the object O_i.

For slow networks, there is a possible delay of $T_{max} + Q$, but in fast networks, this leads to a substantial reduction in capacity.

4   Eliminating Access Conflict of Sources to the MS

If the sources are sending signals to the *MS* without synchronization, conflict appears at the entrance to a repeater and must be eliminated.

- **Method of Fixed Scales [1].** Sources use the moment of detection of the conflict as a synchronization command going from the *MS*, and then sources transmit the message to the *MS*-logical scale, which is a sequence of bit positions. The source that transmits a message to *MS* at the so-called temporary time scale, where one of the positions of the scale is assigned to each source, is entitled to send a message to the given *MS*. The conflicting source inserts one in its position. The logical scale arrives to the *MS* and is returned to all conflicting sources that determine the ordinal number of their transmission. The messages are transmitted sequentially as a single message, which eliminates the conflict.

In some cases, it is advisable to renounce the fixed number of scale positions. For example, the receiver may access an unknown number of sources. In these cases, apply the following uses of the scale with an element of randomness.

- Priority Method. A scale is generated. Different binary priority codes are assigned to the sources. The source randomly selects a scale position and places into it the value of the highest bit of its priority code.

The scale is sent into the *MS* and returned to the sources. Now, the well-known method of conflict elimination is used [2]. The sources may send in the bit position of the scale zero and one bits simultaneously. In this case, the sources that sent zero bits have finished struggling for the right to transmit a message. The rest of the sources operate with the next position of the priority code until exhausting all positions of the priority code.

As a result, the scale will have only conflict-free sources. Then, the positions where no struggle occurred are ignored. If the struggle is won by more sources than the receiver can service, part of the source is rejected.

- Random Method. Assume that only a minor part of objects needs to access the receiver simultaneously. The source selects randomly a scale position and writes into it a bit equal to one. The scale length is selected to be sufficiently large to minimize the probability of more than one source selecting the same position. The actions over this scale are the same as those over the deterministic one. Under this condition, the probability of conflict of the source messages is small.

5 Distributed Computing, Group Commands

5.1 Distributed Computing

This section describes network tools that perform two types of data processing for the messages sent over the network.

- Type 1. Operations: logical sum, logical product, the determination of maximum or minimum are realized in a network for two or a large group of operands at the same time. The result of operation is accessible to all participants of the operation simultaneously.
- Type 2. Calculations are done for the numbers included in the transmitted message, which passes through a chain of series-connected objects. Carrying out calculation requires no delay in the transmission of a message. All logical operations, finding of *max*, *min*, arithmetic addition, subtraction and multiplication are performed.

Operations are Type 1. A group source simultaneously sends messages to the *MS* module so that they overlap each other bit by bit, forming a common message. Each digit in the message is represented by two bits: 10 for unit and 01 for zero.

All objects watching this *MS* will receive a message from the *MS*. For the operation "logical addition," the imposition of a pair of bits 10 and 01 must be interpreted as a unit. For the operation "logical multiplication," the imposition of a pair of bits 10 and 10 must be interpreted as a unit.

When the maximum and minimum values of sources are determined, they send a message to the *MS* as in the previous case, but only one bit is allocated for the number position in the message. In such messages transmitted to the *MS* from sources, each such number position is allocated for a separate subgroup of the group of sources. The sources of all sub-groups will calculate a maximum (minimum) value in all subgroups simultaneously. These subgroup sources send the messages containing numbers in *MS* such that similar bits of the numbers coincide. Next, every source sends in *MS* the high-order digit of a number, which must be compared. The *MS* returns a signal to the source, and if this source has sent to the *MS* a digit zero and received from the *MS* a digit one, then this source is stopped. This operation continues for all other positions of the compared numbers. As a result, the sources of all subgroups will compute the maximum number from the numbers sending their sources. These actions do not differ from the actions with the priority process scale (Sect. 5). By inverting the message representation of bits one and zero, we may calculate minimum in a similar way.

This result created the *MS* without the participation of any logic devices and is very fast. For example, let us require comparing N numbers distributed between N objects. Usually, this requires many operations. In our case, a calculation requires simultaneous sending of N numbers to the *MS*, and the result is available simultaneously for all objects.

Performing Operations of Type 2. Let objects be connected in the chain: the first object sends a message to the R of the second object, the second object receives the message and sends it to the R of the third object, and so on. The numbers in the message are transmitted by signals of two types: f_a for a bit value 1 and f_b for a bit value 0. All steps of the operation are carried out without delay in the message received by the object, as shown in Fig. 2.

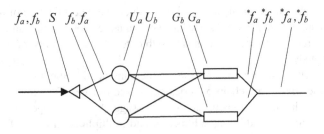

Fig. 2. Computation in the chain

The device in Fig. 2 receives signals f_a or f_b. Separator S directs the incoming signal, depending on its value, into one of the two paths. The object in the chain sends electrical control signals to the switches U_a and U_b. To perform the computation, the object controls the switches before the arrival of the next bit in the message. For this, the object uses only the value of the bit number stored in the object and a type of operation. After that device receives an input signal, it goes to a switch and then goes to one of two sources G_a or G_b, which generate signals $*f_a$ or $*f_b$, respectively. The signal from the switch turns it and the signal from (G_a, G_b) is directed to the next object in the chain.

Objects are realized via the following four steps, using U_a and U_b.

Action M1. The signal f_a (f_b) is transformed into output signal $* f_b$ $(*f_a)$. Action M2. The signals f_a and f_b are transformed into $*f_a$. Action M3. The signals f_a and f_b are transformed into $* f_b$. Action M4. The signals f_a and f_b are transformed into $*f_a$ and $*f_b$, respectively.

These actions do not analyze the value of the incoming signal; therefore, the output signal has no time delay.

Actions M1–M4 are sufficient for performing the referred above operations with the number in the message and with the number stored in the object [1, 2]. Examples of performing distributed computations without delay are given in [3].

The signal $*f_a$ formed instead of f_a, and signal $*f_b$ formed instead of signal f_b because the MS, receiving a signal 2 from the source, converted this signal to signal f_1. Therefore, in Fig. 2, the inlet signal and outlet signal should be different types.

We emphasize that the use of pairs of signals f_a and f_b for the computations in the chain requires using the corresponding pair of signals instead of the above signals f_1 and f_2.

Here are two simple examples that show the absence of delay in operations of type 2.

Assume that chain objects must perform the bitwise logical multiplication. Each object before the operation analyzes the value of the bit stored in the object and prepares the following step. If the value of the bit is 0, it is necessary to perform action M3 differently from M4. Then, a message is transmitted into the chain, and the time needed for it to move on the chain while performing logical multiplications is the same as the time of moving without multiplication.

Now, we add the current bit of the binary number stored in the object and the bit that comes from the chain. If the object must be added to a zero, then it selects action M4; otherwise, action M1 is selected. Because each object in the chain selects the previous action that comes from the bit from the chain, the switching time in the chain of objects is not cumulative.

Note for operation 2. To execute the cyclic operations, the last object in the chain must be connected to the first object. To perform the branching, which requires restructuring the network structure, the initiator of the branching must send a branch instruction. Some branching may be performed locally by any object in the chain, thus changing its actions based on previous results.

5.2   The Command for a Group of Objects

The command for a group of objects (GC) is a message moving along a chain of objects and carrying data and instructions to the objects to process the data indicated in GC, to change the instructions in the GC and perform local actions in the object [2]. Regarding changes in the GC content, the object analyzes part of the GC passing through it and replaces the remaining part with its information without delaying GC via this transformation. As a result, the GC moving along the chain of objects captures the information about the objects and varies itself along the way. Its effect on the object depends on the actions of the object's predecessors in the chain. A message can also be a group program consisting of a sequence of GC_s generated by several objects. In the application of computational operations of type 1, GC can come to many objects simultaneously.

6   Examples of Using the Network Capabilities

6.1   Barrier Synchronization

The barrier synchronization is a laborious problem for programming. Let a group of P sources transmit messages to the receivers waiting for these messages. Transmission is possible after preparation of messages by all sources, which requires different times. In a group of P sources, a representative of the entire group is assigned. Its repeater MS_p is known for all sources and receivers that watch MS_p.

After preparing a message, the sources transmit to MS_p a continuous signal $*f$ forbidding return of signals f_2. Having prepared a message, the source removes this signal. Upon readiness of all sources, MS_p begins to return the signal f_2. Having received f_2, the objects transmit the messages synchronously with delays $*T_i$ to the repeater MS_p, and all receivers get it as a united message.

6.2   Synchronization of Messages Sent from the Source to a Group of Receivers

The message source sends a request to receivers regarding the adoption of the message. In response to the request, each receiver sends a signal $*f_1$ to its module MS that forbid him to return signals f_1. When the receiver is ready to receive messages, the receiver sends the signal $*f_2$ to the MS and the ban will be lifted.

The source monitors the modules MS of all receivers, sending to all MS simultaneously signals f_1; the disappearance and then appearance of clock signal f_1 serves as a transmission opportunity.

Using signals $*f_1$, $*f_2$ and instead $*f$ (Sect. 6.1), the receiver eliminates the need for a continuous transmission of signals to the MS.

6.3   Networking in MPI

This important topic is touched upon briefly; examples of the impact of the network properties on the implementation of the MPI functions will be given, and the possibility of creating new functions will be shown.

1. The use of channel switching (Sect. 2) establishes a connection between the source and the receiver for the required time, which simplifies the implementation of a number of functions of MPI.
2. Suppose that a group of processes, each of which is located in a separate object of a network using the MPI_ALLTOALL function, must transport a message to all processes—receivers also located on the individual objects. Usually, this function requires long-term action of the program. With the use of barrier synchronization (Sect. 6.1), it is quickly carried out in the hardware of the network controller of the object.
3. The receivers are located as in paragraph 2. The source of command MPI_BCAST must send copies of a message to a group of receivers. The source previously must ascertain their willingness to receive the message. To check it, the source uses rapid synchronization from Sect. 6.2. Then, copies of the messages are sent simultaneously.
4. The function MPI_GRAPH_CREATE displays virtual topology relations in the program into the topology of the real system. The direct neighbors in a virtual topology in a real system can be connected through a long chain of links.

The proposed system has a real short connection between the devices for each long virtual link in the program. These links are created rapidly in real time. The established connection is maintained for an arbitrary length of time.

New functions MPI creation. The above examples relate only to the implementation of MPI functions, but the network also allows developing new MPI functions. For this, the network has new features: fast-change topology network, channel switching, synchronization, method of conflict resolution, and methods of distributed computations.

6.4   Evolutionary Computation

In many evolutionary algorithms, a group of objects, acting in parallel, repeatedly finds private solutions, which are also repeated, that are compared to find the maximum or minimum. The methods of calculating the *max* or *min* from Sect. 5.1 work very quickly. There is the next variant of such actions. Initially, each object finds the best local solution. This may require different times for them. To determine the point in time when these calculations will all be completed, is used barrier synchronization is used. After synchronization, the max must be determined via an operation of type 1 or 2.

6.5   Fighting with Failures in the Network and in the System

Consider two issues: the elimination of network damages which violate the integrity of the system and the search for the quantity and location of faulty system objects. The only type of network components whose damage affects the integrity of the system is the *MS*. If the object-receiver detects failure in its module *MS*, it uses a spare module. Possible access conflict must be resolved, as done in Sect. 4. If failures are greater than the available number of spare *MS* modules, the object will be connected to a module that is already occupied by other objects and will use the module in conjunction with them. Thus, if there is at least one serviceable module *MS*, the system remains operable. To determine the quantity and location of faults in the system, objects are connected together into a chain. *GC* is moved into the chain. To identify the faulty objects in a chain, each active object puts into the *GC* its name or coordinates in the system.

7 Network Amdahl's Law for the TDP Network

According to Amdahl's Law, if a program contains a portion that is run in parallel on n computers and a sequential part, then the time the program is run is reduced (2).

$$T = T_s + T_p/n \tag{2}$$

Here, T, T_s, T_p are the execution time of the program and of its serial and parallel portions, respectively. With increasing n, the contribution of the second term is reduced.

Taking into account transfer of the messages through the network, this law takes the form of a network Amdahl's law (3).

$$T = T_s + T_p/n + T_{nw} \tag{3}$$

Here, T_{nw} is time spent on transfer in the network. Typically, these expressions are converted, but this form will be sufficient.

For the TDP network, the runtime in network Amdahl's law is different (4).

$$^*T = k_1 T_s + k_2 T_p/n + T_{pnw} + k_3 T_{nw} \tag{4}$$

Here, k_1, k_2, k_3 are the numerical coefficients, and T_{pnw} is part of the computing facilities running on the network. Due to the emergence of T_{pnw}, the values of the constituent expressions can change, which is taken into account by the coefficients. Times T_s and T_p may be reduced due to the transmission of the data for processing to the network. The T_{nw} value may be reduced due to the emergence of T_{pnw}; thus, $k_3 T_{nw}$ becomes less than T_{nw}. Let us determine the possibility of $^*T < T$.

For the operations in the article, this ratio is realized, as demonstrated, for example, for the logical multiplication in Sect. 6.1.

This operation requires only one transfer of the number from objects to the *MS*, as the calculation time is not spent and $^*T = T_{max}$ (transfer requires 3 ns for a distance of 1 m). As a result, all the coefficients k equal zero. Thus, we excluded all successive operations; the result was obtained in the shortest possible time, without the involvement of computational tools. Objects use only their own network channels without affecting the other objects. Messages can have a small length. The complex exchange protocol is excluded. Time of the operation is irrespective of the quantity of operands.

These features are specific also for other operations in the paper.

Conclusions of Amdahl's law for the TDP network show that this network has new properties.

8 The Technical Implementation of Network Resources

To implement the proposals described in this paper, the components that perform functions of the demultiplexer, the repeater and the systems informer must be developed. Analysis of the literature shows the presence of devices that are close to those required. We refer, in particular, to the following results.

Demultiplexer - stored in every object and connects an object to any *MS* or, simultaneously, to any group of *MS*. The required organization of the demultiplexer is described in [4, 5], where for a laser lattice was used. Articles [6, 7] are examples of realizations of laser lattices.

Repeater. The main components of the repeater are the retroreflector, the light modulators, and the photo detectors [1, 8, 9]. As an example of a work where all of these components are used, we give present the results of paper [10] in a condensed form.

In Fig. 3, 1 = the light signal coming from a remote laser source; 2 = the lens; 3 = the plate, with many modulators/photo detectors; 4 = the mirror located in the lens focal surface. Items 2 and 4 are components of the retroreflector "cat's eye".

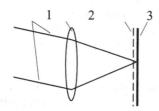

Fig. 3. Repeater with retroreflector

Signal 1 is focused on mirror 4 and returns to the source. Signals from other sources fall into other parts of the mirror and return to the sources. In the forward and reverse paths, each signal passes through one corresponding element 3.

Element 3 is used as a photo detector to receive a signal. It is used as a light modulator to return a signal to the source. The signal of the source carries the message received by photo detector 3. The source transmits a continuous signal after the transmission of the message, using 1. This signal, modulated by element 3-modulator, returns the message of the receiver to the source. Thus, each source has an independent operating channel. Currently, there are similar approaches of other authors.

Repeaters in the present paper require similar components with the following modifications.

– Simplification: modulator/photo detector (3) should be common to all objects.
– Increasing complexity: the modulator must be selective in frequency. However, it is possible to have only the means of paper [10]: the device, according to Fig. 3, allows you to split the stream after the lens into two streams and send them to two devices (3, 4).

Systems Informer. Recently, developments of devices transmitting signals with modulation speeds above 10 GHz have appeared [11].

There have been publications on the development of systems on a chip using optical wireless communication nodes [12]. Some solutions discussed in our paper are also applicable in such systems. As in [12], in our case, we can create wireless links in dust proof construction comprising only optoelectronic components of the repeaters and the communication modules.

9   Conclusion

The most distinguishing features of the considered network are as follows.

– The topology of the offered wireless optical network can be changed during the execution of a single program command.
– Messages sent by many sources in the network are delivered to the receiver (or group of receivers) as a united message without temporary pauses between the individual messages.
– In the content of transmitted messages, the network tools can perform calculations without spending extra time for calculation.
– The network tools significantly accelerate the implementation of a number of complex functions (for example, MPI) and enable new types of applied algorithms to be constructed.

One of the founders of the modern theory of complex networks, A.L. Barabási, emphasized [13] that networks have become the focal point of all areas of research in the XXI century.

This is true for SCs, in which a network connects individual devices to form a single complex system. The paper shows that the impact of networks in this area may be beyond their use as a means of transporting messages.

References

1. Stetsyura, G.: Fast execution of parallel algorithm on digital system with dynamically formed network structures. Large-Scale Syst. Control **57**, 53–75 (2015). Institute of Control Sciences of Russian Academy of Sciences, Moscow. (in Russian). http://ubs.mtas.ru/upload/library/UBS5703.pdf

2. Stetsyura, G.: Basic interaction mechanisms of active objects in digital systems and possible methods of their technical realization. Control Sci. (5), 39–53 (2013). (in Russian). http://pu.mtas.ru/archive/Stetsyura_13.pdf, doi:10.1134/S000511791504013X

3. Stetsyura, G.: Combining computation and data transmission in the systems with switches. Autom. Remote Control **69**(5), 891–899 (2008). http://www.mathnet.ru/php/archive.phtml?wshow=paper&jrnid=at&paperid=664&option_lang=rus

4. Stetsyura, G.: Extending functions of switched direct optical connections in digital systems. Large-Scale Syst. Control. **56**, 211–223 (2015). Institute of Control Sciences of Russian Academy of Sciences, Moscow. (in Russian). http://ubs.mtas.ru/upload/library/UBS5610.pdf

5. Patent RU 2580667 C1 10.01.2015 Stetsyura, G

6. Maleev, N.A., Kuzmenkov, A.G., Shulenkov, A.S., et al.: Matrix of 960 nm vertical-cavity surface-emitting lasers. Semiconductors **45**(6), 836–839 (2011). (in Russian). http://journals.ioffe.ru/ftp/2011/06/p836-839.pdf

7. Bardinal, V., Camps, T., Reig, B., et al.: Collective micro-optics technologies for VCSEL photonic integration. Adv. Opt. Technol. (2011). doi:10.1155/2011/609643

8. Stetsyura, G.: Organization of switched direct connections of active objects in complex digital systems. Autom. Remote Control **77**(3), 523–532 (2016). (in Russian). doi:10.1155/2011/609643, http://ubs.mtas.ru/upload/library/UBS4906.pdf

9. Patent RU 2538314 C1 10.04.2016 Stetsyura, G

10. Rabinovich, W.S., Goetz, P.G., Mahon, R., et al.: 45-Mbit/s cat's-eye modulating retroreflectors. Opt. Eng. **46**(10), 1–8 (2007)

11. Gomez, A., Shi, K., Quintana, C., Sato, M., et al.: Beyond 100-Gb/s indoor wide field-of-view optical wireless communications. Photonics Technol. Lett. **27**(4), 367–370 (2015). IEEE

12. Savidis, I., Ciftcioglu, B., Xu, J., et al.: Heterogeneous 3-D circuits: Integrating free-space optics with CMOS. Microelectron. J. **50**(4), 66–75 (2016)

13. Barabási, A.L.: The network takeover. Nat. Phys. **8**(1), 14–16 (2012)

Analysis of Processes Communication Structure for Better Mapping of Parallel Applications

Ksenia Bobrik[(✉)] and Nina Popova

Moscow State University, Moscow, Russia
bobrik.kp@gmail.com, popova@cs.msu.su

Abstract. We consider a new approach to the classical problem of allocating parallel application processes to nodes of a high-performance computing system. A new algorithm which analyzes the communication structure of processes is presented. The obtained communication structure can be used to recommend mapping for a high-performance computing system with a given topology. The input for the proposed algorithm is the data representing the total length of messages sent between every two processes. A set of processes is analyzed as a system of particles which evolve under the influence of attractive and repulsive forces. The identified configuration of the particles reflects the communication structure of the underlying parallel application and can be used for effective mapping heuristics.

Keywords: Mapping · Supercomputers · Parallel applications

1 Introduction

Current development of large-scale parallel architectures is closely associated with a rapid growth in number of computing nodes per system: top-10 systems from The Top500 list [1] comprise from hundreds of thousands to millions of cores. At the same time, a number of direct communication links between nodes in most cases remains comparatively constant at about several dozens due to physical and technical constrains. Therefore fully interconnected networks are not practically feasible as increase in number of nodes implies quadratic growth in number of links per system and linear growth in number of links per node [2].

Total execution time of a parallel application to a large extent depends on time consumed by exchange of data between parallel applications' processes (communication time). Increase in the number of nodes leads to a larger average distance between nodes, and, therefore, larger communication costs for processes allocated to these nodes, which may substantially inflate communication time required to run an application.

Communication patterns may vary significantly. There are two base types of communications, which can even use different hardware implementations within a single computing system: all-to-all communications, where a process sends a message to every other processes, and point-to-point communications, where a process sends a

© Springer International Publishing AG 2016
V. Voevodin and S. Sobolev (Eds.): RuSCDays 2016, CCIS 687, pp. 264–278, 2016.
DOI: 10.1007/978-3-319-55669-7_21

message to particular another one. All-to-all communications are essentially featured with low scalability, whereas communication time required for point-to-point communications can be reduced. This reduction among other means can be achieved by allocating the processes communicating heavily with each other to the closest possible nodes. Advantageously prevalent number of practically significant parallel applications have sparse communication patterns, where point-to-point communications prevail and, moreover, each process communicates with relatively small amount of other processes [3]. Beyond that several recent works offered a potential to replace all-to-all by point-to-point communications [4].

In this paper we focus on the mapping problem, which is to allocate the processes communicating heavily with each other to the closest possible nodes of the parallel system. The found placement is referred as the mapping. The mapping problem has been proven to be NP-complete [2].

Recent results in this field are represented primarily by heuristic algorithms to find a mapping which minimizes a chosen metric, e.g. hop-byte metric, congestion, average dilation, and, eventually, communication and execution time. For example, [5] provides a combination of greedy heuristics resulting in reduction of parallel application execution time by 25% compared to standard mapping. The referred result was achieved running the parallel application with a recursive doubling communication pattern on a 3D-torus system (Cray XT5 Kraken at NICS).

Most commonly used and investigated mapping algorithms are based on graph bisection algorithms (please refer for description to [5]), various greedy approaches (partly summarized in [5] as well), including greedy heuristics originally developed for mathematically equivalent problems like heuristics searching for local optimal solution for quadratic assignment problem [6], modifications of greedy heuristics for graph embedding [3]. Large number of authors are implementing the pairwise interchanges algorithm [7] for mapping optimization. In majority of works topology-aware algorithms are developed which receive system topology as an input and use this information on each step. Comprehensive survey of previously introduced methods and heuristics is provided in [10].

In this paper we introduce a new approach to solve mapping problem: we propose to generate mappings based on the preliminary analysis of parallel application communication patterns. We suggest performing this analysis using application's communication matrix. The result of such analysis may serve as an input for various newly created topology-aware mapping algorithms and improve the results of the existing ones. More precise explanation and detailed recommendations for practical application are provided in Sects. 3, 7 and 8 of this paper.

This paper is structured in the following way: we formally define the mapping problem (Sect. 2), we validate relevance and practical importance of communication structure analysis (Sect. 3), we introduce an algorithm to perform communication structure analysis (Sect. 4), we present an implementation of the algorithm (Sect. 5) and test results (Sect. 6), and then provide with practical applications of the proposed algorithm to the mapping problem (Sect. 7). Key results, conclusions and directions for future work are summarized in Sects. 8 and 9.

2 Mapping Problem Formalization

Let a parallel application and its processes be denoted as $S = \{s_1, s_2, \ldots, s_{N_s}\}$. We assume that N_q of identical nodes of a parallel system $Q = \{q_1, q_2, \ldots, q_{N_q}\}$ are available to run this parallel application and that only one process will be allocated to each node, i.e. target mapping is a permutation:

$$q_i = q(s_i). \tag{1}$$

During execution of a parallel application each process communicates with other processes by sending messages. We denote the total length of all messages sent from process s_i to process s_j plus total length of all messages sent from process s_j to process s_i as $f_{i,j} = f(s_i, s_j)$.

The communication matrix $F = \{f_{i,j}\}$ is symmetrical and is defined by the parallel application and its input data. The communication matrix does not depend on the mapping or the parallel system topology. Matrix F may correspond to the entire execution time or only part of it. The latter may be of interest if the goal is to analyze a particular phase of the parallel application.

Various system resources are utilized when data is sent from one process to another. Therefore, communication costs should be taken into account. We denote the communication costs of sending a message with a length of 1 from node q_i to node q_j as $E = \{e_{i,j}\}$. In this paper we assume $e_{i,j}$ to be constant and not dependent on processes allocated to nodes q_i and q_j. Thus, E is defined only by the parallel system topology. In the simplest case (a homogeneous system) communication costs may be approximated by a distance between nodes. The communication time required to send a message from process s_i to process s_j can be approximated as a product of the communication costs and length of the message. Therefore, the total communication time required for data exchange between process s_i and all other processes given mapping (1) is

$$E_i = \sum_{j, i \neq j} e_{q(s_i)q(s_j)} f_{i,j}. \tag{2}$$

The problem of finding a mapping minimizing

$$\mathrm{E} = \left(\sum_{i=1,\ldots,N_s} E_i \right) \tag{3}$$

is called a mapping problem with function E being the total communication cost function.

3 Generating a Mapping Based on the Communication Structure Analysis

Apparently, an obvious idea is that optimal mapping for point-to-point communications should be achieved by allocation processes which communicate with each other most to nodes which are closest to each other in the given parallel system architecture.

Various researches [5] have explored this idea and performed experiments proving that even simplest greedy algorithms employing this principle are effective. By effectiveness in this case we understand practically meaningful result which is an execution time reduction achieved for various parallel applications by running them with recommended mappings instead of standard ones. It is worth to mention though that previously proposed algorithms ignored substantial part of information provided as input (communication matrix or equivalent), which implicates that the results may be further improved. Additionally precise knowledge of target parallel system architecture is used on every iteration step in majority of studied approaches, which makes these results inapplicable to even slightly different topology without repetitive calculations.

For example, algorithms based on recursive bisection do not take into account communications between clusters obtained on previous iterations meaning very generally that half of information gets lost on each step. Researches have already pointed out the possibility of obtaining poor quality results using this approach [9]. Same construction is used in hierarchical mapping algorithms (e.g. [11]), which aim to reduce the size of a problem. Negative consequences caused by such construction are sometimes party mitigated by authors (for example, a fine-tuning step is introduced in [11] as a final stage of the algorithm) but cannot be completely eliminated. The greedy heuristics proposed in [3] and [5] (and similar algorithms are frequently proposed in many different works) are picking up next process to place from a set consisting of every process which communicates to already placed ones. Communication between processes in this case is considered only to determine an order of consideration. Such greedy approaches as well as algorithms based on recursive bisection are thus not considering general communication structure. If these algorithms are applied to a parallel application with a block diagonal communication matrix (and, to be precise, we assume that block size is not power of 2), this communication structure will not be recognized and will be only partially captured by an output mapping.

Researchers from Berkley have studied parallel applications and algorithmic methods used in different fields taking into account its short- and long-term practical significance [8]. They outlined thirteen practically important classes of algorithms which have similar computation and communication patterns ("Thirteen Dwarves"). In particular, Spectral Methods (e.g. FFT) Dwarf and Structured Grids Dwarf were outlined. Typical communication matrices for those Dwarves are block diagonal matrices. Moreover, a block structure is remained when application is scaled, i.e. when number of processes running for application increases.

Thus, it is necessary to develop a method to better analyze communication patterns. To highlight the broader meaning of the desired results we would refer to the output of such analysis as communication structure instead of communication patterns. Understanding communication structure of a parallel application is important to ensure more suitable (and, thus, effective) mapping strategies are proposed. In addition, such communication structure analysis will provide opportunity to generate scalable mapping algorithms for parallel applications with scalable communication matrices. This is especially relevant given already mentioned [3] deterioration in quality of greedy algorithms when applied to larger size problems.

Summarizing the above, we propose to decompose the mapping problem and to solve it in two steps:

- Step (problem) 1: an analysis of communication structure. Step 1 does not depend on a target parallel system topology.
- Step (problem) 2: a mapping generation. Step 2 is performed based on Step 1's output and target parallel system topology and does not require parallel application itself or any other information on parallel application.

It is important to keep in mind that Step 1 is to analyze overall communication structure and should not be limited to a cluster analysis of processes. As shown before the latter approach may miss critical information needed to recommend suitable mapping.

This approach also results in ability to generate mappings for a given parallel application and for various target topologies without repetitive calculations once Step 1 is performed.

In the following section we present an algorithm to analyze communication structure. The proposed algorithm is based on simulation of attractive and repulsive forces similar to intermolecular forces or gravity/repulsion between planets.

4   Attraction & Repulsion Algorithm for Communication Structure Analysis

4.1   Idea and Description

We represent each process as a particle on plane. Let's assume certain forces act on particles defining the motion laws. The idea is to define forces and, subsequently, motion laws so that particles corresponding to processes communicating with each other more extensively are moved closely to each other. If the mapping strategy allocates neighboring particles to nodes close to each other, then the proposed forces and motion laws should lead to minimization of the communication cost function (3).

Let's assume there are two forces acting on particles:

- Attractive Force. The attractive force arises between two particles independently of the distance between them. The greater the communication $f_{i,j}$ between two processes, the greater the attractive force acting between the corresponding particles is.
- Repulsive Force. The repulsive force between two particles increases when particles move closer to each other. While the distance between particles is at an acceptable level the repulsive force between them is insignificant or close/equal to 0.

We propose the following formalization of the idea.

If two processes i and j communicate with each other, i.e. if $f_{i,j} > 0$ then we assume that the attractive force acts on corresponding particles with its module linearly dependent on communication:

$$\left|\overrightarrow{F_{i,j}^{att}}\right| = c_{att} * f_{i,j} \tag{4}$$

The attractive force is non-zero and $c_{att} > 0$ is constant for all pairs of particles.

On the contrary, the repulsive force does not depend on communication and does not equal zero only if the distance between the particles is less than a certain threshold.

$$\left|\overrightarrow{F_{i,j}^p}\right| \xrightarrow{dist(i,j)\gg 1} 0 \tag{5}$$

At the same time the repulsive force rapidly increases to infinity when two particles move closer to each other.

$$\left|\overrightarrow{F_{i,j}^p}\right| \xrightarrow{dist(i,j)\to 0} \infty \tag{6}$$

Various functions may be chosen to define the repulsive force as long as criteria (5) and (6) are satisfied. As an example we propose using the inverse squared distance between particles.

We denote coordinates of particle i at time t as $\vec{r}_i(t)$. The motion law for particles can be written as:

$$\vec{r}_i(t+h) = \vec{r}_i(t) + h \sum_j \left(\overrightarrow{F_{i,j}^{att}} + \overrightarrow{F_{i,j}^p}\right) \tag{7}$$

Initial coordinates of particles can be set randomly or, if any prior knowledge is available, predefined.

Additional forces may be introduced to achieve better visualization of the simulation process. These forces may include an attraction to the center or repulsion from predefined boundaries. These additional forces are optional and should not affect the result to a significant extent.

4.2   Iterative Simulation

The proposed algorithm is an iterative simulation of particle movements driven by attractive and repulsive forces. It receives the communication matrix of a parallel application as the input and returns the final coordinates of the particles on the plane to use in mapping generation strategies.

1. Create particles s_k with random coordinates $\vec{r}_s$ corresponding to each process of parallel application S
2. Until the stopping criteria is satisfied, for each iteration i do:

   For each particle s_k of parallel application S:
   Calculate the sum of attractive and repulsive forces acting on this particle
   Update the coordinates of the particle as per the motion law:

$$\vec{r}_{s_k}(i) = \vec{r}_{s_k}(i-1) + \overrightarrow{F_{s_k}^{att}} + \overrightarrow{F_{s_k}^{p}}$$

3. End of iterations
4. Return particle coordinates $\vec{r}_s(i)$ ("configuration of particles")

5   Attraction & Repulsion Algorithm Implementation

We have prepared test implementations of the Attraction & Repulsion Algorithm in Visual Studio 2010 and Matlab using matrix operations. Experiments for the applications with up to 2048 processes were performed. The number of iterations was chosen either empirically or using the stopping criteria based on the average position for the last several iterations, i.e.:

- The average coordinates for the last n and m iterations ($p_n(\overrightarrow{r_{s_k}})$ and $p_m(\overrightarrow{r_{s_k}})$ respectively) are calculated for each particle s_k corresponding to the processes of the parallel application S;
- If $\sum_s \left| p_n(\overrightarrow{r_{s_k}}) - p_m(\overrightarrow{r_{s_k}}) \right| < \varepsilon$, then the stopping criteria is satisfied.

6   Attraction & Repulsion Algorithm Results

6.1   Test Example

Let's consider the following test problem. Let's assume a parallel application has 20 running processes. Processes 1–10 are communicating with each other and not communicating with processes 11–20. Similarly, processes 11–20 are communicating only between each other. The total size of the messages sent between each two processes (if not zero) is a positive constant.

It is obvious that communication-wise the processes are split into two equal groups. Let's refer to processes 1–10 as Group 1 and to processes 11–20 as Group 2. The particles of Group 1 are represented by lighter unfilled circles, the particles of Group 2 are represented by darker filled circles.

This obvious conclusion should be expressed via the communication structure analysis and reflected in the recommended mapping, and thus we should expect the Attraction & Repulsion Algorithm to reflect it in the final particle configuration.

6.2   Step-by-Step Execution on a Test Example

We set the initial coordinates so that particles are spread across a circle with the center at (0, 0) as shown in Fig. 1. Based on our experiments these initial coordinates represent the worst-case scenario for the proposed algorithm with the most number of iterations needed to achieve convergence. Lines originating from the center of each particle illustrate the direction in which this particle moves.

Figures 1, 2 and 3 show the first iterations of the algorithm. As elements of the communication matrix are non-negative numbers, particles are attracted to each other and thus move closer to the center until repulsive forces start to prevail. The velocities of particles decrease during this phase. In Fig. 3 we can see the particles moved as close as possible to each other.

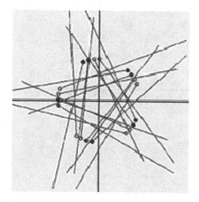
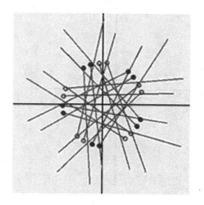

Fig. 1. Initial position

Fig. 2. Moving towards the center

Fig. 3. Equivalence of repulsive and attractive forces

Figures 4, 5 and 6 show reconfiguration of the particles after the whole system gets disturbed by an increased effect of the repulsive forces. The velocities of particles increase during this phase leading to rapid reconfiguration. In Fig. 5 the particle configuration begins to reflect the group structure of the communications: particles within Group 1 and Group 2 move closer to each other. Group 1 starts to appear in Fig. 6.

Figures 7, 8 and 9 show the last iterations of the algorithm. The formation of Group 1 is evident in Fig. 7, it generated a repulsive force towards the particles of Group 2. Driven by these forces, Group 1 moves away from the particles of Group 2 in Fig. 8. The final configuration is presented in Fig. 9. The velocities of particles at the end of the algorithm's execution are close to zero and the groups are clearly structured.

Fig. 4. Repulsive effect

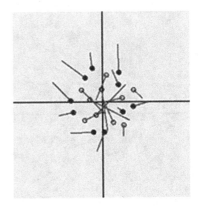

Fig. 5. Start of reconfiguration

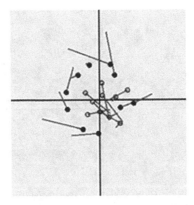

Fig. 6. Separation of the groups

6.3 Analysis of Communications Typical to Structured Grid Class of Algorithms

As pointed out in Sect. 3, one of the most practically important classes of algorithms is the Structured Grid class [8]. A typical communication matrix for this class is presented in [8]. We took this matrix as a base for our experiments and analyzed communication structure via Attraction & Repulsion Algorithm. In Fig. 10 we present the

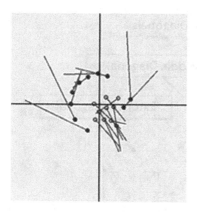

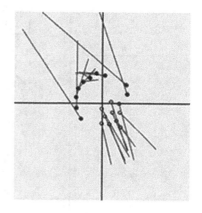

Fig. 7. Group 1 is formed   **Fig. 8.** Groups diverge further

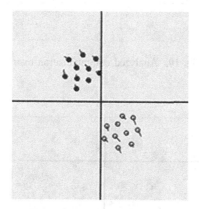

Fig. 9. Final configuration

communication matrix we analyzed, which was created based on the corresponding matrix in [8]. We define Internal, Middle and External Diagonals as shown in Fig. 10 and present results achieved by Attraction & Repulsion Algorithm in three cases:

(1) Communications on Internal Diagonals prevail: Fig. 11
(2) Communications on Middle Diagonal prevail: Fig. 12
(3) Communications on External Diagonal prevail: Fig. 13

Figures 11, 12 and 13 show final configurations of the particles in each case. A number of a corresponding process is written nearby the respective particle. These experiments were performed with an attraction to center force to achieve better visualization.

As shown in Fig. 11 if values on Internal Diagonal are larger than on Middle and External Diagonals, particles form groups of size 4. At the same time communications on Middle and External Diagonals are not neglected: in fact, we can easily see that groups of size 4 also form larger groups of size 32.

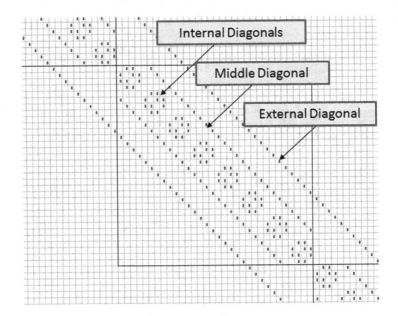

Fig. 10. Analyzed communication matrix

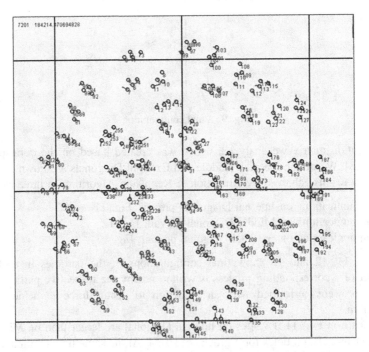

Fig. 11. Final configuration received by Attraction & Repulsion Algorithm for the analyzed communication matrix in case communications on the Internal Diagonals prevail

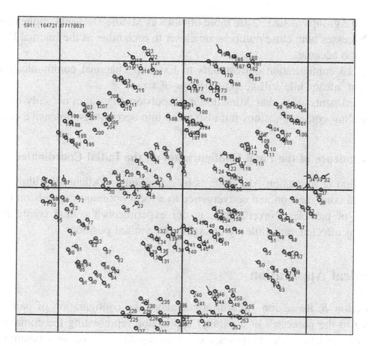

Fig. 12. Final configuration received by Attraction & Repulsion Algorithm for the analyzed communication matrix in case communications on the Middle Diagonal prevail

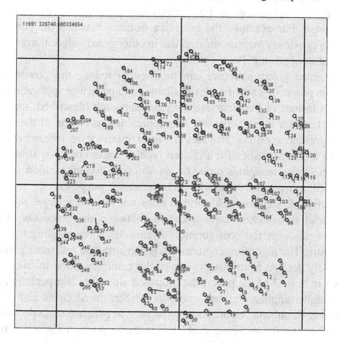

Fig. 13. Final configuration received by Attraction & Repulsion Algorithm for the analyzed communication matrix in case communications on the External Diagonal prevail

In Fig. 12 groups of size 32 are more obvious as Middle Diagonal implies blocks of size 32. Processes with close numbers are closer to each other as the Internal Diagonals are taken into account.

In Fig. 13 configuration corresponds to External Diagonal communications to a larger extent, meanwhile still keeping groups of size 4.

All experiments show that Attraction & Repulsion Algorithm not only catches the most prevailing communications but also takes into account less intensive ones.

6.4   Dependence of the Final Configuration on the Initial Coordinates

The Attraction & Repulsion Algorithm is by definition a gradient algorithm, meaning that the final configuration and convergence to a local optimum depends on the initial coordinates of particles. Nevertheless, in our experiments the final configuration of particles was affected very little by changing their initial positions.

7   Practical Application

The Attraction & Repulsion Algorithm calculates the configuration of particles corresponding to the processes of a parallel application representing the communication structure of the latter. The particles corresponding to the processes communicating intensely with each other are close to each other in the output configuration.

After the final configuration of particles is obtained, the determined patterns, groups and mutual spatial arrangement of the particles may be used to generate effective and suitable mappings. For example, the euclidian distance between the particles can be considered as a proximity measurement to use in other greedy algorithms (in particular, in algorithms described in [5]).

The determined group structure can be used to reduce the power of a set of processes to the power of a set of nodes available for a particular application on a given parallel system instead of the recursive bisection approach (described, for example, in [3, 7]) widely used for these purposes. A graphical interpretation of the Attraction & Repulsion Algorithm's results provides a visual interface to researchers giving an opportunity to create heuristics for a chosen type of communication structure or even manually define/chose mappings based on an observed communication structure. The proposed algorithm may also be used for clusterization purposes.

The Attraction & Repulsion Algorithm does not rely on a target parallel system topology. Instead it analyzes communications between the processes of a parallel application and produces the configuration of particles corresponding to the communication structure. The algorithm captures the group structure and other patterns which are not evidently derived from observation of the communication matrix.

As shown in Sect. 3 of the paper, the proposed algorithm is especially relevant for widespread parallel applications belonging to the Structured Grids and FFT classes. The Attraction & Repulsion Algorithm may be used to generate scalable mappings and be applied to solving the mapping problem via decomposition (Step 1 referred to in Sect. 3), thus saving resources utilized for repetitive calculations for various parallel system topologies.

8   Conclusions

The evolution of modern high-performance parallel computing systems has resulted in an increasing number of nodes, while the number of direct links between nodes is increasing far less rapidly. This leads to inflation of the average distance between nodes and thus potentially extends communication time.

Therefore, optimization of time spent on communication becomes critically important to ensure proper performance of parallel applications. The mapping problem, or the problem of allocating processes to nodes in a way that minimizes communication time, is one of the directions for improving the overall performance and execution time of a parallel application.

The mapping problem has been proved to be NP-complete. Current studies are searching for better heuristics for the mapping problem but in most cases ignore the overall communication structure of a parallel application.

We have demonstrated in Sect. 3 that proper communication structure analysis may open the gates to scalable mapping generation techniques. In the same section we proposed considering the mapping problem as a composition of two sub-problems: a communication structure analysis (does not depend on target topology) and mapping generation based on the communication structure for a given topology.

We have introduced a new algorithm for the analysis of the communication structure of a parallel application. The analysis is performed based on the communication matrix of a parallel application.

We envisage a wide range of practical applications of the proposed algorithm as its output reflects the communication structure in a most general way, creating an opportunity to:

- Determine proximity of the processes based on the final configuration received by the proposed algorithm: several algorithms for mapping rely on proximity for decision-making at every iteration, while simple metrics do not capture the overall communication structure which is identified by the Attraction & Repulsion Algorithm;
- Cluster a set of processes based on the final configuration: clusterization may be used instead of bisection to split a graph into a number of sub-graphs or to generate mapping;
- Create more suitable heuristics based on an understanding of the communication structure.

Additional advantages of the Attraction & Repulsion Algorithm include:

- Graphical visualization of the results: a researcher may choose heuristics from a given set or define their own based on the communication structure;
- Independence from topology: this algorithm can be used as an algorithm for Step 1 of the proposed mapping problem decomposition.

9 Future Work

We envisage the following as the most promising areas of further research:

- A study and analysis of the communication structures for different classes of parallel applications using the Attraction & Repulsion Algorithm;
- Generation of effective mapping strategies for various parallel system topologies using the particle configurations obtained by the Attraction & Repulsion Algorithm;
- An experimental study of performance of the existing mapping algorithms on various topologies if modified to account for the particle configurations of the Attraction & Repulsion Algorithm for a particular parallel application.

Acknowledgements. This work was supported by 14-07-00628 and 14-07-00654 RFBR grants.

References

1. Top-500: The List. http://www.top500.org/. Accessed 14 July 2016
2. Bokhari, S.H.: On the mapping problem. IEEE Trans. Comput. **30**(3), 207–214 (1981)
3. Hoefler, T., Snir, M.: Generic topology mapping strategies for large-scale parallel architectures. In: ICS 2011 (2011)
4. Pekurovsky, D.: P3DFFT – highly scalable parallel 3D fast fourier transforms library. Technical report (2010)
5. Sudheer, C.D., Srinivasan, A.: Optimization of the hop-byte metric for effective topology aware mapping. IEEE (2012)
6. Li, Y., Pardalos, P.N., Resende, M.G.C.: A greedy randomized adaptive search procedure for the quadratic assignment problem. DIMACS Series in Discrete Mathematics and Theoretical Computer Science (1991)
7. Sadayappan, P., Ercal, F.: Cluster-partitioning approaches to mapping parallel programs onto a hybercube. In: 1st International Conference on Supercomputing, Athens, Greece (1987)
8. Asanovic, K., Bodik, R., Demmel, J., Keaveny, T., Keutzer, K., Kubiatowicz, J., Morgan, N., Patterson, D., Sen, K., Wawrzynek, J., Wessel, D., Yelick, K.: A view of parallel computing landscape. Commun. ACM **52**(10), 56–67 (2009)
9. Simon, H.D., Teng, S.H.: How good is recursive bisection? SIAM J. Sci. Comput. **18**, 1436–1445 (1997)
10. Hoefler, T., Jeannot, E., Mercier, G.: An overview of process mapping techniques and algorithms in high-performance computing. In: Jeannot, E., Zilinskas, J. (eds.) High Performance Computing on Complex Environments, pp. 75–94. Wiley, Hoboken (2014)
11. Chung, I., Lee, C., Zhou, J., Chung, Y.: Hierarchical mapping for HPC applications. In: IEEE International Parallel and Distributed Processing Symposium (2011)

Experimental Comparison of Performance and Fault Tolerance of Software Packages Pyramid, X-COM, and BOINC

Anton Baranov[✉], Evgeny Kiselev, and Denis Chernyaev

Joint Supercomputer Center of the Russian Academy of Sciences - Branch of
Federal State Institution «Scientific Research Institute for System Analysis of the
Russian Academy of Science», Moscow, Russia
{abaranov,kiselev,dchernyaev}@jscc.ru

Abstract. The paper is devoted to the experimental comparison of performance and fault tolerance of software packages Pyramid, X-COM and BOINC. This paper contains the technique of carrying out the experiments and the results of these experiments. The performance comparison was carried out by assessing the overhead costs to arrange parallelization by data. In this case special tests simulating typical tasks of parallelization by data were designed by the authors. The comparison of fault tolerance was performed by simulating various emergency situations that arise during computations.

Keywords: Parallelization by data · BOINC · X-COM · Software package pyramid · Overhead costs to arrange parallelization · Fault tolerance of parallel computations

1 Introduction

Among various application tasks solved with the help of supercomputers an important group is formed by the tasks of parallelization by data, when one and the same computation sequence (application algorithm) is performed at all members of the set (pool) of input data. In this case, the computation algorithm may often be implemented in the form of a single sequential program (SSP), for which the size of the input data is defined by the value of one or several parameters. Such amounts of computing job form the pool of the input data that is eventually defined by the set of all values of SSP parameters and their all possible combinations.

Modern cluster computing systems consist of integrated nodes, where a separate control host can usually be distinguished. During parallelization by data one or several instances of SSP with different values of input data are performed on each node. To arrange the parallel computations means to launch SSP instances on the whole set of the available resources, in this case maximal performance and fault tolerance have to be provided.

During parallelization by data the computing rate is usually directly proportional to the capacity of the computing resources, CPU cores in particular. Modern cluster computing systems can have up to several thousands of nodes with the total number of

V. Voevodin and S. Sobolev (Eds.): RuSCDays 2016, CCIS 687, pp. 279–290, 2016.
DOI: 10.1007/978-3-319-55669-7_22

cores 10^5–10^6 or more. Longtime execution of the application tasks in such system is associated with high probability of fault on one or several nodes in the cluster during computation. Organization of fault tolerant parallel computations involves both recurrent saving of checkpoints and eliminating faulty nodes with automatic redistribution of the computing load to non-faulty nodes. In the case computations continue with some rate degradation.

To arrange parallel computations with parallelization by data, a number of technologies can nowadays be applied, including the one realized by the software package (SP) Pyramid [1]. The paper [2] contains experimental comparison of the Pyramid with the technologies MapReduce and MPI. The authors continued the experiments started in [2] on the comparison of the Pyramid with the alternatives, in this paper the software packages X-COM [3] and BOINC [4] are considered as such.

Thus far, experimental comparisons of the mentioned SPs for handling the tasks of parallelization by data have not been performed. This paper is meant to make up for this deficiency and to provide the users with information for further choice of the software package.

2 Software Packages for Arrangement of Parallel Computations with Parallelization by Data

The software package (SP) Pyramid [1] is intended for operation at a hierarchically structured computer system. The computer system includes the central server, the cluster management servers, and nodes as parts of the clusters. The Pyramid was initially designed for application programmers to get rid of parallel programming problems. The applied computing algorithm is realized in the form of a SSP, while launching a massive set of SSP instances on the cluster nodes and distributing the computing job is the care of the Pyramid software.

The Pyramid package provides computation reliability and fault tolerance. Failure of one or several nodes as well as one or several clusters does not lead to stopping the computations, it only slows them down. As soon as the disabled node is restarted (after recovery) the Pyramid automatically begins to distribute the computing job to it. Furthermore, the Pyramid saves checkpoints, which allows restoring the computations after the entire computer system failure.

Let us consider the alternatives of the Pyramid package: the SPs X-COM and BOINC.

The system X-COM is designed by the specialists of Research Computing Center of Moscow State University and is written in Perl programming language, which makes it one of the most lightweight means of parallelization by data. X-COM is based on client-server architecture. X-COM server is in charge of dividing the original task into blocks (jobs), distributing the jobs to the clients, coordinating all the clients, checking the result integrity and accumulating the results. Any computational unit (workstation computer, cluster node, virtual machine) able to perform an instance of an application program can act as the client. The clients are computing the blocks of the application task (jobs received from the server), requesting jobs from the server, and transferring results to the server.

BOINC implements one of the types of distributed computing – volunteer computing. Their specific feature is using the idle resources of PCs, workstations, computer clusters, when the resource is not used by its owner. At idle time, BOINC can use the resource to perform an application SSP developed as a part of a BOINC project. An example of BOINC project is SETI@HOME [5], a scientific experiment that uses Internet-connected computers in the Search for Extraterrestrial Intelligence (SETI). Project members participate by running a free program that downloads and analyzes radio telescope data.

Same as X-COM, BOINC consists of the server that distributes jobs, and a number of clients that carry out the distributed jobs. The BOINC server contains at least one web-server for client calls receiving and processing, database server that stores the assigned jobs status and the complying results, the scheduler that assigns the jobs to the clients. In the last few years, BOINC has gained widespread use; this software is constantly progressing, being supported by many millions of volunteer users and programmers.

We would like to highlight the following major differences of the SPs X-COM and BOINC from the Pyramid package:

1. X-COM and BOINC are designed for arranging parallel computations in distributed metacomputing system, though they can also be used in a single computing cluster. The Pyramid does not work in the distributed environment.
2. To solve an application task, X-COM and BOINC require developing the server and the client management programs, whereas for the Pyramid it is enough to prepare the job passport.
3. With X-COM and BOINC, saving checkpoints is the care of the management programs developer. The Pyramid saves checkpoints automatically.

The purpose of this research is to define the performance level of using the SPs X-COM and BOINC for arrangement of parallelization by data in the undistributed computing environment, as compared to the SP Pyramid.

3 Methods of Experimental Comparison of the Software Packages for Parallelization by Data

The authors carried out the comparison of the software packages Pyramid, X-COM and BOINC according to the factors of performance and fault tolerance, with a dedicated method applied to define the value of each factor experimentally. The study of the published papers on the research topic revealed that assessment of performance of software packages for parallelization by data is carried out by measuring the time of performing test programs simulating the real computing jobs. In the papers [6–9], the word count example is used to assess the performance of the software packages. In the paper [7] the authors use the test that counts all the occurring RGB colors in a given bitmap image and the test that computes a line that best fits a file containing coordinates. In [9] the test that counts user visits to Internet pages is used for performance assessment.

The fault tolerance of the software for arrangement of parallel computing is assessed by simulating a failure of one or several nodes at different stages of the computing process. In [8] the authors carry out a series of experiments simulating loss of connection with the host in the moment of the processed job uploading, in the middle of the computations, and during the results acquisition.

As the test rig for research the SPs Pyramid, X-COM and BOINC, the authors used the computing cluster consisting of 7 nodes, with each node containing two 4-core CPUs Intel Xeon L5408, thus making the total number of CPU cores – 56. The cluster is equipped with a virtualization platform, which allowed installing the researched SPs as a set of virtual machines under the control of the guest OS Linux Debian version 8.

3.1   Methods of Comparison by Performance

To compare the performance of the software packages the authors defined the overhead costs for parallelization as the time spent by each SP to arrange the computations. The overhead costs were defined as follows.

Let the single sequential program (SSP) process the whole input data on one CPU core for a certain time T. If parallelization by data is applied at p CPU cores, the time T of processing the same size of input data will ideally be reduced by factor of p. This will not happen in reality because of the overhead costs to arrange the parallelization: time costs for data transfer among the cluster nodes, for requesting the services (DBMS, web-server, scheduler, and etc.), delays between receiving the data and starting its processing, between the end of processing and the start of results transfer.

Let us introduce the definition of an elementary computing job, by which we will mean the processing of an indivisible (atomic, elementary) chunk of the input data. An elementary job can be the processing of a string generated from input arguments or read from a text file, or exhaustive search of values at a certain minimal range of the input data. It is important that an elementary job cannot be divided in smaller parts and, consequently, cannot be parallelized.

Let an elementary job be done for the time τ, and the entire size of the input data make N elementary chunks. Consequently, at one CPU core the entire size of the input data will be processed for the time $T_1 = N \cdot \tau$. In case of using p CPU cores the ideal time of the processing T_p will be

$$T_p = \frac{T_1}{p} = \frac{N \cdot \tau}{p} \tag{1}$$

Let the researched SP perform the processing of the input data at p CPU cores for the time $T_{exp}(p)$. Then the share of the overhead costs μ introduced by the SP will make

$$\mu = \frac{T_p}{T_{\exp}(p)} = 1 - \frac{N \cdot \tau}{p \cdot T_{\exp}(p)} \tag{2}$$

As we can see, the share of the overhead costs depends on the parameters N, τ and p. By varying the values of one of those parameters, with other two values stable, we can assess the dynamic pattern of the overhead costs to arrange the parallelization by data for each of the researched SP.

3.2 Test Cases

For experimental definition of the overhead costs the SSP test cases should be chosen. The cases should meet the following requirements:

1. The test SSP should be able to process from one to any number of elementary chunks of input data, i.e., SSP should be able to perform any number N of elementary jobs.
2. For the test SSP, the feature of setting the time τ for performing an elementary job should be considered.
3. If the test SSP performs N elementary jobs for the time τ for each, then the time of performing the SSP at one CPU core should equal $N \cdot \tau$.
4. The data chunk processed by the test SSP should completely be defined by the values of its parameters.
5. Cases when the data chunk processed by the SSP is defined by the value (value range) of single parameter are quite common. It is therefore reasonable to consider such cases separately using the suitable test case.
6. We should separately consider the case when the data chunk processed by the SSP is defined by the combination of values (value ranges) of several parameters, which drastically increases the complexity of organizing the parallel computations. Notice that this case in particular is typical with the SP Pyramid practical using because this software implements exhaustive search of all possible combinations for several SSP parameters.
7. We should separately consider the case when the data chunk is defined by a string (string range) of the input text files. This case is typical for X-COM and BOINC application.

To perform the experimental comparison the authors developed three test applications that meet the stated requirements.

The test case *Opp_one* simulates the execution of a SSP with single input parameter. The value of this parameter sets the range of the input data as the triple "a b s", where a is the beginning of the range, b is the end of the range and s is the step. For example, the triple "10 20 3" sets the sequence of the searched numbers 10, 13, 16, 19, each of the numbers defines the elementary chunk of data. The test *Opp_one* has the feature of setting the data chunk processing time τ. E.g., if set $\tau = 1$ s, the time of processing the data chunk set by the input parameter "10 20 3" will make 4 s.

The test case *Opp_three* simulates the execution of a SSP with three input parameters. The values of the first two parameters define the ranges of the input values as triples "a b s", where a is the beginning of the range, b is the end of the range and s is the step. The value of the third parameter defines the list of the searched strings. E.g., if the value of the first parameter is "10 15 3", the value of the second is "1 2 1" and the value of the third is "str1 str2", then the input data chunk is defined by 8 possible combinations of the three parameters values:

"10 1 str1" "10 1 str2" "10 2 str1" "10 2 str2" "13 1 str1" "13 1 str2" "13 2 str1" "13 2 str2"

Same as the test *Opp_one*, the test *Opp_three* provides the opportunity of setting the time τ for processing one combination of the input parameters values, i.e. the time of processing of one elementary data chunk. For our example with $\tau = 1$ s, the time of processing the eight combinations will make 8 s.

The test case *Opp_file* simulates the processing of the strings read from a text file sent to it as the parameter. The program receives two parameters at the input – the file name with the strings and the time τ of processing one string from the file. E.g., if the file contains 20 strings, and the time τ of processing one string is 2 s, the execution time of the test case *Opp_file* at one CPU core will be 40 s.

3.3   Methods of Comparison by Fault Tolerance

Considering the structure of the experimental rig the following method was used for testing the fault tolerance of the compared SPs. All potential failures can be divided into three groups: failures related to the cluster node rebooting, failures related to emergency power off of one or several nodes, and failures related to loss of connection in the interconnect. To simulate the failures of the first group, one of the cluster nodes was rebooted during the computations (by virtual machine restarting at the hypervisor control panel). The failures of the second group were simulated by powering off the node during the computations (by switching off the virtual machine at the hypervisor control panel). Besides the simulation of one node failure, the failure of all nodes of the cluster and the control host was emulated. After switching off the nodes were restarted, with further attempt to proceed with the interrupted computations. To simulate the failures of the third group, the virtual switch was switched off.

4   Results of the Performed Experiments

4.1   Results of the Experiments for Defining the Overhead Costs

During the experiments, for each of the compared SPs the elapsed run time of the three test cases – *Opp_one*, *Opp_file*, and *Opp_three* – was measured, then the overhead costs to arrange the parallelization by data were calculated by the formula (2). Time of carrying out each test case was measured in three stages:

- with variable number of CPU cores p and constant input data size in N elementary chunks and time τ of processing one elementary data chunk;
- with variable τ and constant p and N;
- with variable N and constant p and τ.

Figure 1 presents the dynamic pattern of the overhead costs to arrange the parallelization of the SPs X-COM, BOINC and Pyramid for test case *Opp_one* with variable number of CPU cores, 10^5 elementary chunks data size, and 1 s time of processing one elementary chunk. Notice the increase of the overhead costs of BOINC from 6% to 26%. X-COM and the Pyramid show a slow increase of the overhead costs from 2% to 6–7%.

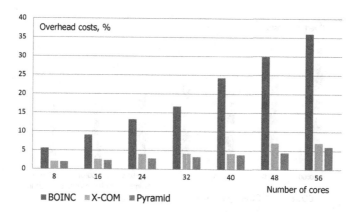

Fig. 1. Overhead costs for test case *Opp_one* with variable number of CPU cores, $N = 10^5$ elementary chunks, $\tau = 1$ s

Figure 2 presents the dynamic pattern of the overhead costs for test case *Opp_one* with variable value of one elementary chunk processing time, 10^5 elementary chunks data size, and 56 CPU cores. With the time increase of processing one elementary chunk, BOINC shows the decrease of the overhead costs from 51% to 20%, with their further stabilization at 18–20%. X-COM and the Pyramid show a slow decrease of the overhead costs share from 12% to 2%.

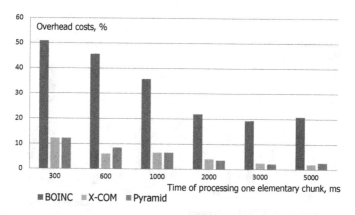

Fig. 2. Overhead costs for test case *Opp_one* with variable value of time of processing one elementary chunk, $N = 10^5$ elementary chunks, $p = 56$

Figure 3 shows the dynamic pattern of the overhead costs for the test case *Opp_one* with variable data size, 56 CPU cores, 1 s time of processing one elementary chunk. BOINC shows the decrease of the overhead costs from 34% to 20%, with their further stabilization at 20%. X-COM and Pyramid show a slow decrease of the overhead costs from 6% to 1%.

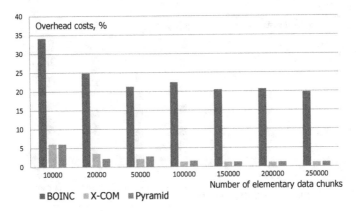

Fig. 3. Overhead costs for test case *Opp_one* with variable input data size, $p = 56$, $\tau = 1$ s

The results presented in Figs. 2 and 3 indicate the performance increase of the compared SPs in the case of coarsening grain of parallelism as in the case of increasing the input data size.

Figure 4 shows the dynamic pattern of the overhead costs for the test case Opp_file with variable number of cores. The size of the input text file was 105 strings, the time of processing of one elementary data chunk (one string from the file) was 1 s. All compared SPs show the increase of the overhead costs. X-COM shows the share of the overhead costs from 2% to 6%, the Pyramid – from 7% to 14%, BOINC – from 5% to 36%.

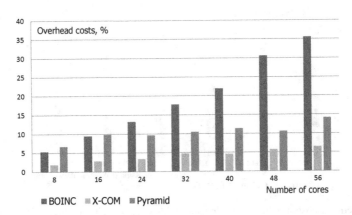

Fig. 4. Overhead costs for test case *Opp_file* with variable number of cores, $N = 10^5$ strings, $\tau = 1$ s

Figure 5 presents the dynamic pattern of the overhead costs for the test case *Opp_file* with variable number of strings in the input text file, 1 s time of processing one string, and 56 cores. X-COM shows the decrease of the overhead costs from 7% to 2%, with the file size of 50,000 strings and more the overhead costs are 2%.

The Pyramid shows the decrease of the overhead costs from 15% to 11%, and with the file size 20,000 strings and more the costs are within 10–11%. BOINC shows a relatively quick decrease of the overhead costs with the increase of the file size up to 100,000 strings, then presenting a slight change of the overhead costs within 20–22%.

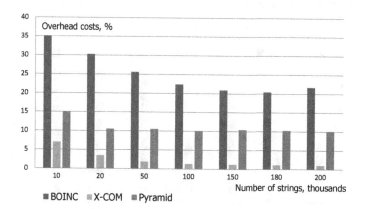

Fig. 5. Overhead costs for test case *Opp_file* with variable number of strings in the file, number of CPU cores $p = 56$, $\tau = 1$ s

Figure 6 presents the dynamic pattern of the overhead costs for test case *Opp_file* with variable time of processing one string, data size 10^5 strings in the file and 56 cores. X-COM shows a quick decrease of the overhead costs from 15% to 5%, with a further slow decrease down to 2%. The Pyramid shows a slow decrease of the overhead costs from 16% to 6%. BOINC shows a quick decrease of the overhead costs with the increase of the time for processing one string up to 2000 ms, then presenting a slight change of the overhead costs within 20–22%.

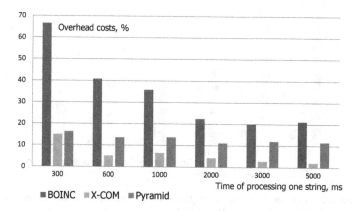

Fig. 6. Overhead costs for test case *Opp_file* with variable time of processing one string, $N = 10^5$ strings, $p = 56$

Figure 7 presents the dynamic pattern of the overhead costs for the test case *Opp_three* with variable number of CPU cores, data size of 10^5 elementary chunks, and 1 s time of processing one elementary chunk. X-COM shows the increase of the overhead costs up to 23%, and then a decrease down to 13%. The Pyramid shows a slow increase of the overhead costs from 2% to 5%. With BOINC, the overhead costs increase quickly from 11% to 33%.

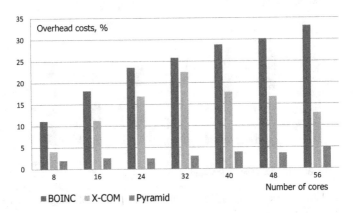

Fig. 7. Overhead costs for test case *Opp_three* with variable number of cores, $N = 10^5$ of elementary chunks, $\tau = 1$ s

Figure 8 presents the dynamic pattern of the overhead costs for the test case *Opp_three* with variable time of processing one elementary chunk, data size 10^5 elementary chunks, 56 cores. X-COM shows the decrease of the overhead costs share from 14% to 11%, the Pyramid – from 9% to 1%, BOINC – from 54% to 24%.

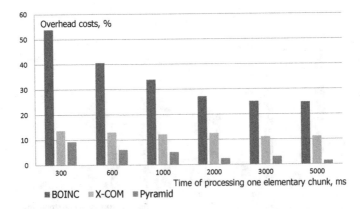

Fig. 8. Overhead costs for test case *Opp_three* with variable time of the processing one elementary data chunk, $p = 56$ cores, $N = 10^5$ elementary chunks

Figure 9 presents the dynamic pattern of the overhead costs for the test case *Opp_three* with variable data size, 56 CPU cores, 1 s time of processing one elementary chunk. With X-COM, the overhead costs share is within the range of 11–12%. The overhead costs of the Pyramid decrease from 4% to 2%, and BOINC shows the decrease from 34% to 20%.

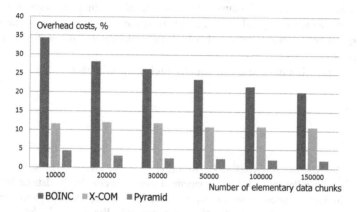

Fig. 9. Overhead costs for test case *Opp_three* with variable input data size, $p = 56$, $\tau = 1$ s

4.2   Results of the Experiments for Fault Tolerance Comparison

During the experimental testing for fault tolerance all compared SPs showed equal reaction to all the emergency situations simulated according to the method:

- after the rebooting of a cluster node, the computing job starts to be distributed to it automatically;
- if a node is off, the computing job is distributed among the living nodes;
- if the control host fails the computations stop;
- if the network interface of the control host is switched off, the control host endlessly expects connection with the hosts;
- if the network interface of a node is switched off, the computing job is redistributed among the other nodes;
- if all nodes are restarted, after the attempt to continue the interrupted job the computations continue from the moment of the last checkpoint saving.

5   Conclusions

1. BOINC has showed many-fold larger overhead costs compared to the Pyramid and X-COM. This is mostly due to the complex structure of the software that contains a number of components including DBMS. BOINC was designed for reliable operation in distributed environment, and using it within single cluster systems is impractical.

2. Overhead costs of all SPs in all test cases increase with the increase of the number of cores, and, as a rule, decrease with the increase of the input data size and the time of processing one elementary data chunk.
3. X-COM has showed many-fold better results with processing the strings read from a text file, with the best performance (overhead costs about 2%) on large files with a coarse-grained parallelism.
4. The Pyramid has showed many-fold better results with processing several SSP parameters, which is logical as this software is the only one of the researched SPs that implements exhaustive search of all possible combinations for several SSP parameters.
5. All researched SPs have showed equally high fault tolerance.

References

1. Baranov, A., Kiselev, A., Kiselev, E., Korneev, V., Semenov, D.: The software package «Pyramid» for parallel computations arrangement with parallelization by data (in Russian). In: Scientific Services & Internet: Supercomputing Centers and Applications: Proceedings of the International Supercomputing Conference, Novorossiysk, Russia, 20–25 September 2010, pp. 299–302. Publishing of Lomonosov Moscow State University, Moscow (2010)
2. Alekseev, A., Baranov, A., Kiselev, A., Kiselev, E.: The experimental comparison of the technologies of parallelization by data «Pyramid», MapReduce and MPI (in Russian). In: Supercomputing Technologies (SCT 2014): Proceedings of the 3-D All-Russian Scientific and Technical Conference, Divnomorskoe, Gelendzhik, Russia, 29 September–04 October 2014, , vol. 1, pp. 77–80. Publishing of the Southern Federal University, Postov-on-Don (2014)
3. Filamofitsky, M.: The system X-Com for metacomputing support: architecture and technology (in Russian). J. Numer. Methods Program. **5**, 123–137 (2004). Publishing of the Research Computing Center of Lomonosov Moscow State University, Moscow
4. BOINC. Open-source software for volunteer computing. http://boinc.berkeley.edu
5. SETI@HOME. http://setiathome.ssl.berkeley.edu
6. Costa, F., Silva, L., Dahlin, M.: Volunteer cloud computing: mapreduce over the internet. In: 2011 IEEE International Parallel and Distributed Processing Symposium, pp. 1850–1857. http://hpcolsi.dei.uc.pt/hpcolsi/publications/mapreduce-pcgrid2011.pdf
7. Naegele, T.: MapReduce Framework Performance Comparison. http://www.cs.ru.nl/bachelorscripties/2013/Thomas_Naegele___4031253___MapReduce_Framework_Performance_Comparison.pdf
8. Tang, B., Moca, M., Chevalier, S., He, H., Fedak, G.: Towards mapreduce for desktop grid computing. http://graal.ens-lyon.fr/∼gfedak/thesis/xtremmapreduce.3pgcic10.pdf
9. Pavlo, A., Paulson, E., Rasin, A., et al.: A comparison of approaches to large-scale data analysis. http://database.cs.brown.edu/sigmod09/benchmarks-sigmod09.pdf

Internet-Oriented Educational Course "Introduction to Parallel Computing": A Simple Way to Start

Victor Gergel and Valentina Kustikova[✉]

Lobachevsky State University of Nizhni Novgorod,
Nizhni Novgorod, Russian Federation
{victor.gergel,valentina.kustikova}@itmm.unn.ru

Abstract. Educational course "Introduction to Parallel Computing" is discussed. A modern method of presentation of the educational materials for simultaneous teaching a large number of attendees (Massive Open Online Course, MOOC) has been applied. The educational course is delivered in the simplest form with a wide use of the presentational materials. Lectures of the course are subdivided into relatively small topics, which do not require significant effort to learn. This provides a continuous success of learning and increases the motivation of the students. For evaluation of the progress in the understanding of the educational content being studied, the course contains the test questionnaires and the tasks for the development of the parallel programs by the students themselves. The automated validation and scalability program evaluation are provided. These features can attract a large number of attendees and pay the students' attention to the professional activity in the field of supercomputer technologies.

Keywords: Parallel computing · Massive Open Online Course · Shared-memory systems · OpenMP

1 Introduction

Parallel computing is a relatively new field of computer science. The problem of developing effective methods of teaching the parallel computing fundamentals is particularly acute when it comes to the need for a rapid transition from theory to practice, from the mathematical foundations of parallel methods to implementation of specific applications.

E-learning is a new stage of developing the educational technologies based on self-paced study of a discipline. The purpose of e-learning is to provide the ability of obtaining the necessary knowledge for maximally possible number of interested parties, regardless of their age, social status and location. At the same time, the basis of the learning process is an intensive self-study, largely due to the presence of a strong motivation.

© Springer International Publishing AG 2016
V. Voevodin and S. Sobolev (Eds.): RuSCDays 2016, CCIS 687, pp. 291–303, 2016.
DOI: 10.1007/978-3-319-55669-7_23

The paper describes the e-learning course "Introduction to Parallel Computing". This course is based on the e-learning system http://mc.e-learning.unn.ru of Nizhni Novgorod State University. The course is aimed at the students who have programming skills in the programming language C/C++ (implementation, compilation, debugging). Along with this, it is assumed that these students have basic mathematical knowledge corresponding to the second or third course of university. A distinctive features of the course are its availability for a large number of persons of different social groups and the possibility of automatic self-control. The course provides both automatic control of theoretical knowledge gained during the study of lecture materials and practical knowledge gained during the development of parallel programs.

The paper is structured as follows. In Sect. 2, we consider similar massive open online courses (MOOC) on the parallel computing. Sections 3 and 4 describes the lecture and practical parts of the course. Section 5 provides a method of studying the course. Section 6 describes the methodology of assessment of students' knowledge. Section 7 provides a brief background on the development tools of the course.

2   Related Works

The field of parallel computing is rapidly growing. With the emergence of new hardware architectures and parallel programming techniques learning technologies and related training courses are developed. The structure of such courses is standardized and described in a number of recommendations [1,2].

During the creation of the System of scientific and educational centers of supercomputer technologies, covering the whole territory of Russia, training curricula of entry-level professionals, retraining and advanced training of the teaching staff are developed as well as education programs of special groups, and distance learning programs based on Internet-University of Supercomputer Technologies [3]. Distinctive features of the Internet-University project are implementation of the classic forms of learning based on new technologies and accessibility of education through the use of the Internet [4]. This project can be considered one of the first steps towards solving the problem of the mass popularization of supercomputer technologies.

At present, more and more popularity gains new training techniques and presentation of educational materials for the training of a large number of students (Massive Open Online Course, MOOC) [5]. This procedure is not spared the field of parallel computing. Next, we consider MOOCs on parallel technologies because it is the subject of this paper.

Among the most relevant courses on the site MOOC [5] which are similar in the subject and the content to the developed course there are such courses as "Parallel Programming Concepts" [6], "Parallel Programming" [7] and "High Performance Scientific Computing" [8]. The first of these courses [6] considers parallel programming models (data parallelism, message passing, the model of functional programming), patterns and some of the best programming practices for shared-memory and distributed-memory systems. This course is a wide overview of parallelism concepts. The course [7] considers parallel programming fundamentals, and demonstrates, how well ideas of functional programming fall to the paradigm of data parallelism. In general, [7] aims to the specialists in the field of functional programming. The course [8] aims to the learning tools and techniques of parallel programming. This one contains technical description of the command line (Unix, Mac OS), revision control systems, base materials on OpenMP and MPI technologies for the programming language FORTRAN, the possibility of parallel programming in IPython, questions of testing software, its verification and validation, as well as other issues that inevitably arise in the process of developing computationally-intensive applications. Among the advanced level courses "Intro to Parallel Programming. Using CUDA to Harness the Power of GPUs" [9] and "Heterogeneous Parallel Programming" [10] can be identified. Both courses involve the study of parallel programming foundations for GPU on the example of NVIDIA CUDA technology.

The developed course is an attempt to explain the basics of parallel computing in the most simple and understandable way with the widespread use of video presentations and materials. This provides assignments for self-development of parallel programs on the example of one of the widely used technologies of parallel programming for shared memory systems OpenMP. In contrast to the courses discussed above, our course provides automatic knowledge control which reduces lecturer labor costs and accelerates obtaining feedback. The automatic control implies theoretical knowledge control through quizzes and practical knowledge control through the automatic validation and scalability evaluation of parallel programs developed for solving well-known mathematical problems. Moreover, the student receives a full test report about program execution.

The course materials are developed based on a textbook [11], as well as previously used in the skill enhancement programs in Nizhni Novgorod State University [12] and training programs of special groups in the field of high-performance computing.

3   Summary of the Lectures

The theoretical part of the course consists of 10 lectures, each lecture is separated into small topics. The lecture is supported by its description, the topic – by a presentation and a short video that is not longer than 15 min in order to improve the quality of material perception and assimilation. Moreover, there is a video containing a discussion of key issues considered in the lecture at the end of the theoretical part of the lecture.

The importance of parallel computing is discussed in the first lecture *Introduction to parallel programming*. The general course structure is stated. The hardware and software requirements to solve practical problems are described.

Lecture 1. Introduction to parallel programming.
Topic 1. Importance of parallel computing.
Topic 2. General characteristics of the course.

The notion of parallel computing is introduced in the second lecture *Basic concepts of parallel programming*. General performance measurements (speedup and effeciency) are considered. The problem of computing the sum of a sequence of numbers is discussed to demonstrate applicability of the stated metrics of a parallel algorithm. By this example, the difficulties in parallelization of a sequential algorithm that is not initially focused on a possibility to organize parallel computations are noted. In order to uncover a "hidden" parallelism, a possibility to convert the origin sequential computational scheme is demonstrated, the cascade scheme obtained as a result of this conversion is described. By the same example, it is shown that a greater parallelism of computations being performed may be achieved if introducing redundant computations.

Lecture 2. Basic concepts of parallel programming.
Topic 1. The notion of parallel computing.
Topic 2. Performance metrics.
Topic 3. Supercomputers today.

The third lecture *Parallel programming with OpenMP technology* provides the overview of the OpenMP technology to form a base for a quick start in the development of parallel applications for shared-memory systems. A number of concepts and definitions that are fundamental for the OpenMP standard are given. The OpenMP directives and their basic clauses are discussed in this lecture. The **parallel** directive is described; an example of the first parallel program using OpenMP is demonstrated. The concepts of *construct, region* and *section* of a parallel program are discussed. The questions of computing load balancing among threads are considered by the example of the data parallelism for loops. The **for** directive is described, the methods of controlling the distribution of loop iterations among threads are stated. The issues of controlling the data environment for threads running are considered. The notions of shared and local variables for threads are introduced. The description of the important *reduction operation* often used when processing shared data is given. The problem of computing load balancing among threads based on the task parallelism is considered and the **sections** directive is described. Some additional information about the OpenMP technology is given in the lecture description.

Lecture 3. Parallel programming with OpenMP technology.
Topic 1. Approach fundamentals.
Topic 2. OpenMP directives. Parallel region definition.
Topic 3. OpenMP directives. Management of data scope.
Topic 4. OpenMP directives. Computation distribution between threads.
Topic 5. OpenMP directives. Reduction.

The purpose of the fourth lecture *Principles of parallel algorithm design* is learning the methodology of parallel algorithm development. This methodology includes the stages of decomposing computations into independent subtasks, analysis of information dependencies, scaling and disributing the set of subtasks among computing elements of certain available computer system. The discussed method of parallel algorithm design is demonstrated by the example of solving the gravitational N-body problem.

Lecture 4. Principles of parallel algorithm design.

Topic 1. Scheme of developing parallel program.

Topic 2. Stages of parallel algorithm design. Decomposing computations into independent subtasks.

Topic 3. Stages of parallel algorithm design. Analysis of information dependencies.

Topic 4. Stages of parallel algorithm design. Scaling the set of subtasks.

Topic 5. Stages of parallel algorithm design. Subtasks distribution among computational elements.

The fifth and sixth lectures provide classic examples of parallelism for the problems of matrix-vector and matrix-matrix multiplication. This lectures cover the possible schemes of matrix distribution among threads of a parallel program oriented on a multiprocessor shared-memory computing system and/or on a computing system with multicore processors. The data partitioning schemes are the general schemes that can be used to organize parallel computations for any matrix operations. Among the discussed schemes there are the methods to partition matrices into *stripes* (vertically or horizontally) or into rectangular sets of elements (*blocks*). Each algorithm is introduced according to the general scheme of parallel method design – first the basic subtasks are selected, then the information dependencies of the subtasks are determined, after that the subtask scaling and distributing among computing elements are discussed. In conclusion the efficiency analysis of the parallel computations for each algorithm is carried out, the results of computational experiments are demonstrated. The possible variants of implementation are given.

Lecture 5. Parallel methods of matrix-vector multiplication.

Topic 1. Sequential algorithm.

Topic 2. Execution time evaluation of the sequential algorithm.

Topic 3. Parallelization principles of matrix computations.

Topic 4. Matrix-vector multiplication in the case of rowwise data partitioning.

Topic 5. Efficiency analysis of rowwise data partitioning.

Topic 6. Matrix-vector multiplication in the case of columnwise data partitioning (self-study).

Lecture 6. Parallel methods of matrix-matrix multiplication.

Topic 1. Sequential algorithm.

Topic 2. Basic parallel matrix multiplication algorithm.

Topic 3. Matrix multiplication algorithm based on block-striped data partitioning.

Topic 4. Block algorithm with efficient cache memory usage.

Topic 5. Matrix multiplication algorithm based on the band data partitioning (self-study).

The seventh lecture *Parallel programming with OpenMP technology (advanced)* is a logic continuation of the third lecture that covers the basic directives of the OpenMP technology. The objective of this lecture is to discuss basic methods to organize thread interactions in a parallel program developed using OpenMP, and also to study OpenMP functions and environment variables, which allow us to control the execution environment for the OpenMP programs. The description of some more directives (**master, single, barrier, flush, threadprivate, copyin**) is also given in this lecture.

Lecture 7. Parallel programming with OpenMP technology (advanced).
Topic 1. OpenMP directives. Synchronization.
Topic 2. OpenMP functions.

The next lecture *Parallel methods for solving partial differential equations* covers the problems of numerical solution of partial differential equations. The Dirichlet problem for the Poisson equation is considered as a training example. The most widely spread approach to numerical solution of differential equations is the finite difference method. The required amount of computations for this method is huge. This lecture discusses the possible methods to organize parallel computations for the grid methods on the shared-memory and distributed-memory multiprocessor computing systems. When discussing the organization of parallel computations in the shared-memory systems, the attention is paid to the OpenMP technology. The problems that arise when applying this technology and the solution of these problems are also considered. The deadlock problems are solved with the help of semaphores; the unambiguous calculations are achieved applying the red/black row alternation scheme.

Lecture 8. Parallel methods for solving partial differential equations.
Topic 1. Problem statement.
Topic 2. Deadlock problem.
Topic 3. Possible ambiguous computations in parallel programs.
Topic 4. Elimination of ambiguous computations.
Topic 5. Wavefront schemes of parallel computations.
Topic 6. Block-wise data representation.

A classification of the computing systems is introduced, the parallelism concepts based on multithreading and multiprocessing are considered in the ninth lecture *Parallel computing organization for the shared-memory systems*. The lecture gives a brief characteristic of multiprocessor computing systems of the MIMD type according to the Flynn's classification. Taking into account the way RAM is used in the systems of this type they are subdivided into two important groups of shared-memory and distributed-memory systems – *multiprocessors* and *multicomputers*. Then, it is noted that the most perspective way to achieve high-performance computing nowadays is to organize multiprocessor computing devices. To ground this statement the main methods to organize multiprocessing are considered in details – *Symmetric Multiprocessor (SMP)*, *Simultaneous Multithreading (SMT)* and *multicore systems*. As a continuation of the theme of multicore processors, analysis of a number of particular multicore processors widely used nowadays and developed by the major software companies is represented in the lecture description. Moreover, to give a complete picture the lecture description covers a number of hardware devices (video cards and computing co-processors) that can be used to achieve a considerable calculation speedup.

Lecture 9. Parallel computing organization for the shared-memory systems.
Topic 1. Parallelism as the basis for high-performance computing.
Topic 2. Classification of computational systems.
Topic 3. Symmetric multiprocessor systems.
Topic 4. Simultaneous multithreading.
Topic 5. Multicore systems.

The tenth lecture *Modeling and analysis of parallel computations* explains the theoretical foundations of modeling and analysis of parallel computing. The model of graph "operation-operands" is introduced, this model can be used for description of the information dependencies existing in the algorithms selected for solving a problem. Representation of computations using the similar models allows to obtain analytically a number of characteristics of parallel algorithms being developed. The execution time, scheme of optimal scheduling, estimates of maximum possible performance of methods for solving stated problems are among such characteristics. To construct the theoretical estimates more quickly the lecture also considers the concept of paracomputer as a parallel system with an unlimited number of processors.

Lecture 10. Modeling and analysis of parallel computations.
Topic 1. Modeling parallel computations.
Topic 2. Model analysis.
Topic 3. Efficiency characteristics of a parallel algorithm.
Topic 4. Examples of efficiency analysis.

4 Summary of the Course Assignments

The practical part of the course consists of the following components:

1. **Quizzes.** It supposes testing the knowledge gained during the lecture study. Each quiz contains several questions with a set of possible answers.
2. **Assignments for self-study.** This is assignments related to the estimating performance measurements or developing parallel programs using OpenMP technology. For example, the course contains the following assignments: develop a model and estimate the speedup and efficiency measurements for the problem of dot product of two vectors; develop a program to search the minimum (maximum) value among the vector elements and etc. Student performs verification and validation of the prepared solution by himself. The material required for solving these problems is represented in the lectures.

3. **Labs.** The course contains four labs, that involve the development of parallel programs for shared-memory systems using OpenMP technology. A system of automatic control conjugated with the e-learning system performs validation and scalability evaluation of the program. This system provides web-interface, where students are able to upload their programs and to track the stage of automatic control. There is a possibility to view compilation and test reports to estimate faults of the developed program.

At the same time, to understand the importance and significance of parallel computing the course contains the demonstration of real projects carried out in Nizhni Novgorod State University, these projects use parallel computing for solving complex practical problems. Essentially these is interviews with project leaders who talk about the importance of the use of parallel computing for modeling the dynamics of the heart, digital medicine and many other fields. The study of this material does not affect the assessment of the student's knowledge, however, it is recommended by the lecturers in order to expand listener's horizons.

5   Learning Methods

The course is available for participating from September to December. Supposed that students study the lecture materials by them-self in accordance with the proposed curriculum. Methods of self-study of the course consists of the following actions (Fig. 1):

1. Study materials of each topic in a lecture (presentation, video). For more information, the student can apply to the lecture description, which is a comprehensive document with dedicated sections for each topic of the lecture.
2. Watching a video with a discussion of key lecture issues.
3. Solving assignments for self-study (if any).
4. Implementing lab (if any).
5. Performing the appropriate quiz to the lecture.
6. Watching a video with an example of a project carried out in Nizhni Novgorod State University, which uses parallel computing to solve complex practical problems (if any).

To access to the materials of the next lecture, you have to complete the lab (if any) and the lecture quiz successfully. The student chooses by him-self the time of learning the lecture materials and performing the practical assignments in the period of availability of the course for registration. Questions arising during the course study, student is able to ask lecturers or other students in a course forum. Based on points obtained for labs and quizzes the student receives a final record of points that represents the level of material digestibility. Points are computed automatically to ensure the objectivity of knowledge assessment.

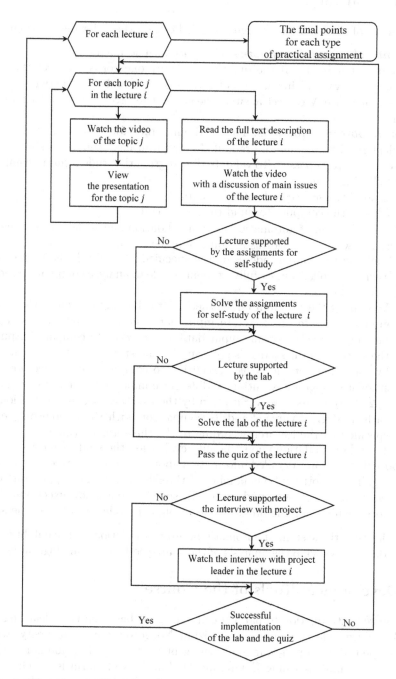

Fig. 1. The scheme of learning the course "Introduction to Parallel Computing"

6 Assessment

Assessment of students' knowledge is carried out on the following information:

1. *Points obtained for quizzes of lectures.* The maximum grade for each quiz question is "1". If the question has only one correct answer then the student receives "1" in the case of correct answer and "0" otherwise. In the case of presence N correct answers the student receives "$1/N$" point for each correct answer.
2. *Points obtained for labs.* The maximum grade for each lab is "100". The student receives maximum grade if the program has been successfully compiled and all tests passed. To validate the program the student has to complete the following steps:
   - Upload the source code of the program.
   - Choose the compiler to build the source code.

   When the system of automatic program validation and scalability evaluation receives new job, it performs a sequence of actions described below.
   - Building the source code using the specified compiler. If the source code wasn't compiled successfully, student is able to analyze compilation report to reveal errors.
   - Validating the program on synthetic data. Lecturer prepares three programs: *generator* to generate input data, *solver* to solve the assigned problem correctly for any input data and *checker* to compare output of the *solver* and the student's program. Each test involves data preparation by the *generator*, the *solver* execution to compute correct decision, the student's program execution on different number of threads (for example, 1, 2, 4, 8), correctness verification by the *checker*, speedup evaluation for each number of threads and its comparison with the reference speedup specified by the lecturer according to the theoretical estimates.

   At the end of validation and scalability evaluation the student receives a test report that contains detailed information about his program execution and its testing. The report contains number of threads created during program execution, verdict (test accepted or failed), speedup, efficiency, execution time, memory required for the program execution, exit code and *checker* message.

Notice that the system of automatic program validation and scalability evaluation does not provide the analysis of the source code for technological defects.

7 Development Tools of the Course

The course "Introduction to parallel computing" is based on the e-learning system http://mc.e-learning.unn.ru of Nizhni Novgorod State University, which is developed on the open-source e-learning platform Moodle (Modular Object-Oriented Dynamic Learning Environment) [13]. This platform is a well-known and it is actively supported by the community. The platform provides a wide range of possibilities for developing and preparing the content of online-courses.

Each object on the course page is a component of a special type listed below.

1. *"Label"* is used to represent plain text on the page or another component of the course.
2. *"URL"* is used to add links to external resources, particularly videos of lectures.
3. *"File"* module enables a lecturer to provide a file as a course resource: course program, presentation and text description of a lecture.
4. **"Page"** module enables to create a web page resource using the text editor. It is helpful for convenient and functional display of videos with discussions of substantive issues of the course lectures and interviews with project managers, which decide practically important problems with the use of parallel computing.
5. *"Quiz"* is used to prepare the test content for each lecture and to determine an evaluation system for each question. The teacher can allow the quiz to be attempted multiple times, with the questions shuffled or randomly selected from the question bank [13].
6. *"Forum"* allows students to communicate with lecturers and other participants. The forum activity module enables participants to have asynchronous discussions i.e. discussions that take place over an extended period of time [13].
7. *"Assignment"* is useful to represent control assignments. The assignment activity module enables a teacher to communicate tasks, collect work and provide grades and feedback [13].
8. *"Glossary"* is used to create a dictionary for the developed course.

If the content of the course is ready for publication, you have fast Internet and proper skills to work with Moodle, preparing the course pages takes no more than one workweek of the course developer. Learning Moodle capabilities is not difficult problem because there is a large number of tutorials in the Internet supported by numerous videos.

Developing the system of automatic parallel program validation and scalability evaluation is a particular technical problem that supposes the development of connected services, applications and web-services. The problem of integrating this system with Moodle is easy to solve because of the existence of documented application programming interface (API). Development of the specified system and organization of its interaction with Moodle is out of the scope of this paper.

8 Employment

The course passed by the test group of students from Lobachevsky State University of Nizhni Novgorod. It is assumed that the course will be employed during International scientific youth school "High-performance computing, optimization and applications" that will take place in Lobachevsky State University of Nizhni Novgorod from November 7 till 11, 2016.

9   Conclusion

This paper represents the course "Introduction to parallel computing" in the development of which a modern method of presentation of the educational materials for simultaneous teaching a large number of attendees has been used. The course curriculum, the method of self-study and assessment methodology are described. The general goal of the course is a self-study of parallel computing fundamentals using the most simple and comprehensible form, implying the presence of a wide range of presentations and video materials. The structuring of the course and the presence of the large number of assignments for self-control help to improve the quality of self-study of parallel computing foundations. A distinctive feature of the course is automatic knowledge control. The control implies theoretical knowledge control through quizzes and practical knowledge control through the automatic validation and scalability evaluation of parallel programs developed for solving well-known mathematical problems. These fundamental features are able to attract a large number of listeners and thereby to pay students' attention to professional work in the field of parallel computing and supercomputing technologies.

In the nearest future, we plan to prepare English version of the course. Moreover, we suppose to extend practical part of the course by the applied problems of global optimization [14,15], image processing, computational geometry and another interesting fields.

Acknowledgments. This research was supported by the Russian Science Foundation, project No 16-11-10150 "Novel efficient methods and software tools for the time consuming decision making problems with using supercomputers of superior performance".

References

1. Computer Science Curricula (2013). https://www.acm.org/education/CS2013-final-report.pdf
2. NSF/IEEE-TCPP Curriculum Initiative on Parallel and Distributed Computing - Core Topics for Undergraduates. http://grid.cs.gsu.edu/~tcpp/curriculum/?q=home
3. Internet-University of Supercomputer Technologies. http://www.osp.ru/os/2009/02/7323236
4. Voevodin, V., Gergel, V., Popova, N.: Challenges of a systematic approach to parallel computing and supercomputing education. In: Hunold, S., et al. (eds.) Euro-Par 2015. LNCS, vol. 9523, pp. 90–101. Springer, Heidelberg (2015). doi:10.1007/978-3-319-27308-2_8
5. Free Online Courses for everyone! List of MOOCs offered by the Best Universities and Entities. https://www.mooc-list.com
6. Parallel Programming Concepts. https://open.hpi.de/courses/2efebcdd-0749-4ca3-8396-58a09e7bf070
7. Parallel Programming. https://www.coursera.org/learn/parprog1?siteID=.GqSdL GGurk-V9YazzHdzg5C8LLm6cpr8A&utm_content=10&utm_medium=partners&utm_source=linkshare&utm_campaign=*GqSdLGGurk

8. High Performance Scientific Computing. https://www.coursera.org/course/scicomp
9. Intro to Parallel Programming. Using CUDA to Harness the Power of GPUs. https://www.udacity.com/course/intro-to-parallel-programming-cs344
10. Heterogeneous Parallel Programming. https://www.coursera.org/course/hetero
11. Gergel, V.P.: Theory and Practice of Parallel Computing, Moscow (2007)
12. Gergel, V., Liniov, A., Meyerov, I., Sysoyev, A.: NSF/IEEE-TCpp. Curriculum implementation at the State University of Nizhni Novgorod. In: Proceedings of the International Parallel and Distributed Processing Symposium, IPDPS, pp. 1079–1084 (2014). Article no. 6969501
13. Moodle - Open-source learning platform. https://moodle.org
14. Gergel, V., Sidorov, S.: A two-level parallel global search algorithm for solution of computationally intensive multiextremal optimization problems. In: Malyshkin, V. (ed.) PaCT 2015. LNCS, vol. 9251, pp. 505–515. Springer, Heidelberg (2015). doi:10.1007/978-3-319-21909-7_49
15. Lebedev, I., Gergel, V.: Heterogeneous parallel computations for solving global optimization problems. Procedia Comput. Sci. **66**, 53–62 (2015)

Parallel Computational Models to Estimate an Actual Speedup of Analyzed Algorithm

Igor Konshin[1,2,3]([⊠])

[1] Dorodnicyn Computing Centre of the Russian Academy of Sciences,
Moscow 119333, Russia
Igor.Konshin@gmail.com

[2] Institute of Numerical Mathematics of the Russian Academy of Sciences,
Moscow 119333, Russia

[3] Lomonosov Moscow State University, Moscow 119991, Russia

Abstract. The paper presents two models of parallel program runs on platforms with shared and distributed memory. By means of these models, we can estimate the speedup when running on a particular computer system. To estimate the speedup of OpenMP program the first model applies the Amdahl's law. The second model uses properties of the analyzed algorithm, such as algorithm arithmetic and communication complexities. To estimate speedup the computer arithmetic performance and data transfer rate are used. For some algorithms, such as the preconditioned conjugate gradient method, the speedup estimations were obtained, as well as numerical experiments were performed to compare the actual and theoretically predicted speedups.

Keywords: Parallel computations · Computational complexity · Communication complexity · Speedup

1 Introduction

During the last decades a parallel computing has been the basic tool for the solution of the most time consuming problems of mathematical physics, linear algebra, and many other branches of modern supercomputer application [1]. The most fitted parallel computational model allows to adequately estimate the numerical algorithms efficiency. It gives a possibility to compare the performance of the analyzed algorithms for the concrete architectures and choose the most successful ones in advance.

There are a lot of parallel computation models (see, for example, [1–3]), however some of them are too superficial to estimate the quantitative speedup values, on the contrary the other requires to take into account too detailed information on the algorithm and a run of code implementation. Moreover, the most of the models do not reflect the peculiarity of the architectures of the computers, on which the implemented algorithms are running. Using only macro-structure of algorithms there would be interesting to decide which algorithm

© Springer International Publishing AG 2016
V. Voevodin and S. Sobolev (Eds.): RuSCDays 2016, CCIS 687, pp. 304–317, 2016.
DOI: 10.1007/978-3-319-55669-7_24

properties are the most important for deriving of practical quantitative speedup estimates and which computation systems characteristics could be applied to get these estimates.

To analyze an algorithm properties for different computer architectures, for example, for computers with the shared or distributed memory fitted computational models may be required. While analyzing algorithms for each computational model a knowledge of the conditions and regions of the model applicability becomes very important.

This paper describes two parallel computation models, presents the efficiency estimations obtained on their base, and defines the conditions of their applications. Specification and analysis of the parallel efficiency upper bounds estimates are performed for some linear algebra algorithms, including the preconditioned conjugate gradient method. The qualitative comparison of estimates obtained and results of numerical experiments are presented.

2 Parallel Computational Model for the Shared Memory Computers

Let the parallel computations be performed on a shared memory computer and programming environment OpenMP be used for parallelization. In the most simple cases, OpenMP can be treated as an insertion of compiler directives for loops parallelization. At some intermediate parallelization stage or due to intrinsic algorithm properties some arithmetic operations can be performed in a serial mode.

Let us consider that the computations are sufficiently uniform by the set of arithmetic operations performed and, in principle, we can calculate the amount of such operations and obtain the total numbers of parallel and serial operations. Let the fraction of serial operation be $f \in [0; 1]$ (this value sometimes is denoted by "s" from "serial" but we are not doing so to separate the usage of "S" for speedup).

If the time T for the arithmetic operations is linear with respect to the number of arithmetic operations, then it is easy to estimate the maximum speedup, which can be obtained for some implementation of the considered algorithm.

2.1 Amdahl's Law

Let denote by $T(p)$ the time of program running on p processors, than the speedup received for computations using p processors will be expressed by classical formula:

$$S(p) = T(1)/T(p). \tag{1}$$

If the fraction of serial operations is equal to f, then using formula (1) we can estimate the maximum achievable speedup by the following way:

$$S(p) = T(1)/(fT(1) + (1-f)T(1)/p)) = p/(1 + f(p-1)). \tag{2}$$

The last formula expresses the Amdahl's law [5]. It can also be treated in more general case as the maximum achieved speedup which can be obtained for arbitrary parallel architecture for the analyzed algorithm or the program code [6].

For the analysis of the obtained formula (2) it can be noted that if the fraction of serial operations is just 1%, then when running the program on 100 processors, the serial part of the code on a single processor will take about the same time as the parallel part of operations on 100 processors. It results in about 50 speedup (or 50% of efficiency). If the number of processors used can be arbitrarily increased, then the maximum achieved speedup will be equal to $S = 1/f$. Another extreme case is the linear speedup $S(p) = p$ achieved for $f = 0$.

In addition, we can define the best conditions of the Amdahl's law applicability for a shared memory computer, that is, when the algorithm actual speedup will be close to the estimated by (2). The basic conditions are:

- the arithmetic operations are quite uniform;
- all the used threads take part in the computations of the parallel part of the code;
- load balancing for all active threads;
- scalability of threads usage, i.e., the performance of the threads does not depend on the threads number (or, in other words, the execution time for the parallel part of the code is actually p times reduced when running on p threads).

Let us analyze the last condition in more detail.

2.2 Actual Efficiency of the Parallel Program

Let us inquire the issue: which maximum speedup can be achieved for the ideally parallel algorithm on some parallel computer cluster.

We consider implementation of DAXPY operation from BLAS1, which with the OpenMP directive can be written as following:

```
#pragma omp parallel for
for (i=0; i<n; i++) y[i] += a * x[i];
```

The numerical experiments were performed on the computer cluster [7] of the Institute of Numerical Mathematics of the Russian Academy of Sciences (INM RAS). The computer nodes specification from x6core queue that has been used for the experiments are:

- Compute Node Asus RS704D-E6;
- 12 cores (two 6-cores processors Intel Xeon X5650@2.67GHz);
- RAM memory: 24 GB;
- Disc memory: 280 GB;
- Operating system: SUSE Linux Enterprise Server 11 SP1 (x86_64).

Table 1. The efficiency of the DAXPY operation within OpenMP and MPI environment.

p	E^*_{omp}	S^*_{omp}	E^*_{mpi}	S^*_{mpi}
1	1.000	1.00	1.000	1.00
2	0.939	1.87	0.948	1.89
3	1.778	5.33	0.994	2.98
4	1.929	7.71	0.987	3.94
5	1.496	7.48	0.994	4.97
6	1.481	8.89	0.986	5.92
7	1.095	7.66	0.997	6.98
8	1.011	8.09	0.977	7.82
9	0.863	7.77	0.988	8.89
10	0.842	8.42	0.933	9.33
11	0.638	7.02	0.960	10.56
12	0.385	4.63	0.985	11.82

The Intel C compiler 4.0.1 with the MPI 5.0.3 support was used.

For comparison, the above mentioned code fragment has been run not only under the OpenMP environment but under MPI environment as well. The values of speedup (1) and the efficiency $E = S/p$ obtained are given in Table 1. Some specific results were defined for the parallelization by OpenMP:

- the expected reduction of the efficiency E^*_{omp} for large number of threads $p = 11, 12$ due to insufficient bandwidth of the memory channel;
- the unexpected superlinear speedup for $p = 3, ..., 8$ threads, that violate the Amdahl's law, probably due to coherent memory access operations and effective compiler processing.

In case of the MPI implementation the computational efficiency E^*_{mpi} for such an "ideal" algorithm has expectedly been very close to 1.

Thus, if we would like to improve the formula of the Amdahl's law (2) in accordance with the specific of the numerical experiment on the certain computer cluster, we should multiply the right-hand side of the Eq. (2) by E^*_{omp}:

$$S(p) = pE^*_{\mathrm{omp}}/(1 + f(p - 1)). \tag{3}$$

The value of E^*_{omp} here would be considered as given in tabular form in accordance with Table 1. Then the possible superlinearity would be included in formula (3), that prevents the failure of the Amdahl's law.

3   Parallel Computation Model for Distributed Memory Computers

The key peculiarity of the parallel algorithm execution on the distributed memory computer is a memory exchange operations and an additional loss of the efficiency connected there with. Issues connected with the memory exchanges can be considered in more detail.

3.1 Message Transmission Rate

To estimate the time spent at the data exchanges the well-known formula can be used:

$$T_c = \tau_0 + \tau_c L_c, \tag{4}$$

where τ_0 is the initialization time for the message transmission, τ_c is the rate of the data exchange (i.e., measured by time of the data exchange of unit length), T_c is the time spent for the transmission of length L_c. Generally, the initialization time τ_0 (latency of the transmission) can be rather long, for example, $\tau_0 = 100\tau_c$, i.e., the time spent for transmission of 100 words can take just only two times more than the transmission of one word. However, if the length of the transmission is large enough, for example, greater than 1000, than the latency can be neglected.

The most effective algorithm implementation would be the implementation of a transmission of great length. Such an algorithms are called the algorithms with "large-grained" parallelism. If this algorithms class is analyzed, the simplified formula can be applied:

$$T_c = \tau_c L_c. \tag{5}$$

In other words, we neglect the latency of the communication network and consider that the rate of data transmission is specified only by network capacity. It should be noted that in the specified propositions the transmission time becomes linear with respect to the data length. Additionally, it means that the total length of all transmissions will define the total transmission time for several successive transmissions. Subsequently, this fact allows us to essentially simplify the efficiency estimation of the parallel algorithms analyzed.

3.2 Estimate of the Algorithm Parallel Efficiency

Let us introduce the same notations as in Subsect. 2.1. Let p be the number of processors used and $T(p)$ be the execution time for the algorithm on p processors. Respectively, the speedup that can be obtained by the algorithm will be expressed by the formula $S = T(1)/T(p)$, while the efficiency of the algorithm will be specified by the ratio $E = S/p$.

To estimate the computation time for the algorithm we need the knowledge of both characteristics of the analyzed algorithm and the parameters of the parallel computer used.

Let L_a be the total number of arithmetic operations of the algorithm and τ_a is the time spent per one such operation. Similar, let L_c be the total transmission length and τ_c be the time of transmission of the unit length. Then, the total time for arithmetic operations can be expressed by the formula $T_a = \tau_a L_a$ and the total time for communications is $T_c = \tau_c L_c$.

Now, everything is ready to speedup estimation, but we introduce two auxiliary values. The first one will describe the general characteristic of the parallel computer properties:

$$\tau = \tau_c/\tau_a, \tag{6}$$

specifying how many arithmetic operations can be performed when transmitting a number from one processor to another (in case of theoretically unlimited fast data transmissions or formally synchronous transmissions, $\tau = 0$; for computers with sufficiently fast transmissions we can expect approximately $\tau = 10$; while on case of slow communications we have about $\tau = 100$).

The second important value is the characteristic of the algorithm parallel properties:

$$L = L_c/L_a, \tag{7}$$

denoting a value being reverse to how many arithmetic operations are actually performed by the algorithm when transmitting a number.

Finally, we can estimate the speedup:

$$S = S(p) = T(1)/T(p) = T_a/(T_a/p + T_c/p) = pT_a/(T_a + T_c) = p/(1 + T_c/T_a)$$
$$= p/(1 + (\tau_c L_c)/(\tau_a L_a)) = p/(1 + \tau L), \tag{8}$$

and, analogously, estimate the efficiency:

$$E = S/p = 1/(1 + \tau L). \tag{9}$$

As a result, we obtain a fairly simple formula for efficiency estimate, depending on two parameters τ and L only, characterizing parallel properties of computer and algorithm, respectively. At first glance, it is surprising that the last formula has no explicit dependence on the number of processors p, but what actually happens is that it implicitly presents in characteristic L via the dependence of all transmissions L_c total length with the given amount of processors p.

Let us summarize the assumptions that has been made during derivation of the upper bound of the speedup and efficiency of the parallel algorithm when running on the shared memory computer:

- in contrast to the Amdahl's law formula, it is considered that all computations are completely parallelizable and sequential part of the algorithm is absent ($f = 0$);
- the delay in computations is due to the data transmissions only, and the algorithms with synchronous communications are mainly suited for this model;
- parallel computations are well balanced, i.e., there is no delay due to imbalance;
- computational nodes are uniform, it means that parameter τ is the same for all nodes (though as it is known MPI can be performed on nonuniform computer systems);
- the arithmetic operations rate τ_a is independent on the number of processors p (for distributed memory computers it is performed more frequently, than on the shared memory computers with the use of OpenMP, see Table 1);
- the data transmission rate τ_c is independent on the number of processors p as well (this less obvious fact means the scalability of communication network).

3.3   Estimation of the Linear Algebra Algorithms Parallel Efficiency

Let us consider the application of the constructed speedup and efficiency estimates for some examples of linear algebra algorithms.

Example 1 (ideally parallel operations).

(a) Sum of two vectors:

$$Z_i = X_i + Y_i, \quad i = 1, ..., n. \tag{10}$$

(b) Vector normalization (multiplication by a constant):

$$X_i = \alpha X_i, \quad i = 1, ..., n. \tag{11}$$

(c) AXPY operation (as a combination of two above mentioned operations, intensively used in numerical methods, implemented in BLAS1):

$$Y_i = \alpha X_i + Y_i, \quad i = 1, ..., n. \tag{12}$$

(d) Multiplication of block-diagonal matrix by a vector, each block corresponds to certain processor, moreover a sparsity structure inside each block does not matter if total amount of nonzero elements inside blocks are about the same.

(e) Solution of linear system with block-triangular matrix when performing forward or backward substitutions. As in the previous case, the block structure can be arbitrary if the number of nonzero elements in each block triangle is about the same.

It is obvious that for these operations it is not necessary to perform the data transmissions ($L_c = 0$, and hence $L = 0$), therefore the speedup will be linear: $S = p$ for any value of τ, and the efficiency will be overall: $E = 1$. The computations are independent, and for cases (a)–(c) it is possible to exploit maximum number of processors $p = n$. It should be noted that in all cases the uniform load balancing is assumed, i.e., vector components amount for the cases (a)–(c) and number of nonzero elements inside the block for the cases (d) and (e).

Example 2 (dot product or inner product).

$$c = \sum_{i=1}^{n} X_i Y_i. \tag{13}$$

Firstly, we should locally compute the partial sum at each processor, and then it is necessary to compute the total summation and send the result to processors. By means of MPI library it can be done by using, for example, the function MPI_Allreduce(). The way of this function implementation is not fixed in MPI standard and is left at the discretion of specific MPI implementation. However, to estimate the speedup we can apply the simplest way by sending the partial sums to a master processor and perform summation on it, and then distribute a result to other processors.

The total number of arithmetic operations (considering the summation with multiplication as a single operation, as well as a separate summation on the master processor) will be equal to $L_a = n + (p - 1)$, but the total length of all data exchanges is $L_c = 2(p - 1)$.

As while calculating L we are interested only in the ratio of these values, it is more convenient to write them down with respect to a local processor, i.e., $L_a = (n + (p - 1))/p$ and $L_c = 2(p - 1)/p$. Further, if not stated otherwise we will mean precisely such estimates.

As a result, the speedup estimate will look like:

$$L = L_c/L_a = 2(p - 1)/(n + (p - 1)), \tag{14}$$
$$S = p/(1 + 2(p - 1)\tau/(n + (p - 1))). \tag{15}$$

For example, for $n = 10^6$ and $\tau = 10$ we can calculate several estimate values:

$$S(p = 1) = 1, \quad S(p = 100) \approx 99.8, \quad S(p = 1000) \approx 980, \tag{16}$$

and for $\tau = 100$ with the same vector dimension we obtain:

$$S(p = 1) = 1, \quad S(p = 100) \approx 98, \quad S(p = 1000) \approx 800. \tag{17}$$

The dot product operation is very important and is frequently used in linear algebra. It is observed that in case the amount of processors increases up to $p = 1000$, there is a drastic fall of operation speed. The estimations provided by this paper indirectly confirm this observation.

Example 3 (multiplication of a dense matrix by a vector). Let us consider the matrix-by-vector multiplication for dense square matrix:

$$Y_i = \sum_{j=1}^{n} A_{ij}X_j, \quad i = 1, ..., n, \tag{18}$$

considering that on each processor the portion of block rows are stored, as well as the corresponding parts of vectors X and Y:

```
[:]    [== == ==]    [:]
---    ----------    ---
[:]  = [== == ==]  * [:]
---    ----------    ---
[:]    [== == ==]    [:]
```

To perform the multiplication, it is necessary to collect on each processor the copy of vector X of full dimension, and then to perform multiplication on the local part of the matrix A located on the processor.

Let n be a matrix dimension, and the matrix rows are distributed by processors equally, then:

$$L_a = n^2/p, \quad L_c = (n/p)(p-1), \quad L = L_c/L_a = (p-1)/n, \tag{19}$$

$$S = p/(1 + (p-1)\tau/n). \tag{20}$$

If $n = 1000$ and $\tau = 10$, then $S(p = 1) = 1$, $S(p = 10) \approx 9$, $S(p = 100) \approx 50$.

Example 4 (multiplication of a transposed dense matrix by a vector). Let us consider the matrix-by-vector multiplication for a transposed dense square matrix:

$$Y_i = \sum_{j=1}^{n} A_{ij}^T X_j, \quad i = 1, ..., n, \tag{21}$$

considering that on each processor the portion of block rows of A (block columns of A^T) is stored, as well as the corresponding parts of vectors X and Y:

```
[:]    [:: :: ::]    [:]
---                  ---
[:] = [:: :: ::] * [:]
---                  ---
[:]    [:: :: ::]    [:]
```

To perform the multiplication, first, it is necessary to compute the local partial sum Z (of full dimension) as the product of the local block columns and the local part of the vector X, then send the parts of the vector Z to the respective processors, and, finally, sum the received parts of the vector Z to obtain the final local part of the vector Y.

Let n be the matrix dimension, then:

$$L_a = n^2/p, \quad L_c = (n/p)(p-1), \quad L = L_c/L_a = (p-1)/n, \tag{22}$$

$$S = p/(1 + (p-1)\tau/n). \tag{23}$$

It is surprising, that, despite of the very different algorithm structure, the obtained estimate is the same as in Example 3. This is due to the same total length of interprocessor communications.

Example 5 (multiplication of a band matrix by a vector). Let us consider the matrix-by-vector multiplication for a band matrix stored by rows considering as it was stated before that each processor stores a portion of matrix rows as well as the corresponding parts of vector X and the resulting vector Y:

```
[:]    [===      ]    [:]
---    ---------     ---
[:] = [   ===    ] * [:]
---    ---------     ---
[:]    [      ===]    [:]
```

The portion of block rows of matrix A is stored on each processor, as well as the corresponding parts of vectors X and Y.

Let n be the dimension and r be the bandwidth of the matrix, then in order to perform the multiplication of the local part of the matrix each processor should additionally receive r components of vector X from two neighbouring processors:

$$L_a = (2r+1)n/p, \quad L_c = 2r(p-1)/p, \tag{24}$$

$$L = (2r/(2r+1))(p-1)/n \approx (p-1)/n, \quad S \approx p/(1+(p-1)\tau/n). \tag{25}$$

The most surprising in this estimate is the fact that it reproduces almost exactly the previous estimates and is almost independent on the half bandwidth r. It means, that although the number of arithmetic operation is reduced, the communication length is reduced in the same proportion.

Example 6 (multiplication of a sparse multi-diagonal matrix by a vector). Let us consider the matrix-by-vector multiplication for a sparse matrix with nonzero elements located on diagonals corresponding some discretization stencil. Let each processor stores a portion of matrix rows as well as the corresponding parts of vector X and the resulting vector Y:

```
[:]   [\\ \    ]    [:]
---   -----------   ---
[:] = [ \ \\\ \ ] * [:]
---   -----------   ---
[:]   [    \ \\]    [:]
```

Let n be the dimension and r be the semi-bandwidth of the matrix, and d be the total number of diagonals in the matrix (or number of vertices in the discretization stencil), then in order to perform the multiplication of the local part of the matrix A each processor (as in the previous example) should additionally receive r components of vector X from two neighbouring processors:

$$L_a = dn/p, \quad L_c = 2r(p-1)/p, \quad L = 2r(p-1)/(dn), \tag{26}$$

$$S = p/(1+2r(p-1)\tau/(dn)). \tag{27}$$

It worth to note, that for two-dimensional problem of size $n = m \times m$ with the use of 5-point discretization stencil the parameters of sparse matrix are equal to $r = m$ and $d = 5$. For three-dimensional problem of size $n = m \times m \times m$ with the use of 7-point discretization stencil we should take $r = m^2$ and $d = 7$.

It is worth to note, that the effective semi-bandwidth of the matrix depends on distribution of the domain to processors, for example, in three-dimensional case it is advantageous to cut the domain by 3D domains but not by slices. It may essentially reduce the total communication length and increase the efficiency of the sparse matrix-by-vector operation.

However, in comparison with the multiplication by a matrix with a dense band, the low efficiency of such an operation is due to the respectively less number of arithmetic operation for the same semi-bandwidth, and consequently for the same communication costs.

3.4   Conjugate Gradient Method

As the final example, we derive the estimate for the preconditioned conjugate gradient (PCG) method [8].

We consider the most simple but frequently used preconditioner: the block Jacobi structure with no overlap and incomplete Cholesky IC0 factorization of each block. The basic operations involved in this algorithm have already been studied in Subsect. 3.3:

- three "AXPY" operations (Example 1c);
- two inner "DOT" products (Example 2);
- multiplication of a sparse multi-diagonal matrix by a vector "MVM" (Example 6);
- solution of linear system with block-diagonal preconditioner matrix "SOL" (Example 1e).

We can write out now the speedup estimate for an iteration of PCG algorithm.

Example 7 (conjugate gradient method). The computational and communicational costs for a single iteration of conjugate gradient method with IC0 preconditioning consist of

$$
\begin{aligned}
L_a &= 3L_a^{\text{AXPY}} + 2L_a^{\text{DOT}} + L_a^{\text{MVM}} + L_a^{\text{SOL}} \\
&= 3(n/p) + 2(n/p) + (dn/p) + (dn/p) = (2d+5)n/p, \quad (28) \\
L_c &= 3L_c^{\text{AXPY}} + 2L_c^{\text{DOT}} + L_c^{\text{MVM}} + L_c^{\text{SOL}} \\
&= 3 \cdot 0 + 2(2(p-1)/p) + (2r(p-1)/p) + 0 = (2r+4)(p-1)/p. \quad (29)
\end{aligned}
$$

After that the "parallelism" characteristic of the algorithm can be expressed as

$$
L = L_c/L_a = (2r+4)(p-1)/((2d+5)n), \quad (30)
$$

while the speedup estimation will be expressed as follows:

$$
S = p/(1 + \tau L) = p/(1 + (2r+4)\tau(p-1)/((2d+5)n)). \quad (31)
$$

3.5   Numerical Experiment and Comparison with the Speedup Estimate

For the numerical experiments INM RAS cluster [7] with already described in Subsect. 2.2 computational nodes from queue "x6core" was used.

First, we compute the "parallelism" characteristic of the computer, which was applied in Subsect. 3.2 when deriving the estimate. Operation DOT over double precision vectors of length 10^6 was used to estimate the arithmetic performance of the cluster, while two simultaneous asynchronous data exchanges with the double precision vectors of the same length was used to estimate the transmission rate. The communications were performed without overlapping with arithmetic operations. The following values were obtained:

$$\tau_a = 3.14 \cdot 10^{-10}, \quad \tau_c = 3.06 \cdot 10^{-8}. \tag{32}$$

It means that the main "parallelism" characteristic of the computer can be set to:

$$\tau = \tau_c/\tau_a = 100. \tag{33}$$

To verify the obtained estimates the developed in the INM RAS parallel program platform INMOST [9] was used. It can be loaded as a source code from [10]. As the model problem we have used the test program solver_test002 developed by the author of the paper, the program is accessible from the same site as well. The linear system matrix were constructed by discretization of 3D 7-point stencil for the domain of size $n = m \times m \times m$. The resulting linear system was solved by PCG method from the external package PETSc [11]. The additive Swartz method with no overlap and IC0 factorization in subdomains was used by setting the following parameters:

```
-ksp_type cg
-pc_type asm
-pc_asm_overlap 0
-sub_pc_type ilu
-sub_pc_factor_levels 0
```

A set of problems with different dimensions was considered, the dimension of domain in each direction was $m = 64, 96, 128, 160$. The total number of unknowns ranged from about 262 thousand to about 4 million, while a number of processors was chosen equal to $p = 1, 2, 4, 8, 16, 32, 64$.

For the final form of the PCG method speedup formula estimated by (31) the following parameters were used $r = m^2$, $n = m^3$, $d = 7$, and $\tau = 100$. For four considered linear systems the actual speedup with respect to the run on a single processor were obtained, and the plots of theoretical estimates by formula (31) were drawn as well. The obtained plots are presented on Figs. 1 and 2. It is worth to note that the plots behavior is qualitatively coincided.

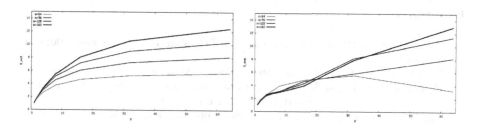

Fig. 1. The speedup estimated by formula (31) and the actual speedup for problems with $m = 64, 96, 128, 160$.

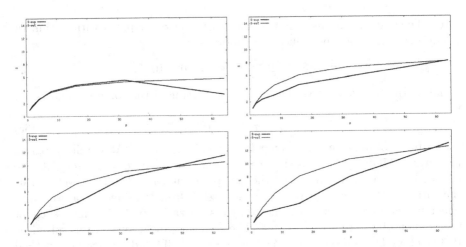

Fig. 2. Comparison of estimated and actual speedup for problems with $m = 64, 96, 128, 160$.

4   Conclusions

Two parallel computation models were presented for computers with both shared and distributed memory. Based on the macro-structure algorithm properties the speedup estimates were obtained for runs on parallel computers. The estimate for shared memory computers is built on the portion of serial computations of the algorithm, while the estimate for distributed memory computer clusters is based on the "parallelism" characteristics of both the considered algorithm and the computer in use.

The numerical experiments demonstrate that the theoretical speedup estimates and the actual experiment results are in qualitative agreement.

Acknowledgements. This work has been supported in part by RSF grant No. 14-11-00190.

References

1. Voevodin, V.V.: Parallel Computing. BHV-Petersburg, St. Petersburg (2002). (in Russian)
2. Bogachev, K.Y.: Parallel Programming. Binom, Moscow (2003). (in Russian)
3. Gergel, V.P., Strongin, R.G.: Parallel Computing for Multiprocessor Computers. NGU Publ., Nizhnij Novgorod (2003). (in Russian)
4. AlgoWiki: open encyclopedia of algorithm properties. http://algowiki-project.org. Accessed 15 June 2016
5. Amdahl, G.M.: Validity of the single-processor approach to achieving large scale computing capabilities. In: AFIPS Conference Proceedings, Atlantic City, NJ, 18–20 April, vol. 30, pp. 483–485. AFIPS Press, Reston (1967). http://www-inst.eecs.berkeley.edu/n252/paper/Amdahl.pdf. Accessed 15 June 2016

6. Antonov, A.: Under the Amdahl's law, No. 430. Computerra (2002)
7. INM RAS cluster. http://cluster2.inm.ras.ru. Accessed 15 June 2016 (in Russian)
8. Saad, Y.: Iterative Methods for Sparse Linear Systems. PWS, Boston (1996)
9. Vassilevski, Y., Konshin, I., Kopytov, G., Terekhov, K.: INMOST - A Software Platform and Graphical Environment for Development of Parallel Numerical Models on General Meshes. Moscow State University Publ., Moscow (2013). (in Russian)
10. INMOST - a toolkit for distributed mathematical modeling. http://www.inmost.org. Accessed 15 June 2016
11. PETSc (Portable, Extensible Toolkit for Scientific Computation). https://www.mcs.anl.gov/petsc. Accessed 15 June 2016

Techniques for Solving Large-Scale Graph Problems on Heterogeneous Platforms

Ilya Afanasyev[1]([⊠]), Alexander Daryin[2], Jack Dongarra[1],
Dmitry Nikitenko[1], Alexey Teplov[1], and Vladimir Voevodin[1]

[1] Lomonosov Moscow State University, Moscow, Russia
afanasiev_ilya@icloud.com, dongarra@icl.utk.edu,
{dan,voevodin}@parallel.ru, alex-teplov@yandex.ru
[2] Yandex.Technology GmbH, Moscow, Russia
shurick@yandex-team.ru

Abstract. The paper introduces techniques for solving various large-scale graph problems on hybrid architectures. The proposed approach is illustrated on the computation of minimum spanning tree and shortest paths. We provide a precise mathematical description accompanied by the information structure of required algorithms. Efficient parallel implementations of several graph algorithms are proposed based on this analysis. Hybrid computations allow using all the available resources on both multi-core CPUs and GPUs. Our implementation uses out-of-core memory algorithms to handle graphs that don't fit in the main memory. Experimental results confirm high performance and scalability of the proposed solutions. Moreover, the proposed approach can be applied to other graph processing problems, which have recently rapidly increased in demand.

Keywords: Hybrid computations · CUDA · GPU · Large-scale graph processing · Graph algorithms · APSP · MST · All pairs of shortest paths · Minimum spanning tree

1 Introduction

Large-scale graph processing problems are recently becoming more and more demanded in various application fields. The most common examples are social networks and web-graphs, containing millions of vertices and billions of edges. Current paper reveals possibility of processing these graphs on server architecture with modern multi-core CPUs and GPUs installed. Such large graph processing presents many challenges, including: limited GPU's memory for graph storage, poor data locality, lot of cache misses and as a result extremely long computation time. Moreover, it's not always clear if it is better to solve graph problem on CPUs, GPUs, or if it is possible to use hybrid approach. This paper introduces supercomputer co-design technology, aimed to help researchers implement specific graph problem on various hardware architectures and platforms.

The proposed techniques are applied to implementation of two important graph processing problems: computation of minimum spanning tree (MST) and all pairs of shortest paths (APSP). For each of these problems a detailed algorithm analysis and its

© Springer International Publishing AG 2016
V. Voevodin and S. Sobolev (Eds.): RuSCDays 2016, CCIS 687, pp. 318–332, 2016.
DOI: 10.1007/978-3-319-55669-7_25

informational structure are provided. As a result, efficient parallel implementations have been developed, accompanied by performance and scalability studies.

2 Target Platform

All algorithm research and implementations are designed for server architecture, with modern multi-core CPUs and Nvidia GPUs installed. This architecture is well suited for solving large-scale graph problems for the following reasons: first, modern servers have significant amount of memory to store real-world graphs, and, secondly, have sufficient parallelism resources to process these graphs in a reasonable amount of time. The ratio of the CPU and GPU memory is an important feature of the selected architecture; usually, servers have significant amount of memory (up to several tens or hundreds GB), while GPU memory is very limited (around 6–24 GB on modern GPUs).

Provided in the current paper, experiment results were obtained on a single node of Russian top-performance supercomputers "Lomonosov" [1] (early stages) and "Lomonosov-2" (final results). The compute node of last one is equipped with the following hardware:

– 14-core CPU Intel® Xeon® E5-2697 with hyperthreading supports;
– 64 GB RAM;
– NVIDIA® Tesla™ K40 GPU with 12 GB device memory.

3 Design Principles

This section describes techniques for graph problems solution. Proposed techniques involve graph problem properties research, which helps to create an efficient parallel implementation. This implementation must be able to solve the problem at hand with any possible configuration of provided hardware. Several fundamentally different server architectures are considered to be the target platform:

1. server with a high-performance CPU and a modern GPU or multiple GPUs;
2. server with a high-performance CPU, but with no GPU (or with outdated models);
3. server with an outdated CPU, but with a modern GPU.

Architecture of the first type will solve graph problems much more efficiently in hybrid computational mode, of the second type — in CPU-only mode, of the third — in GPU-only. In ideal case, any implementation aimed for further usage in different application fields must be able to process graphs efficiently on any types of architecture. That's why before implementing specified graph problem it is necessary to answer the following questions:

– Is it better to use CPU or GPU mode on the available hardware?
– Is it possible to implement hybrid computational mode?
– Is it possible to use multiple GPUs?
– Is it possible to implement the algorithm for any graph size?

To answer the questions listed above, current paper suggests the following technology. While researching selected graph problem from the theoretical point of view, it is necessary to:

1. formulate a precise mathematical description of the selected graph problem;
2. review existing algorithms;
3. select most suitable graph storage format;
4. research information structure of reviewed algorithms (using information graphs), to select the algorithm with the largest parallelism resource or most suited for current hardware;
5. create a modified algorithm for current hardware specifications.

During the implementation stage, from the programming point of view it is important to:

1. implement different computation approaches, most common are: CPU, GPU, hybrid;
2. support out-of-core computational mode at least on GPUs (that allows processing graphs with a size larger than current memory available by parts);
3. accurately implement selected algorithm, while using all available parallelism resource;
4. perform different tests on various graph types and sizes, as a result presenting performance and scalability results.

The following sections show how the described technology was applied to different graph problems — minimum spanning tree and all pairs of shortest paths computation. The application of described technique allowed creation of flexible high-performance parallel implementations for these problems.

4 Minimum Spanning Tree

4.1 Mathematical Description and State-of-Art

Minimum spanning tree (MST) problem was firstly described in Boruvka paper [2]. Given a connected undirected graph $G = (V, E)$ with vertices $V = (v1, v2, \ldots, vn)$ and edges $E = (e1, e2, \ldots, em)$, with assigned weights $w(e)$ to each $e \in E$, it is required to compute a tree $T * \subseteq E$ connecting all the vertices and having a minimal possible weight of all such trees.

If the graph G is not connected, such tree doesn't exist. In this case it is required to compute MST for each connected component of graph G. The set of such trees is called minimum spanning forest (MSF). These notions are considered to be the same later in the article.

There is an important MST property, called edges associativity, later used in the proposed algorithm. This property can be formulated in the following way:

$$MST(E_1 \cup E_2 \ldots \cup E_n) = MST(MST(E_1) \cup MST(E_1) \cup \ldots \cup E_n))$$

There are three most common algorithms for solving MST problem: Boruvka, Kruskal [3], and Prim [4] algorithms. All these algorithms have sequential complexity of $O(m \log n)$ operations for graphs with n vertices and m edges. However, the largest parallelism resource has Boruvka algorithm, that's why current implementation of MST problem is largely based on this algorithm.

Boruvka's algorithm is introduced in paper [2]. First attempts to implement this algorithm on multi-core processors are described in paper [5]. Also, several implementations on GPUs are described in papers [6, 7]. Paper [7] offers several optimizations, such as GPU primitive operations like scan and prefix sum usage.

4.2   Graph Storage Format

It is important to determine graph storage format before implementing any graph algorithm. The chosen format influences performance and even computational complexity of corresponding algorithm heavily. Work edges list format is selected. This decision has been made for the following reasons:

- edges lists can be efficiently merged and divided;
- MST operation doesn't require traversals of adjacent vertices to a specified one; moreover, each iteration of algorithm traverses all the edges in parallel, that's why edges list usage doesn't lead to significant performance losses;
- the graph is undirected, so in edges list format it is possible to store only half of the edges (for example with source vertex ID less that destination vertex ID).

Each edge is represented by three values (source vertex ID, destination vertex ID, edge weight) in current implementation. The edge weight can be represented by any characteristic or number (with a certain order relation <and>). During all of the tests 32-bit floating point numbers have been used.

4.3   Modifications to the Classical Algorithm

The chosen Boruvka algorithm was implemented with both CPU-based and GPU-based approaches. Hybrid implementation is a simple generalization of GPU version for large-scale graph processing in combination with CPU one. Medium-sized graphs (which fit into host or device memory) are processed with standard Boruvka algorithm. Sometimes during the computations it is necessary to check to which components belong a specified vertex. Union-Find data structure [8] is used for this purpose. Implementations of both CPU and GPU algorithms are very similar, so the description, provided below, suits for both cases.

Algorithm 1. Boruvka algorithm for medium-sized graph processing

1. In the beginning each vertex belongs to separate component.
2. On each step:
(a) For each component an incident edge with minimal weight is found.
(b) Minimal edges are added to MST, while corresponding components are merged.
3. Algorithm stops when only single component remains, or there is no incident edge
   in any component

In edge list format atomic operations are required on 2.a step where minimal edges updates occur. But this doesn't result into a huge problem, since provided tests demonstrate that atomic operations don't cause significant performance losses either on CPUs or GPUs.

For large-scale graph processing on GPUs slightly different algorithm is used:

Algorithm 2. Modified algorithm for large-scale graph processing

1. Graph edges list is divided into parts (fragments); each part fits into device memory.
2. Each part is loaded to device memory.
3. For each loaded part a separate MST computation is performed (using algorithm 1).
4. The edges, which belong to MST for current fragment are loaded back to CPU and merged into common edges list.
5. For the new edges list (from step 4) MST is calculated. The resulted tree is the required answer for the whole algorithm.

All minimum spanning trees for the separate fragments on step (3) can be computed in parallel; the only requirement for efficiency is balanced workloads (so computation of each MST can be performed for the same amount of time).

The same way hybrid computation approach works: for example, if graph doesn't fit into device memory, half of its edges can be processed on CPU in step (3), while the other part — on GPU. The computation of final MST can be performed either on CPU or GPU. Same idea can be used to compute MST on multiple GPUs too. This approach results into significant performance boost on really large graphs.

Same idea can be applied to process graphs with a size larger than host memory. Described algorithm requires only to store $O(|V|)$ data, which is significantly less than $O(|E|)$ in standard algorithm.

4.4   Resource of Parallelism and Information Graph of the Algorithm

To evaluate how proposed implementations use inner parallelism of described algorithm, information graphs of algorithm are used. Information graphs can be defined as a directed graph with vertices corresponding to algorithm operations and edges corresponding to data dependences between them (some operation output used as an input by different operations). Information graph is introduced in book [9].

There are two levels of parallelism in modified Boruvka algorithm: parallelism in standard Algorithm 1 (bottom), and parallelism while processing different fragments in Algorithm 2 (top). More specific description of these levels are listed below:

- minimal incident edge search can be performed for each fragment in parallel that allow creating efficient implementations of both CPU and GPU algorithm on the bottom parallelism level;
- MST computation for each fragment can be performed in parallel on the top parallelism level. This allows using multiple GPUs or a hybrid approach.

Informational graphs are provided in the following figures with detailed descriptions. Figure 1 corresponds to standard Boruvka (Algorithm 1), Fig. 2 — to a large-scale graph (Algorithm 2).

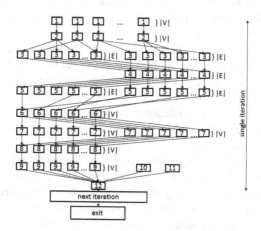

Fig. 1. Informational graph of Algorithm 1

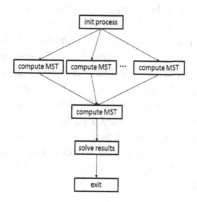

Fig. 2. Informational graph of Algorithm 2

The bottom level parallelism (Fig. 1) is represented on levels {3, 4, 5}, which correspond to operations of minimal edges search, and on levels {6, 7, 8}, corresponding to merge trees operations. Different copy and initialization operations {1, 2, 8, 9} can be also executed in parallel. On each main loop iteration the remaining trees count value is being checked {12}. If it hasn't changed during current iteration — loop break occurs.

The top level parallelism (Fig. 2) is represented by parallel computations of minimum spanning trees {compute_mst} for different edges list fragments. These computations are followed by final MST computation that uses results of previous computations; afterwards final results are saved to output arrays {save_results}.

4.5   Performance Analysis

Most of the performance results are provided for synthetic RMAT [10] graphs, since these graphs have very similar structure with real-world social networks and web graphs. It's important to notice that graph structure may strongly influence the performance, so in addition to RMAT, SCCA2 [11] and random uniform graphs have been used during the testing process.

In this work performance metric is used to compare different versions (CPU, GPU and hybrid). Performance is defined as TEPS, and is equal to amount of traversed edges per second. In addition to different computational modes comparison, the performance can be compared for different graph sizes and types. This comparison helps to understand how graph processing efficiency changes with graph size variation; usually performance goes down with graph size increase because graph data becomes larger than cache or device memory, etc.

The results provided have been obtained for graphs with vertices count in range from 2^{20} to 2^{27}. All graphs have RMAT structure and average connections count equal to 32. Thus, the maximum size of graph, which has been processed during program testing was 48 GB of the total 64 GB RAM in current system.

Figure 3 demonstrates performance comparison of different computational modes. First let's examine GPU version performance: on graph scale 20–25 a slight performance drops can be noticed. The reason is that graph size grows, resulting to a less frequent cache usage. The drop between 25–26 sizes is much more significant and has different cause — graph of 226 size doesn't fit into device memory and requires processing by parts. The reason why hybrid computational mode starts giving significant performance improvements at the same size is because one part can be processed now on CPU. The bigger graph is — the better computational balance can be achieved, that's why on 27 scale hybrid modes gives even better speed up.

Figure 4 shows CPU version scalability for different threads count (from 1 to 28). Current testing environment has 14-core CPU, so 28 threads are running in hyperthreading mode, which gives significant performance boost, as shown on figure.

The result of performance comparison for different graph types is provided on Fig. 5. All tests provided on this figure have been obtained on graph with fixed 2^{22} size and average degree equal to 32.

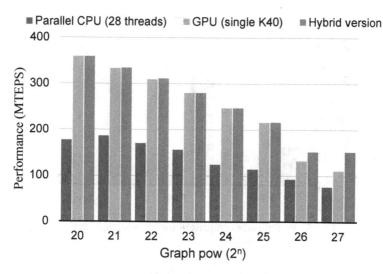

Fig. 3. Performance comparison of various computational modes: CPU, GPU, hybrid

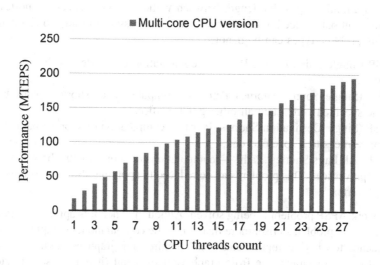

Fig. 4. Scalability of parallel CPU version

5 Shortest Paths

5.1 Mathematical Description and State-of-Art

Given an undirected graph $G = (V, E)$ with vertices $V = (v1, v2, \ldots, vn)$ and edges $E = (e1, e2, \ldots, em)$ with weights $w(e)$ assigned for $e \in E$. Edges sequence $\pi u, = (e1, \ldots, ek)$ from vertex u to vertex v, where all edges distinct from one another and has the same direction is called path between vertices u and v. Path length is defined as total weight of edges, which belong to it:

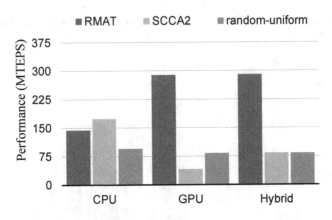

Fig. 5. Performance comparison for various graphs types

$$w(\pi u, v) = \sum w(ei). \; i = 1, k$$

Path with a minimal possible length between vertices u and v is called shortest path. Depending on pairs selection between which it is necessary to find shortest paths, the following problem types can be denoted:

1. SPSP (Single Pair Shortest Path) — computation of shortest paths between two vertices;
2. SSSP (Single Source Shortest Path) — computation of shortest paths between single specified vertex and all other graph vertices;
3. SDSP (Single Destination Shortest Path) — computation of shortest paths from all graph vertices to a specified one;
4. APSP (All Pairs Shortest Path) — computations of shortest paths between all pairs of graph vertices. This problem can be reduced to applying SSSP algorithms to each graph vertex.

In this work APSP problem is being solved. Although, when it is applied to large-scale graphs, several problems arise: more than quadratic computational complexity and as a consequence too long computational time, required for graph processing. Moreover, output data has quadratic size from graph vertices count (it is necessary to store $|V|$ distances with $|V|$ length). That's why the following generalization of this problem has been created: instead of computing pairs from all $|V|$ graph vertices, distances just from several source-vertices have to be calculated. With only single source vertex current generalization can be reduced to SSSP problem, while with $|V|$ different source vertices it can be reduced to APSP.

Shortest paths problems are studied in papers [12–14], where most common Dijkstra's, Bellman-Ford and Floyd-Warshall algorithms are introduced. Among these algorithms Dijkstra's has the best sequential complexity $O(m + n \log n)$. In fact, it asymptotically the fastest known sequential algorithm for this class of problems. The problem with Dijkstra's algorithm is that it is purely sequential, although some researchers attempted to implement it on parallel CPU and GPU [15] architectures.

In contrast to Dijkstra's, Belman-Ford algorithm has great parallelism potential, but worse sequential complexity —$O(mn)$ in the worst case and $O(m)$ in best. The implementation of this algorithm on GPU architecture is described in paper [16].

Floyd-Warshall algorithm also has large resource of parallelism; approaches for its implementation on GPUs are presented in paper [17]. But drawback of this algorithm is that it requires storing matrix of distances between all graph vertices, what is impossible even in distributed memory of largest clusters.

5.2   Graph Storage Format

Edges list storage format is selected again for APSP problem, since it allows processing graphs efficiently, which doesn't fit into device memory. But edges list format isn't suitable for Dijkstra's algorithm, because it requires adjacent vertices traversal on each iteration. Since the further described implementation requires execution of Dijkstra's algorithm on CPU, an extended edges list format is suggested. List of edges is stored in sorted order (by source vertex ID), with an additional array with pointers to the incident edges for each vertex. This approach is quite similar to compress adjacency list format, described in paper [7].

5.3   Modifications to the Classical Algorithm

Modified APSP problem requires computation of shortest paths form several source vertices to all other graph vertices. This computations are independent and can be performed in parallel, what gives a great parallelism potential; moreover, they can be performed by different algorithms and on different hardware (for example on CPU and GPU). Algorithm selection is influenced heavily by graph size and computational mode, which can be done in the following ways:

1. Dijkstra's algorithm is used on CPU (single copy on every core for each source vertex);
2. Bellman-Ford algorithm is used on CPU (optionally, if RAM size doesn't allow storing distance arrays for each core);
3. Bellman-Ford algorithm is used on GPU.

Hybrid computations are performed on "top" level of algorithm, where independent calculations of source-vertices are performed. A shared task queue is created for this purpose, where each task contains information about single source-vertex, required to be processed. Each CPU thread might have several behavior options listed below:

1. computation coordination for single GPU (multiple threads required for multiple GPUs);
2. its own computations, using Dijkstra algorithm;
3. participating in collective problem solution together with other threads, using Belman-Ford algorithm.

```
function APSP (in: Graph graph, array source_vertices[];
out: distances_file)
{
    Queue tasks = create_task_queue()
    for each vertex in source_vertices
        add vertex to tasks
    parallel section
    {
        while tasks.not_empty()
        {
        in atomic do
        {
            cur_vertex = tasks.pop()
        }
        if thread.algorithm = CPU_DIJKSTRA
            call cpu_dijkstra(cur_vertex)
        else if thread.algorithm = CPU_BELLMAN_FORD
            call cpu_bellamn_ford(cur_vertex)
        else if thread.algorithm = GPU_BELLMAN_FORD
            call gpu_bellamn_ford(cur_vertex)
        }
    }
}
```

Here cpu_dijkstra and cpu_bellamn_ford functions are standard implementations of sequential Dijkstra's and parallel Bellman-Ford algorithms; gpu_bellman_ford function have several modifications over a standard algorithm:

- large-scale graph can be processed on GPU even if it doesn't fit into device memory;
- instead of single source vertex the function processes multiple vertices (usually 8 of them), since it optimizes memory accesses and data movements to device memory.

Next paragraph describes GPU algorithm for large-scale graphs processing. In the initial part of computations all edges are divided into equal (or almost equal) parts: $E = E1 \cup E2 \cup \ldots \cup Ek$. Then the following steps are executed:

Algorithm 3. Bellman-Ford GPU algorithm for large-scale graph processing

---

1. Initialization: for each vertex initial distance $(v) = \infty$ is set; the only exception is source vertex, with $t(u) = 0$ distance.
2. Big iteration:
(a) Next edges portion i is loaded to device memory.
(b) Small iteration: standard Bellman-Ford algorithm is performed for edges Ei.
(c) Step (b) is repeated until on current small iteration distances array stops changing
(d) Algorithm is repeated from step (a) until all graph parts are processed
3. Algorithm is repeated from step 2 until all distances stop changing on big iteration

---

5.4   Resource of Parallelism and Information Graph of the Algorithm

In this section information graph will be used again to evaluate how current implementation uses parallelism resources of described algorithms. As already mentioned in
previous section, proposed algorithm has two levels of parallelism: in standard
Bellman-Ford algorithm (the bottom level), and during different source-vertices processing (the top level) (Fig. 6).

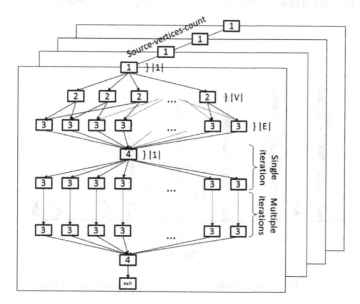

Fig. 6. Informational graph of APSP algorithm

The bottom level parallelism is represented on levels {2 and 3}; these operations
correspond to initialization of distances array (2) and updating those using edges weights
(3). Operation (4) is checking, if there were any changes during current iteration.

The top level parallelism is represented on parallel planes and corresponds to
parallel computation of shortest paths for different source vertices.

5.5   Performance Analysis

Performance metrics will be used again to compare different versions (CPU, GPU,
hybrid and sequential). Provided results are obtained for graphs with vertices count in
range from 2^{20} to 2^{27}. All graphs have RMAT structure and average connections count
equal to 32. Also, CPU computations can be performed with different algorithms —
Dijkstra's and Bellman-Ford, what is also reflected in the following diagrams.

Figure 7 demonstrates performance comparison of different computational modes.
Since processed graph is directed now, it requires twice more memory compared to
MST case, so graph with size 2^{25} doesn't fit into GPU memory. That's the reason of

significant performance drop between 24–25 graph sizes. Also, since Dijkstra's algorithm usage on CPU outperforms Bellman-Ford, hybrid approach with Dijkstra's algorithm most of the time significantly outperform its counterpart.

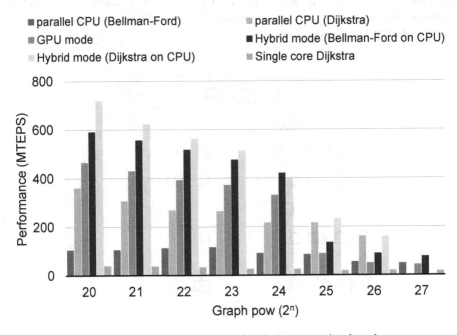

Fig. 7. Performance comparison of various computational modes

Figure 8 demonstrates performance comparison for different graph structure types: RMAT, SCCA2, random-uniform. All tests provided on this figure have been obtained on graph with fixed 2^{22} size and average degree equal to 32.

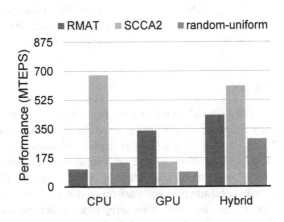

Fig. 8. Performance comparison for various graphs types

6 Conclusions and Future Work

The paper introduces techniques for solving large-scale graph problems on modern server architectures with different hardware configurations. Proposed technology was successfully applied to research and implementation of two important graph problems — computation of minimum spanning tree and all pairs shortest paths. Three different computational modes were discussed and implemented during problem analysis: CPU, GPU and hybrid.

Parallel high-performance implementations were developed as a result, capable of processing graph with vertices count up to 2^{27} and average connections count 32. The implementation of APSP operation achieves 800 MTEPS performance, while MST operation — 350 MTEPS. Also, hybrid computation offers significant performance boost in several situations. As a result of provided analysis it becomes possible to select computational mode and other implementation preferences according to the size problem and the platform hardware to maximize the performance on the selected platform.

Future plans are facing research on other important large-scale graph problems, including strongly connected components, transitive closure and bridges computation.

Acknowledgements. This study was financially supported by the Russian Science Foundation, agreement N14-11-00190 except mathematical problem statement (Sects. 4.1 and 5.1).

References

1. Voevodin, V., Shumaty, S., Sobolev, S.: Usage practices of "Lomonosov" supercomputer. Open Syst. (2012). (in Russian), http://www.osp.ru/os/2012/07/13017641/
2. Borůvka, O.: On a certain minimal problem. Práce Moravské Přírodovědecké Společnosti **III**(3), 37–58 (1926)
3. Kruskal, J.B.: On the shortest spanning subtree of a graph and the traveling salesman problem. Proc. Am. Math. Soc. **7**(1), 48–50 (1956)
4. Prim, R.C.: Shortest connection networks and some generalizations. Bell Syst. Tech. J. (1957). http://doi.org/10.1002/j.1538-7305.1957.tb01515.x
5. Badera, D.A., Cong, G.: Fast shared-memory algorithms for computing the minimum spanning forest of sparse graphs. J. Parallel Distrib. Comput. **66**, 1366–1378 (2006)
6. Harish, P., Vineet, V., Narayanan, P.J.: Large graph algorithms for massively multithreaded architectures. Technical Report IIIT/TR/2009/74 (2009)
7. Vineet, V., Harish, P., Suryakanth, P., Narayanan, P.J.: Fast Minimum Spanning Tree for Large Graphs on the GPU. HPG (High Performance Graphics) (2009). Report No: IIIT/TR/2009/197
8. Anderson, R.J., Wall, H.: Wait-free parallel algorithms for the union-find problem (1991). http://citeseerx.ist.psu.edu/viewdoc/download?doi=10.1.1.56.8354&rep=rep1&type=pdf
9. Voevodin, V.V.: Parallel Computing, 608 p. BHV, St. Petersburg (2002). (in Russian)
10. Chakrabarti, D., Zhan, Y., Faloutsos, C.: R-MAT: a recursive model for graph mining (2006). http://www.cs.cmu.edu/∼christos/PUBLICATIONS/siam04.pdf

11. Bader, D.A., Madduri, K.: Design and implementation of the HPCS graph analysis benchmark on symmetric multiprocessors. In: Bader, D.A., Parashar, M., Sridhar, V., Prasanna, V.K. (eds.) HiPC 2005. LNCS, vol. 3769, pp. 465–476. Springer, Heidelberg (2005). doi:10.1007/11602569_48

12. Bellman, R.: On a routing problem. Q. Appl. Math. **16**, 87–90 (1958)

13. Dijkstra, E.W.: A note on two problems in connexion with graphs. Numer. Math. **1**, 269–271 (1959). http://doi.org/10.1007/BF01386390

14. Floyd, R.W.: Algorithm 97: shortest path. Commun. ACM (1962). http://doi.org/10.1145/367766.368168

15. Nepomniaschaya, A.S., Dvoskina, M.A.: A simple implementation of Dijkstra's shortest path algorithm on associative parallel processors. Fundam. Inf. **43**(1–4), 227–243 (2000)

16. Hector, O., Torres, Y., Llanos, D.R.: A new GPU-based approach to the shortest path problem (2013). http://www.infor.uva.es/ ~ hector/papers/hpcs_2013.pdf

17. Katz, G.J., Kider, J.T.: All-pairs shortest-paths for large graphs on the GPU (2008). http://repository.upenn.edu/cgi/viewcontent.cgi?article=1213&context=hms&sei-redir=1& referer=https%3A%2F%2Fscholar.google.ru%2Fscholar%3Fq%3Dcuda%2Bshortest% 2Bpaths%2Bfloyd%26btnG%3D%26hl%3Dru%26as_sdt%3D0%252C5#search=% 22cuda%20shortest%20paths%20floyd%22

The Elbrus Platform Feasibility Assessment for High-Performance Computations

Ekaterina Tyutlyaeva(✉), Sergey Konyukhov, Igor Odintsov,
and Alexander Moskovsky

ZAO RSC Technologies, Kutuzovskiy Av., 36, Building 23, 121170 Moscow, Russia
{xgl,s.konyuhov,igor_odintsov,moskov}@rsc-tech.ru

Abstract. This paper examines the prospects of the Elbrus computing platform for high-performance computations. The results of the most representative HPC benchmarks (HPCC, NPB, HPCG) and their analysis were presented. The testbed node was equipped with four MCST Elbrus-4C processors and DRAM DDR3 with total capacity 48 Gb. Different factors affecting the performance of FT and MG tests from NPB benchmark suite were analyzed by using Paraver tool, hardware performance counters (HPC) and MPI communications data. The scalability of geological application implementing the double-square-root (DSR) prestack migration method was investigated. Benchmark results show that the code customization to reveal platform-specific optimizations is required for the best performance. Nevertheless, the scalability analysis demonstrates that most tests are linearly scalable within a certain range of processor numbers.

Keywords: HPC Benchmarking · The Elbrus Platform · Scalability analysis · HPL · HPCC · NPB

1 Introduction

The main goal of this work is the initial assessment of the computational potential of the Elbrus [1] platform for HPC, and particularly the feasibility of its usage in geophysics.

A simpler way to examine the total performance of the architecture is to use of benchmarking suites for measuring host CPU performance, memory transfer rates, support libraries, drivers, and compilers effectiveness.

In this research, the HPCC, HPCG, NAS parallel benchmarking suites have been chosen for understanding the performance of scientific applications using Elbrus architecture in HPC. The benchmark reports for test runs with different configurations have been analyzed. We have studied not only peak results, but total performance, portability, programmability, and potential built-in capabilities of the architecture.

The execution time measurement is, in fact, the preliminary assessment of the system performance. The tracing of the program execution with the subsequent analysis could be used for the further architecture characterization. This

© Springer International Publishing AG 2016
V. Voevodin and S. Sobolev (Eds.): RuSCDays 2016, CCIS 687, pp. 333–344, 2016.
DOI: 10.1007/978-3-319-55669-7_26

method allows to gather the performance information by monitoring the application run via the microprocessor behavior and intra-node communications. The tracing method can be used for static and dynamic analysis of the application runtime and computation pattern. In this work, we have used the instrumentation package for profiling and tracing of two application with an illustrative internal structure.

Finally, it is well known that the software from different scientific disciplines varies significantly in program structure and dynamic behavior. So the architecture cannot be accurately characterized with the typical benchmarks only. In this work we have measured the scalability of the real application that processes synthetic seismic reflection data by a depth migration method.

2   Features and Specifications of the Computing Node

Features and specifications of the Elbrus Compute Node are summarized in the Table 1.

Table 1. Specifications of the Elbrus Compute Node

Characteristics	Specifications
CPU	4 × Elbrus-4C (Architecture e2k (VLIW); 800 MHz, 4 cores, Theoretical Peak Performance - 4 × 25.6 GFlops (double precision))
Total	16 cores, 102.4 GFlops (double precision)
RAM	12 × Micron DDR3 4GB, 1600 MHz. 48 GB Total
Energy consumption	300 Watt (Under the HPL benchmark)

3   Standard Benchmarks

There are a lot of benchmarks designed to help evaluating the performance of the compute node and to address the interactions among hardware components, system software on the one side and application scientific software on the other side. Benchmark suites are developed for particular programming and parallelization models and implemented by special libraries, so careful selection of the appropriate benchmarks is required for accurate performance prediction.

The chosen benchmarks with results description are shown below. Wide configuration range for problem sizes, blocks, number of threads and processes has been used in all test runs to evaluate not only the peak performance, but the computational effectiveness under different settings. Tuning optimization flags in an architecture-dependent manner has been done for all benchmarks. The mpich-3.1.4 library has been used for communications.

3.1 HPCC

The HPCC benchmark suite [2] is one of the most popular benchmarks suites, it also includes the High Performance Linpack (HPL) benchmark used in the Top500 list. Tests of the HPCC benchmark suite exploit a range of memory access patterns. It consists of four local (matrix-matrix multiply, STREAM, RandomAccess and FFT) and four global (High Performance Linpack—HPL, parallel matrix transpose—PTRANS, RandomAccess and FFT) kernel benchmarks. The maximal HPL benchmark results obtained for the computing node under the test are listed in the Table 2.

Table 2. The HPL benchmark results

Attribute	Value
Problem size (N)	36000
Block size (NB)	176
Process grid ($P \times Q$)	2×8
Time, sec	378.66
Performance, GFlops	82.15

The native EML mathematical library has been used for basic linear algebra calculations. The directory [3] for the hardware support of memory coherence was turned on.

The state-of-the-art scientific application are handling the big data, so the memory measurements is one of the priority areas for benchmarking. The results for the StarSTREAM benchmark that measures sustained memory bandwidth to/from memory are listed in the Table 3.

Table 3. The StarSTREAM benchmark results

Test	Bandwidth, GB/sec				
	1 thread	2 threads	4 threads	8 threads	16 threads
Copy	14.020	8.528	3.990	3.995	3.884
Scale	14.273	8.824	4.245	4.235	4.177
Add	15.412	8.799	4.078	4.063	3.976
Triad	15.865	9.229	4.315	4.313	4.168

3.2 NAS Parallel Benchmark

The NAS Parallel Benchmark suite (NPB) [4] consists of five parallel kernels and three simulated application benchmarks. The benchmarks are derived from computational fluid dynamics (CFD) applications. Although these applications are not typical for all areas of science and technology, it covers a wide range of problem types.

The observations made in the study are described below:

1. The best results using GFLOPs metric are obtained at the BT benchmark (Block Tri-diagonal solver).
2. MPI version demonstrates good scalability as the number of cores increases from 1 up to 16. For example, the IS benchmark demonstrates sub-linear scalability for all versions (MPI, OMP) and all used classes (B, C).
3. The BT and LU benchmarks show similar performance results. Performance results show a small increase with increasing the benchmark class from B to C.

The FT, MG and LU benchmarks was of special interest to us, because these benchmarks are partly reflecting the typical computational pattern, mathematical and computing techniques, and interdependency of computational blocks used in seismic imaging and fluid dynamics fields. The testing results for MPI version are listed in the Table 4, where *process* is MPI-process.

Table 4. The NPB benchmark results

Benchmark	Class	Performance, GFlops				
		1 process	2 processes	4 processes	8 processes	16 processes
FT	B	0.886	1.816	3.588	6.621	10.525
FT	C	-	1.748	3.447	6.402	10.366
MG	B	1.663	3.388	6.506	10.487	16.04
MG	C	-	3.958	7.617	12.25	13.751
LU	B	1.621	3.107	6.295	11.179	19.784
LU	C	-	3.162	6.321	11.390	20.872

As the results of these benchmarks given in the Table 4 show we have observed good scalability despite the different structure of the conducted benchmarks. In many of the benchmarks, the increasing of the problem class slightly affects the total performance what represents balanced share of the workload and good potential for increasing number of nodes in computational cluster.

3.3   The Fourier Transform

The Fourier transform is one of the most common methods used in virtually all areas of engineering and science. There are plenty of seismic image processing algorithms, particularly, that are based on Fourier transform.

We have used special benchmarks to assess the optimization impacts during fast Fourier transformation using FT benchmark from NPB test suite. The "non-optimized" test run has been done with the only "-O3" optimization flag. The "optimized" test run has been done with optimization flags listed below:

Optimization Flags

```
FFLAGS = -O3 -fwhole -mcpu=elbrus-4c -fcache-opt -ffast -ffast-math
```

Moreover, the directory for hardware support of memory coherence was turned on.

The results of comparative benchmarking are listed in the Table 5.

Table 5. The NPB-FT comparative benchmarks results

	Performance, GFlops				
	1 process	2 processes	4 processes	8 processes	16 processes
Non-optimized version	0.515	0.809	1.439	2.589	3.600
Optimized version	0.886	1.816	3.588	6.621	10.525
Acceleration	1,720	2,245	2,493	2,557	2,924

While the source codes have remained unchanged the almost 3-times performance speedup have been gained. To attain the next level of FFT calculations performance we use architecture-optimized programming libraries. There is Fast Fourier transformation library specialized for the Elbrus microprocessors. Tables 6 and 7 show the results of a Fourier transformations benchmarking implemented by specialized library. In all runs one core of the Elbrus-4S CPU was simulated. The *100*Th./Exp.* column contains the Theoretical to Experimental performance ratio percentage. It should be taken into account that basic Fourier Transformation (so called butterfly in the context of the Cooley-Tukey FFT algorithm [5]) takes 10 flops or 8 fmuladd combined operations.

Table 6. Direct fourier transform

Benchmark	Performance, GFlops		100*Th./Exp.
	Theoretical	Experimental	
Complex-to-Complex (16-bit)	8	6.723	84
Complex-to-Complex (32-bit)	8	6.9	86
Complex-to-Complex (64-bit)	4	4.08	97.56
Real-to-Perm (16-bit)	8	5.67	70.9
Real-to-Perm (32-bit)	8	6.299	78.7
Real-to-Perm (64-bit)	4	3.33	83.3

Note: In the tables there are benchmarks with the value of experimental performance higher than the value of the theoretical performance. This effect is possible due to decreasing of arithmetical operations number in the different

Table 7. Inverse fourier transfrom

Benchmark	Performance, GFlops		100*Th./Exp.
	Theoretical	Experimental	
Complex-to-Complex (16-bit)	8	6.723	84
Complex-to-Complex (32-bit)	8	8	100
Complex-to-Complex (64-bit)	4	3.98	99.5
Real-to-Perm (16-bit)	8	5.16	64.5
Real-to-Perm (32-bit)	8	6.67	83.3
Real-to-Perm (64-bit)	4	3.47	86.9

algorithm stages (multiplication by -1.0 is not executed, by it counts as real multiplication in theoretical performance computations). The *type1-to-type2 (N-bit)* notations in the tables refers to the Input and Output data formats. Although the results shown in the Tables 6 and 7 reflect the Fourier Transform performance using one core they can be scaled linearly for small data ($<$2Mb, local L2 cache size), so we have reason to expect e.g. $6.299 * 4 = 25,196$ GFlops performance for the direct Fourier Transform for Real-to-Perm (32-bit) data.

The overheads should affect the performance with data size and number of used CPU cores increasing, nonetheless according to this benchmarking the architecture have significant potential.

4   Study of Parallel Executions Characteristics on the Elbrus Architecture Using Trace-File Analysis

A trace-file analysis allows to collect performance metrics at known points in source code what provides with information about the correlation between performance and the dynamic behaviour of application. In this work we have used the Paraver [6] performance analyzer based on traces generated by the Extrae [7] package.

4.1   FT

The FT benchmark solves a three-dimensional partial differential equation (PDE) using the fast Fourier transform (FFT). In the performance analysis the MPI-version of test application and class C for test problem have been used. There is the linear performance increase with increasing the MPI-processes number from 1 up to 16, so it is not possible to assess the optimal processes number for the FT benchmark in this case (Table 8).

The idle and running time periods are distributed symmetrically by computation processes during the runtime and correlate to the time to communication synchronizations (Fig. 1).

Table 8. The Idle/Running periods for FT benchmark, class C

Number of processes	Idle time, %	Running time, %
2	11.39	88.61
4	5.80	94.20
8	8.99	91.01
16	11.95	88.05

According to The Elbrus architecture specificities the information about *VLIW instructions per cycle* were gathered instead of the standard IPC metric (Instructions per Cycle). The VLIW IPC distribution (Fig. 2) have some kind of symmetry reflecting the good balance of computations and communications during the test execution.

Moreover, we can assume that workload becomes more equally distributed as the number of used computational cores increases (as it is shown on the Fig. 2 the label color of computation intensity shifts from dark-blue to the light-blue and green).

Because of the good balance of computations and communications for FT application in perspective we may expect its further scalability on the Elbrus platform.

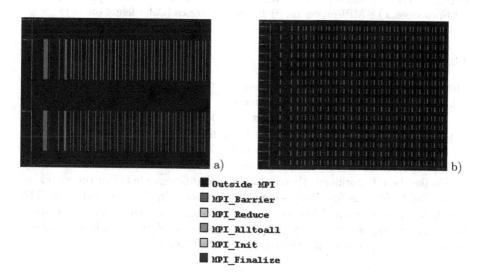

a)   b)

■ Outside MPI
■ MPI_Barrier
☐ MPI_Reduce
■ MPI_Alltoall
☐ MPI_Init
■ MPI_Finalize

Fig. 1. MPI-communication during the FT test runtime, class C; a) 2 MPI-processes, b) 16 MPI-processes

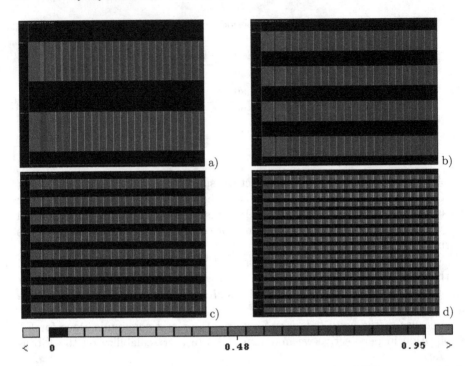

Fig. 2. VLIW IPC during the FT test runtime, class C: a) 2 MPI-processes, b) 4 MPI-processes, c) 8 MPI-processes, d) 16 MPI-processes (Color figure online)

4.2   MG

The MG test approximates the solution to a three-dimensional discrete Poisson equation using the V-cycle multigrid method.

This benchmark requires well structured local and distant communications. We have used MPI-version, class B for trace-file analysis. According to the MPI communication patterns (Fig. 3), it's difficult to distinguish different process groups. The MPI calls and computations period are distributed quite uniformly.

Unlike the FT workload, there is significant difference between the workload level (Fig. 4) for the first and second MPI-process in the first (a) case. The workload equalizes with the processes number increasing what can be explained by the increasing of CPU usage efficiency in the case of 16 cores. It also reflects good scalability potential for the tested architecture.

5   The WEMIG2DMPI Seismic Module

The WEMIG2DMPI module is a part of the GEOLAB [8] software for seismic data processing used in the oil and gas exploration industry. The WEMIG is a 2D-seismic migration method using reverse-time wavefield continuation in frequency/space domains and depth imaging.

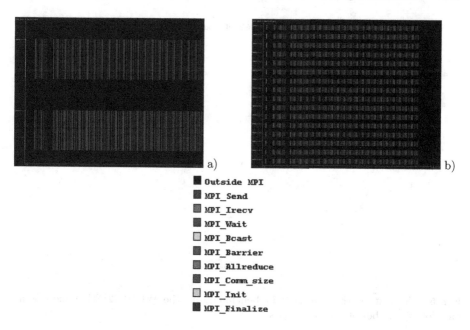

Fig. 3. MPI-communication during the MG test runtime, class B; a) 2 MPI-processes, b) 16 MPI-processes

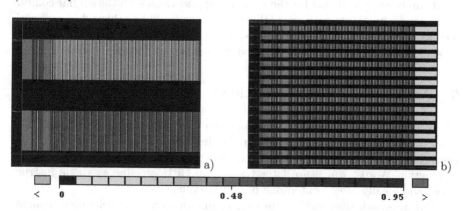

Fig. 4. VLIW IPC during the MG test runtime, class B: a) 2 MPI-processes, b) 16 MPI-processes

This WEMIG2DMPI module has been compiled with following optimization flags:
Optimization Flags

```
-O3 -mcpu=elbrus-4c -mptr64 -ffast -ffast-math
```

The runtime dependency for WEMIG2DMPI module (Fig. 5) shows its good scalability on the Elbrus architecture. Meanwhile the performance decreasing has been observed for the number of MPI-processes higher than the half a number of used CPU cores ($N_{proc} \equiv 8 - 9$).

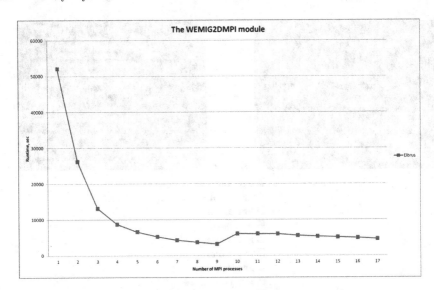

Fig. 5. The test runtime (in sec) in benchmarking the WEMIG2DMPI module as function of number of MPI-processes

It can be supposed that for the scientific applications with the similar computation/communication patterns high performance could be achieved as well with increased number of computation cores. Nevertheless, primarily the architecture-dependent optimization required including specialized libraries.

6   Architecture Differentiation

The Elbrus microprocessor family belongs to the VLIW architecture class. The main VLIW feature is explicit instruction-level parallelism. Different instructions that allow simultaneous execution, are statically scheduled into Very-Long Instruction Words, which are later treated by CPU pipeline as single instructions, and ideally are executed one per CPU cycle.

This approach simplifies the process of microprocessor development comparably with the superscalar architecture, especially supporting out-of-order execution, but it transmits the task of parallelism detection to the compiler (or to the qualified assembler developers).

The execution performance in terms of IPC (Instructions Per Cycle) for the Elbrus (and other VLIW architectures with explicit speculative instructions execution) for one session is defined as follows:

$$IPC(program) = \sum_{WI \in program} \frac{[\sum_{oper \in WI} prob(oper)] * C_{exec}(WI)}{Cycles(program)}, \quad (1)$$

where WI – Very–Long Instruction Word;

$C_{exec}(WI)$ – The number of Very-long Instruction Word executed in one session;

$Cycles(program)$ – The total number of cycles during execution;

$prob(oper)$ – the operation "profitableness"– a complex numeric characteristic that reflects the level of usefulness of operations. It can roughly be described as probability of speculative operation result not being thrown away as unneeded in data flow. For non-speculative operations prob(oper) = 1.0.

Consequently, the effective performance depends on:

1. Width (number of operations) of the Very-Long Instruction Word
2. High operations profitableness in the Very-Long Instruction Word
3. Presence of the pipeline stalls (bubbles) that increase Cycles(program).

The optimizing compiler is responsible for providing positive factors for effective performance, but it still requires additional efforts from the developers.

Combining the developers and compiler efforts one could achieve high performance on VLIW architectures for wide range of algorithms.

7 Conclusion

In this work we have studied the computation behavior of different benchmarks including pseudo-real applications. The applications with different patterns demonstrates good scalability on the Elbrus platform. The presented results especially for FFT calculations allow comparison with the modern CPUs and systems.

The trace analysis of some standard benchmark tests has shown effective CPU utilization during the runtime and tendency to workload equalization with increasing of number CPU cores.

The architecture principles has been analyzed to better understand of the benefits and the pitfalls of the usage of the Elbrus platform. The Fourier transform example illustrates the importance of architecture-dependent optimization.

Proper optimization of some benchmark tests brings out the architecture performance potential of the Elbrus platform.

In further work We are planning to apply architecture-dependent optimization to seismic imaging applications.

Based on the results of our benchmark tests We can presumably expect that the increasing of number of computation cores can lead to further applications scalability and performance growth.

Acknowledgments. This research was supported by the Common State Scientific and Technological Programme "SKIF-Depths" with funding from Ministry of Education and Science of the Russian Federation.

We thank our colleagues from MCST/INEUM, particularly Murad Iskender-Ogly, Alexander Breger, and Sergey Zotov for the consultations.

References

1. The Elbrus CPU technical specification (in Russian). http://www.mcst.ru/mikroprocessor-elbrus4s
2. Dongarra, J., Luszczek, P.: HPC challenge: design, history, and implementation highlights. In: Jeffrey, V. (ed.) Contemporary High Performance Computing: From Petascale Toward Exascale. Taylor and Francis, CRC Computational Science Series, Boca Raton, FL (2013)
3. Petrov, I., Sherstnev, A.: Realization of a directory for hardware support of coherence in the computer system on the basis of microprocessor "Elbrus-2S". Questions Radio-Electron. (3) (2011). (in Russian)
4. Bailey, D., et al.: The NAS parallel benchmarks: Technical report, Moffett Field, NASA Ames Research Center, p. 79 (1994)
5. Cooley, J.W., Tukey, J.W.: An algorithm for the machine calculation of complex Fourier series. Math. Comput. **19**, 297–301 (1965)
6. Paraver, Performance Analysis Tools: Details and Intelligence. http://www.bsc.es/computer-sciences/performance-tools/paraver
7. Paraver, Trace-generation package. https://www.bsc.es/computer-sciences/extrae. Accessed 22 Aug 2016
8. Technology and the GEOLAB Software. http://geolab-it.ru/en/technology-en.html

Using Machine Learning Methods to Detect Applications with Abnormal Efficiency

Denis Shaykhislamov[(✉)]

Lomonosov Moscow State University, Moscow, Russian Federation
sdenis1995@gmail.com

Abstract. At the moment a lot of supercomputing applications are inefficient in terms of the usage of available resources. To decrease the number of such inefficient applications, a tool for supercomputer task flow analysis and detection of inefficient application runs is needed. In this paper several supervised machine learning methods are considered to solve this issue. The classification performed by these methods is based on system monitoring data (e.g. CPU load, network usage etc.). The experiments on real data show that the Random Forest algorithm is currently the best option to accomplish given goal. At the moment the resulting classifier model is being tested on the "Lomonosov" supercomputer. The experiment results demonstrating the efficiency of the resulting model are also included in this paper.

Keywords: Supercomputer · High performance computing · Task flow · Anomaly detection · Program efficiency · Machine learning

1 Introduction

It is very essential that supercomputer's resources are used efficiently. Inefficient usage leads to partially idle resources that could have been used much more efficiently by other users. Also users get results later than expected due to slower execution of their programs. Considering all this, the detection of inefficient applications is essential. In order to solve this problem, a tool that analyzes the supercomputer task flow, detects inefficient application runs and notifies users about these applications is being developed.

There are different tools that help to analyze the behavior of a single program (e.g. Scalasca, Vampir [12], HPCToolkit), and these tools can be used to detect different kinds of efficiency issues. But the problem is that users very often don't even realize that their applications can run inefficiently, so these tools are rarely used. There are also tools like NuPIC [1] and Rocana [2] that detect anomalies in the data flow in real time, but they can only analyze the data flow presented by {time:value}. So these tools can be used to analyze only one dynamic characteristic (e.g. CPU load, memory usage) and they can only point to the fact that something is unusual. But in future we also want to know why these unusual events occur. Moreover, some anomalies can be detected only by analyzing multiple dynamic characteristics. Taking all this into account, these tools are not suitable for solving given problem.

V. Voevodin and S. Sobolev (Eds.): RuSCDays 2016, CCIS 687, pp. 345–355, 2016.
DOI: 10.1007/978-3-319-55669-7_27

I searched for papers about supercomputer task flow analysis and real time anomaly detection using machine learning techniques but no researches on this topic were found. Although some related papers about the usage of machine learning methods in program efficiency analysis can be found [3–5]. For example, in the paper [5] supervised machine learning methods were used to classify launched applications. Authors tested the machine learning algorithms on the problem of finding the inefficient programs in the large set of programs to select a suitable algorithm for their task. But they used their own criteria of what should be called an "inefficient program", e.g. CPU load <30, clock ticks per instruction <2. This kind of anomaly detection is very simple and detects only obvious types of anomalies.

2   Definition of Anomaly

First step was to define the term "anomaly". An anomaly is the application that uses the resources of supercomputer very efficiently or inefficiently so it stands out from the general supercomputer task flow. Finding the exact criteria of application abnormality is not possible at the moment because there is a vast amount of applications on supercomputer that are very different by their properties and the problems they solve. That was one of the main reasons for using machine learning.

There are different kinds of anomalies:

1. Very efficient and very inefficient programs. Very efficient program makes good use of the supercomputer resources and runs a lot more efficiently than an average program, i.e. it optimally distributes work between nodes, minimizes cache misses, etc. Very inefficient program usually wastes resources due to its inefficient implementation of the algorithm or an error occurred in the program, i.e. deadlock, infinite loop, etc. It would be better to detect not only inefficient programs but also efficient ones because these programs could be analyzed to make a recommendation list for other users on how to write efficient programs.

2. Anomalies in the supercomputer task flow, in the user task flow and within a single program. Anomalies in the supercomputer task flow are efficient or inefficient programs that differ greatly in comparison with an average program in the task flow. Anomalies in the user task flow are programs that are different from the other programs of the particular user. This kind of anomaly is interesting because it can be assumed that usually user launches similar programs that are comparable. If there is a program that is drastically different from the others then there is a possibility that this program behaves not as planned and should be treated as suspicious. Also these kinds of programs can be very similar to many other programs in supercomputer task flow and often are labeled as normal while analyzing the overall flow, but they are considered as abnormal to the particular user. Example of the outliers in the user task flow is given in Fig. 1. An anomaly within a single program is found not by comparing it to other programs but by analyzing its dynamic characteristics' behavior during the execution. This can lead to discovery of dependencies between dynamic characteristics for different types of inefficient programs.

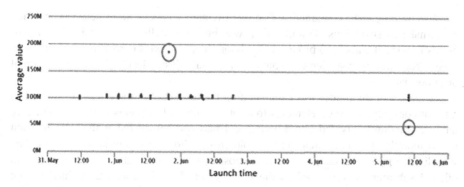

Fig. 1. Example of the anomalies in the user task flow. Each dot represents the user task with given average number of memory writes per second.

In this work it was decided to focus on detecting anomalies in the overall supercomputer task flow.

3   The Use of Machine Learning Methods

All tested machine learning techniques required applications to be presented in a unified format. In this case all supercomputer application data is collected by the monitoring system of the "Lomonosov" supercomputer. It gives access to the following dynamic characteristics:

- level 1 cache misses per second (cache_1),
- level 3 cache misses per second (cache_3),
- system load (Loadavg),
- CPU load (cpu_load),
- memory load operations per second (mem_load),
- memory store operations per second (mem_store),
- amount of received bytes per second via Infiniband (ib_rcv_data),
- amount of received packets per second via Infiniband (ib_rcv_pckts),
- amount of sent bytes per second via Inifiniband (ib_xmit_data),
- amount of sent packets per second via Infiniband (ib_xmit_pckts).

Every dynamic characteristic is represented by time series and every application has different amount of elements in series due to different execution times. That's why it was decided to use feature based classification. Two features where chosen to represent time series: median and oscillation rate. Other features like average, quantile, dispersion were also considered, but they gave less accurate results.

It was decided to compare the following supervised machine learning classifiers: Random Forest [8], Linear Discriminant Analysis [6] and Decision Tree. For Decision Tree, the CART [7] algorithm was used. Other classifiers like Naïve Bayesian or AdaBoost were also tested but they showed worse results than tested methods.

The training data contained 300 programs and 3 classes were selected: normal, abnormal and suspicious. Training data was built manually and contained 110 programs in normal class, 130 programs in abnormal class and 60 programs in suspicious class. Program was considered suspicious if it couldn't be clearly classified as normal or abnormal.

Accuracy of the model was calculated on the test set. Accuracy was considered as ratio of correctly classified elements to a total number of elements in the set. Training data was randomly divided into 4 parts and 1 part was considered as the test set. Then the method that is very similar to the 4-fold cross-validation was used. Remaining data was divided into 4 subsamples, a single subsample was retained as validation set, and other 3 subsamples were used as the training set. The classifier was trained on the training set and accuracy on validation set was calculated. It was done 4 times, for every possible subsample as the validation set. The best out of 4 trained classifiers is chosen based on the accuracy on corresponding validation set. Then the accuracy of the chosen best model on the test set is calculated. This accuracy of the classifier model was considered as its accuracy on test set. Chosen classifiers show these results:

- Linear Discriminant Analysis: 0.74 on the training data.
- Decision Tree: 0.75 on the test set.
- Random Forest: 0.82 on the test set.

Linear Discriminant Analysis was considered not suitable for the problem because this algorithm showed very poor results even on the training data. Decision Tree and Random Forest achieved better results, but Random Forest always showed better results than Decision Tree classifier, and that lead to choosing Random Forest as the final classifier for the system. Also because of the fact that Random Forest is an ensemble of trees and classification is done by voting, the information about how many trees voted for each class can be retrieved. This information can be useful: while classifying the element, if all trees voted for a particular class then this element is considered as normal and most probably classified correctly; but if trees voted for different classes then this could mean that this element has unique properties and should be analyzed more properly.

One of the main parameters of Random Forest classifier is the number of trees in the ensemble. Multiple numbers of trees were tested. The accuracy increased until the number of trees reached 256 and didn't increase afterwards. That's why in this work the parameter was set to 256.

Next step was to find out if the accuracy of the classifier can be increased. ROC curve graphs for Random Forest and Decision tree classifiers were built (Figs. 2 and 3) that show how well classifiers predict different classes.

The area under the curve also shows how well the classifier works – is predicts better if the area under the curve is higher. As can be seen, both Random Forest and Decision Tree classify suspicious programs a lot worse than normal or abnormal programs. After analyzing distributions of dynamic characteristics it was found out that suspicious program's behavior sometimes very alike to the normal or abnormal programs. Due to this reason the classifier sometimes mislabels the program. The confusion matrix (Table 1) that was built using the test set also shows that suspicious programs are very difficult to classify correctly.

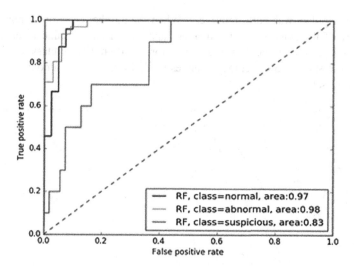

Fig. 2. ROC curve for Random Forest classifier

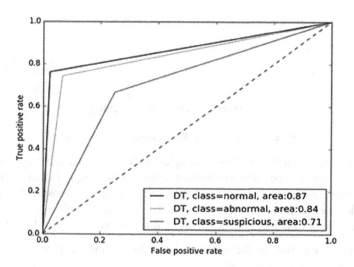

Fig. 3. ROC curve for Decision Tree classifier

Table 1. Confusion matrix for Random Forest classifier

		Classified as		
		Normal	Abnormal	Suspicious
Actual classes	Normal	25	0	2
	Abnormal	1	35	1
	Suspicious	3	4	6

The training data was built iteratively. When the number of samples reached 100, the accuracy of the classifier didn't increase and stayed on the same level. It is best shown on the graph of the learning curve for Random Forest classifier in Fig. 4 (learning curve shows how accuracy changes while increasing the amount of elements in the training set).

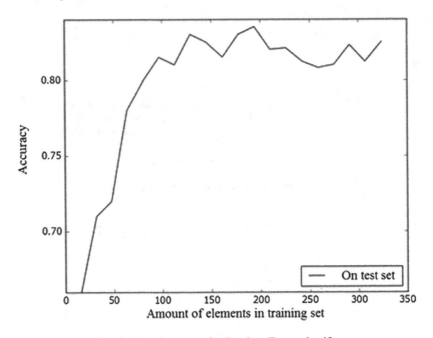

Fig. 4. Learning curve for Random Forest classifier

It is clear that when the number of samples is above 100 the accuracy stays on the same level. There is a possible explanation related to the misprediction of suspicious programs that are alike with normal and abnormal programs. Increasing the amount of samples does not solve the problem of similarity of classes. That's why other methods were tested trying to increase the accuracy. These two methods have been tried:

- Dimension reduction of the samples, i.e. usage of only a part of the dynamic characteristics that are available.
- Usage of new dynamic characteristics derived from available characteristics.

Linear Discriminant Analysis was used for the dimension reduction, but the method showed that all dynamic characteristics are important and dimension reduction only decreases the accuracy. For example, the accuracy dropped from 0.82 to 0.76 when dimension had been halved.

Then the method with derivation of new dynamic characteristics was tried out. First of all, it is unclear which dynamic characteristics should be used. Two methods were tried: (1) manually select some dynamic characteristics that can presumably increase the accuracy; and (2) automatically search for the dynamic characteristic with defined formula that gives the best accuracy.

Five derived characteristics were manually added and it increased the accuracy of the Random Forest classifier up to 0.835 with the same parameters. Following dynamic characteristics were used:

- ln(ib_rcv_pckts_median/ib_rcv_pckts_oscil)
- ib_rcv_pckts_median/ib_xmit_pckts_median
- cpu_user_median/cpu_user_oscil
- ib_xmit_data_median/ib_xmit_pckts_median
- ln(cache_3_median/cache_3_oscil)

Also the best dynamic characteristic described as the ratio of two dynamic characteristics was searched. While using one or two best dynamic characteristics found by brute force, the accuracy increased up to 0.83 and 0.845 correspondingly. It is very difficult at the moment to search for 3 best dynamic characteristics because the complexity of the algorithm grows exponentially. It was decided to stick to the manually selected dynamic characteristics because we understand well what they indicate. Alternatively, derived dynamic characteristics found by brute force seem random, and currently it is unclear for us why they increase the accuracy (e.g. ib_xmit_data_median/loadavg_oscil).

At the moment the Random Forest classifier is used with the accuracy of 0.835 on the test set.

4 Proposed Method Implementation and Evaluation

The diagram of developed system is shown in Fig. 5.

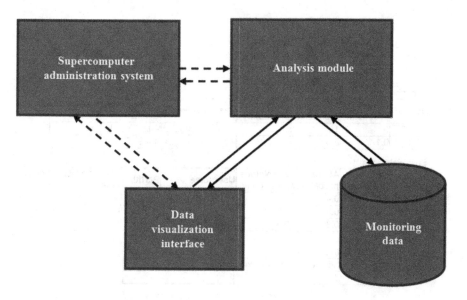

Fig. 5. System's diagram. Solid lines show implemented links, dashed lines show currently not implemented links.

Analysis module and data display interface were developed. In this work super-computer administration system is Octoshell [9], a tool also being developed in the research center in Moscow State University. Link between analysis module and supercomputer administrations system will be essential in future when system will have to promptly alert users about detected programs.

This project is implemented using Python 2.7 language. Scikit-learn [10] library was used because of its variety of optimized machine learning algorithm implementations.

The overall classification algorithm is arranged as follows:

1. The "Lomonosov" supercomputer has monitoring system that collects the data on every program that is being executed at the moment. Monitoring system has its own database where the basic information about the programs is stored (i.e. amount of cores, launch time, partition name). When program finishes its execution, new record in database is created. All the data on dynamic characteristics is held in raw format, i.e. <timestamp>; <node>; <mean value>; <max value>; <minimal value>. Monitoring system saves data every 5 min which means it collects data for 5 min and then calculates mean, maximum and minimum values for every node.
2. Every 30 min the tool described in this paper scans the monitoring system database to detect finished programs.
3. Raw monitoring system data is processed and mean, median, maximum and minimum values are calculated for each program.
4. After the data is collected, trained model classifies the program.

At the moment this application is running on the "Lomonosov" supercomputer and classification is done in real time. Because of the fact that data is collected and programs are classified every 30 min, the amount of programs is small which means that classification works very fast (fractions of seconds).

Abnormal jobs

Abnormal jobs :

Job ID	Normal probability	Abnormal probability	Suspicious probability	Username	Partition	Execution time	Number of cores
13142	0	1	0		regular4	2228.9333	488
13157	0.1	0.5	0.4		regular4	290.9000	8

Suspicious jobs :

Job ID	Normal probability	Abnormal probability	Suspicious probability	Username	Partition	Execution time	Number of cores
13134	0.1	0.4	0.5		regular4	4107.7000	904
13158	0.1	0.3	0.6		regular4	61.5000	256
13158	0.1	0.3	0.6		regular4	61.1333	256

Basic info (.2016) :

Total number of jobs	Number of abnormal jobs	Number of suspicious jobs
54	2	3

Fig. 6. Example of the report sent by email

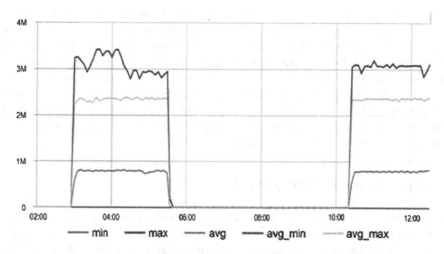

Fig. 7. Number of level 3 cache misses per second

Every day the report is formed based on the detected abnormal and suspicious programs. It is then sent to the supercomputer administrators by email. The example of the report is shown below (Fig. 5). Normal, abnormal and suspicious probabilities are the portions of trees in the classifier that have selected given class.

An example of the detected anomaly is shown below. Figures 6 and 7 show the graph of the number of level 3 cache misses per second and the number of writing to a memory per second correspondingly. These graphs were given by the system of report generation JobDigest [11] that also uses monitoring data. This system was very helpful in building training set for the classifier.

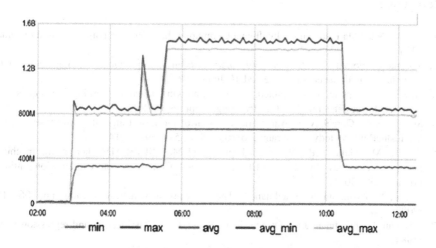

Fig. 8. Number of memory writes per second

All other dynamic characteristics have the same behavior as amount of level 3 cache misses per second. As can be seen the application's behavior is not usual. First hour of the execution dynamic characteristics have very low values but it can be explained by data loading and distribution of the work between the nodes. The behavior of the program is normal for the next few hours but then it becomes very suspicious due to complete lack of level 1 and 3 cache misses. There is a possibility that the data perfectly fits in the cache memory, but this possibility is very slim and it much more likely indicates an error in the program. This program behavior is at least suspicious and the user must be warned about it (Fig. 8).

5   Conclusion and Future Work

In this work, the tool that analyzes "Lomonosov" supercomputer task flow and detects inefficient programs was developed. At the end of the day the report is formed that contains all information about the programs labeled as suspicious or abnormal. The report is sent to administrators. This system is currently being tested on the "Lomonosov" supercomputer and it shows very accurate results (e.g. all detected programs are in fact abnormal or suspicious).

This system only detects if a program has abnormal behavior or not. Future works may involve detecting the reason why program behaves that way and testing other methods to analyze the monitoring data. Also this system detects anomalies only in the overall task flow; it is planned to learn to detect anomalies in the user task flow as well.

Acknowledgements. I thank my research supervisor Vadim Voevodin for helping me with this work. This research is supported by RFBR grant №16-07-00972 and Russian Presidential study grant (SP-1981.2016.5).

References

1. NuPIC: Numenta Platform for Intelligent Computing. http://numenta.org. Accessed 06 June 2016
2. Rocana: Anomaly Detection. https://www.rocana.com/products/technology/advanced-analytics-anomaly-detection. Accessed 10 June 2016
3. Aleem, S., Capretz, L.F., Ahmed, F.: Benchmarking machine learning techniques for software defect detection. Int. J. Softw. Eng. Appl., 6(3) (2015)
4. Sidnev, A.A., Gergel, V.P.: Automatic selection of the fastest algorithm implementations. Vychislitel'nye Metody i Programmirovanie 15(4), 579–592 (2014)
5. Steven, M., Gallo, J.P.: White analysis of XDMoD/SUPReMM data using machine learning techniques. In: 2015 IEEE International Conference on Cluster Computing, pp. 642–649 (2015)
6. Hastie, T., Tibshirani, R., Friedman, J.: The Elements of Statistical Learning, pp. 106–119 (2008)
7. Breiman, L., Friedman, J.H., Olshen, R.A., Stone, C.J.: Classification and regression trees (1984)
8. Breiman, L.: Random forests. Mach. Learn. 45(1), 5–32 (2001)

9. Nikitenko, D.A., Voevodin, V.V., Zhumatiy, S.A.: Octoshell: large supercomputer complex administration system. In: CEUR Workshop Proceedings on Russian Supercomputing Days International Conference, Moscow, Russian Federation, 28–29 September 2015, Moscow, vol. 1482, pp. 69–83 (2015)
10. Scikit-learn, machine learning in Python. http://scikit-learn.org/stable/. Accessed 15 Sept 2015
11. Adinetz, A.V., Bryzgalov, P.A., Voevodin, V.V.: JobDigest – approach to jobs dynamic properties investigation on supercomputer systems. Vestnik UGATU **17**(2), 131–137 (2013)
12. Nagel, W.E., Arnold, A., Weber, M., Hoppe, H.-C., Solchenbach, K.: VAMPIR: visualization and analysis of MPI resources. Supercomputer **1**, 69–80 (1996). SARA Amsterdam

Using Simulation for Performance Analysis and Visualization of Parallel Branch-and-Bound Methods

Yury Evtushenko, Yana Golubeva, Yury Orlov,
and Mikhail Posypkin$^{(\boxtimes)}$

Dorodnicyn Computing Centre, FRC CSC RAS, Moscow, Russia
evt@ccas.ru, golubeva.yana.v@gmail.com,
justice1786@gmail.com, mposypkin@gmail.com

Abstract. The Branch-and-Bound (B&B) is a fundamental algorithmic scheme for a large variety of global optimization methods. For many problems B&B requires the amount of computing resources far beyond the power of a single-CPU workstation thus making parallelization almost inevitable. The approach proposed in this paper allows one to evaluate load balancing algorithms for parallel B&B with various numbers of processors, sizes of the search tree, the characteristics of the supercomputer's interconnect. The proposed approach was implemented as a special tool that simulates the process of resolution of the optimization problem by B&B method as a stochastic tree branching process. Data exchanges are modeled using the concept of logical time. The user-friendly graphical interface can render both real traces and ones produced by the simulator. It provides efficient visualization of the CPU's load, data exchanges and progress of the optimization process.

Keywords: Performance analysis and simulation · Parallel computing · Global optimization · Branch-and-Bound methods · Load balancing

1 Introduction

The Branch-and-Bound method (B&B) is one of the main approaches to the resolution of mathematical programming problems [1, 2]. In contrast to heuristic and stochastic methods, B&B ensures the accuracy of the found solutions and, in some cases, can solve the problem exactly. For realistic problems B&B can consume computational and time resources, significantly exceeding the available capacity. Parallel computing can be used to speed up and reduce the memory requirements for B&B implementation. Balancing computational load between processors plays an important role in the parallel implementation of global optimization methods [3–5]. Typically load balancing means transmission of jobs from one processor (core) to another along the computations.

Today most powerful supercomputers contain 10^6 computational cores and this number continues to grow thus making load balancing a very challenging problem. There is a clear demand for deep and systematic study and comparison of various load

V. Voevodin and S. Sobolev (Eds.): RuSCDays 2016, CCIS 687, pp. 356–368, 2016.
DOI: 10.1007/978-3-319-55669-7_28

distribution strategies. Performing such evaluation on a real multiprocessor computing system requires multiple runs on the very expensive equipment. We propose to use the simulator for these purposes. The simulator allows one to study performance of load balancing algorithms with various numbers of processors, sizes of the search tree, the characteristics of the supercomputer's interconnect. The process of resolution of the optimization problem by B&B method is replaced by a stochastic branching process. Data exchange and computations are modeled using the concept of logical time.

Another important problem is an adequate visualization of the algorithm performance. To address this issue we developed a user friendly graphical interface that enables convenient performance analysis through processor load charts communication tables and aggregate statistics.

2 Related Works

Tools for automated performance analysis and visualization play and important role in parallel application lifecycle. Monitoring systems Ganglia [6], Nagios [7], DiMMon [8] etc. are aimed at collecting and presenting to the user the overall information on supercomputers' performance, including CPU load, memory consumption and data exchange traffic. Such tools are indispensable for assessing the performance of computational clusters or grid systems.

Another family of tools such as TAU [9], HPCToolkit [10], Paraver [11], Vampir [12], HOPSA [13] focus on the performance of individual parallel applications. They can trace and analyze profiling information, including pipeline stalls, cache misses, inter-cache communication in multi-core and multi-socket configurations. For distributed memory system they also collect data exchange information, analyze and visualize the performance of message passing. A good survey of parallel application performance measurement tools can be found in [14].

The tools outlined above can be very helpful for application performance analysis of virtually any parallel application. Unlike this general purposed approach we focus on a particular class of applications: parallel branch-and-bound solvers. Though the visualization part has a lot in common with the mentioned software the data collection part is completely different. Our tool simulates the process of resolution of the optimization problem by B&B method as a stochastic tree branching process. Data exchanges are modeled using the concept of logical time. Such an approach enables rapid evaluation of various load-balancing algorithms without expensive runs on a real supercomputer. The developed tool could serve as a good problem specific addition to other performance analysis software.

To the best of our knowledge the only tool for parallel B&B simulation was proposed in [15]. This tool used Unix fork/exec mechanism to simulate multiprocessing and pipes to simulate message-passing. This approach works well for moderate number of processes. However this approach has a very poor scalability since the overhead of running thousands of independent processes in modern OSes can remarkably impact the performance thereby affecting the accuracy of simulation.

3 Distributed Memory Branch-and-Bound Implementation

The goal of global optimization (GO) is to find an extreme (minimal or maximal) value $f^* = f(x^*)$ of an objective function f (x) on a feasible domain $X \subseteq R^n$. The value f^* and feasible point $x^* \in X$ are called optimum and optimal solution respectively. Without loss of generality one can consider only minimization problems:

$$f\ (x) \to \min, x \in X \qquad (1)$$

The Branch-and-Bound (B&B) is a general name for methods to split an initial problem into subproblems which are sooner or later eliminated by bounding rules. Bounding rules determine whether a subproblem can yield a solution better than the best solution found so far. The latter is called the incumbent solution. Bounding is often done by comparing lower and upper bounds: a subproblem can be pruned if the lower bound for its objective is larger or equal to the current upper bound, i.e. incumbent solution.

Numerous Branch-and-Bound algorithms were developed for different global optimization problems. Some of them were very successful for particular problem kinds, e.g. Travelling Salesman or Knapsack problems. However for many problems Branch-and-Bound methods require the amount of computing resources beyond the power of a single-CPU workstation. Fortunately Branch-and-Bound is highly suitable for parallel and distributed computing: after splitting the parts of the solution space can be processed independently and simultaneously.

Another great advantage of B&B methods is that the general scheme does not significantly vary from one problem to another. The splitting and bounding rules may differ while keeping the general scheme almost intact. The direct consequence of this is the possibility to separate problem-independent and problem-specific parts. Such separation saves a lot of efforts when implementing a new problem or a new method. This is especially true for tools targeted at parallel and distributed environments because the "parallel" part is reused for different optimization problems. We follow this approach in our tools: the computing space management, the work-distribution and communication among application processes is problem-independent.

Our parallel library for global optimization BnB-Solver [16] is built on top of MPI [17] which implies that parallel processes communicate via message-passing. Each process do three basic kinds of activity: performing steps of B&B method, sending data and receiving data. Transmitted data consists of sub-problems and/or incumbent solutions and commands. Exchanging sub-problems performs computations redistribution among processes in order to make the load more or less even. Sending incumbents ensures fast error propagation among parallel processes.

According to the aforementioned concepts managing the resolution process including data exchanges can be encapsulated in a special component called the *scheduler*. The problem-specific part is managed by another component – the *solver* that provides methods to solve the problem, read its state (the number of subproblems in a queue) fetch and extract subproblems. Sending and receiving of subproblems is implemented by the *communicator* component. These parts are composed together by a special bridge class that invokes respective methods of the scheduler, the solver or communicator (Fig. 1).

The proposed approach separates the managing part from implementation details part thereby providing an opportunity for an independent schedulers testing and verification.

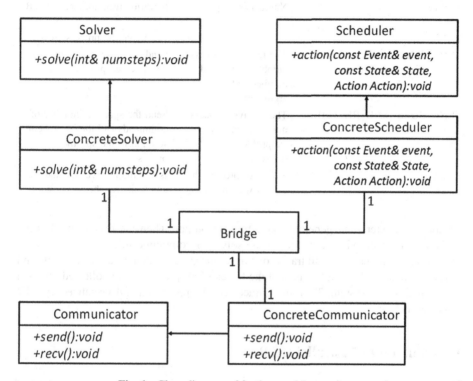

Fig. 1. Class diagram of fundamental interactions

The scheduler is a finite state machine that accepts events and issues actions. Possible events and actions are listed in Tables 1 and 2 respectively. The bridge invokes method action() of the scheduler class that accepts an event and the solver state

Table 1. Event types

Event type	Arguments	Description
ERROR	Error code	An error occurred
START		The beginning of computations
DONE	The real number of steps done	The requested number of steps done
SENT	The number of transmitted items	The requested sending message action done
DATA_ARRIVED	The process that sent the data	The receive command finished and the requested data received
COMMAND_ARRIVED	The process that sent the data	The command arrived

Table 2. Action types

Action type	Arguments	Description
SOLVE	Number of steps	Perform given number of B&B steps
EXIT		Terminate the process
SEND_COMMAND	The receiver process number, command number and arguments	Send the given command to the specified process
SEND_SUBS_AND_RECORDS	The receiver process number, number of subproblems to transmit	Send the specified number of commands and the incumbent solution to the specified process
RECV	The id of process the data is waited from	Issues the receive command and waits for the message

as input parameters and generates an action on output. Then the bridge invokes the methods associated with the action of the solver or communicator.

The scheduler can trace all transitions from one state to another, actions, events and their arguments. If logging is enabled the traces of all processes are collected, merged and written to file system. Then these traces can be processed and visualized by GUI described below.

4   Simulation of Parallel B&B

The simulator was designed for convenient fast and efficient performance testing of parallel schedulers. The simulator uses the real scheduler which is taken intact from the library and provides 'fake' implementations of the solver and the communicator. This approach enables the rapid testing of the schedulers on large trees and thousands of processors because the time consuming resolution steps and communications are substituted by formal actions which take nearly zero time.

The parallel processing is simulated serially. For each simulated process the instance of the scheduler is created. The simulator cyclically iterates through these instances and invokes *action()* methods. If the action is SOLVE then the specified number of steps is simulated and the logical clock is increased according to the modelled time. The B&B method is substituted by a random branching process where the node generates two new nodes with a probability decreasing with distance between the tree root and the node. When the node reaches the maximal tree depth the probability becomes zero. Thus the maximal tree depth controls the size of the whole tree. The time of solving is modelled using the simple formula:

$$t = n\,t_s, \tag{2}$$

where n is the number of performed steps and t_s is the time of one step.

The data transmission is simulated using the concept of logical clock [18]. When the SEND_SUBS_AND_RECORDS command is issued the communicator object stores the message and its timestamp obtained by increasing the current time on a process by the modelled time of a message transmission. The time required to transmit the message is computed by the following formula:

$$t = S t_p + L + S/B, \tag{3}$$

where S is the size of the message, t_p is time needed for packing a unit of data at a sender process, L is the network latency, defined as the time needed to transfer the minimal amount of data throughout network and B is the bandwidth – the amount of data transmitted through the network in a unit of time.

When the RECV command is issued by a scheduler the recipient process the communicator looks up for available messages for this process and if one is encountered it compares the logical time on a recipient tR with the message time stamp tS. The logical time on a recipient is adjusted to the maximum of these values and the obtained value is increased by time required to unpack the message:

$$t_R = \max(t_R, t_S) + S t_u, \tag{4}$$

where S is the size of the message, t_u is time needed for unpacking a unit of data on the recipient.

During the simulation all events and actions are logged. The log files contain all information about logical time of various simulated events. This information is used by graphical user front-end described in the next section.

5 Graphical Front-End

The log files are not suitable for direct analysis by a human. The graphical front-end is aimed at user-friendly graphical visualization and performance analysis of traces produced by either simulator or the real solver. Based on the collected traces the GUI performs the following activities:

- visualizes processors' loads;
- visualizes data exchange among processors;
- computes aggregated performance information such as speedup and efficiency.

Figure 2 shows the window demonstrating processor load plots for individual processors. At the bottom of the window there is a slider similar to one used in multimedia players. It allows an easy and natural navigation throughout the trace. Such representation is convenient for a moderate number of processors. However for hundreds and thousands of processors it can be very inefficient. For such cases BNB-Visualizer provides the processor grid (Fig. 3) which scales well. Blue color is used for depicting computations, red color marks processors blocked in the receiving state. Green color means the processor is sending data.

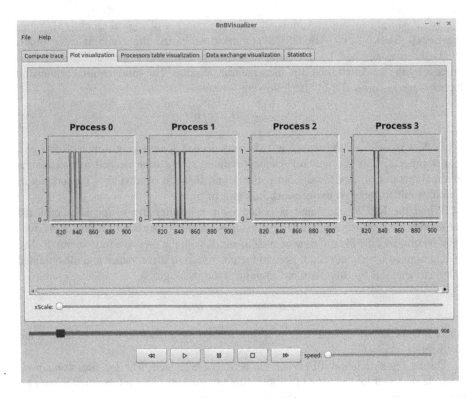

Fig. 2. Processors' load plots

Communications are visualized using two-dimensional chart where processors are aligned along horizontal (senders) and vertical (receivers) axes. The receive actions are visualized by a horizontal blue line and the send action is represented by a vertical green line (Fig. 4). At the Figure lines (1) and (2) correspond to a successful message transmission from the process 9 to the process 0. Line (3) depicts the unsatisfied send issued by the process 0.

The cumulative information about the processors' usage and performance metrics is shown in a separate tab (Fig. 5). This performance chart shows the number of processors occupied at the given moment of time (blue color) and the number of free processors (green color).

6 Experiments

6.1 Case Study I: Selecting Best Parameters for Adaptive Load Balancing

The simulator was used to study the comparative performances of a family of load balancing algorithms working as follows. At the initial phase the 1st (*master*) processor generates some number of sub-problems. At the second stage each of remaining processors (*slaves*) gets a sub-problem from the master and starts its resolution. The

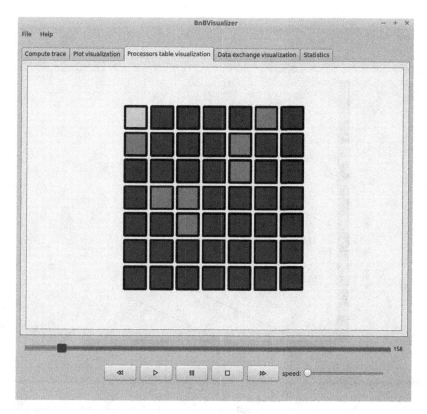

Fig. 3. Processors' grid (Color figure online)

solution process on a slave is interrupted each T iterations and then the slave sends S sub-problems or less to the master. If there are remaining sub-problems on a slave it resumes B&B method. The master processor stops receiving sub-problems from slaves when the number of sub-problems in its pool exceeds M and resumes receiving when it drops below m. This is done by setting parameter S to 0 or to its original value.

Figure 5 shows the performance chart for small values of T. The very intensive data exchange among parallel processes doesn't yield good performance because of large communication expenses.

For moderate values of T the performance is better but we can see significant performance losses at the final stage of the algorithm (Fig. 6). In the middle of the computational process the load balance is good but at the terminal stage it is quite bad.

The natural solution to avoid such performance losses is to introduce dynamic adaptation: when the number of subproblems on the master drops below the number of free processors the parameter T is decreased in 10 times. Thus at the middle of computations when the demand for load redistribution is small T is kept relatively large. At the final stage T decreases in order to provide good load balancing among process through intensive exchange of subproblems. This leads to a better performance (Fig. 7).

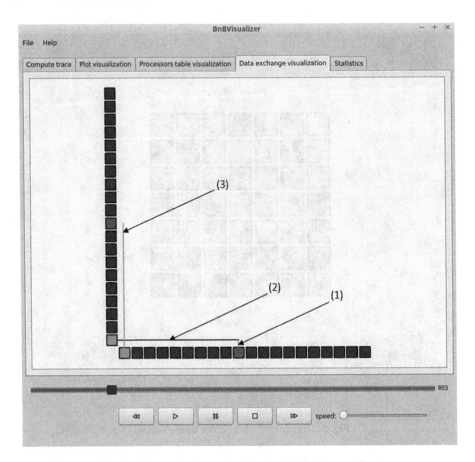

Fig. 4. Communications visualization (Color figure online)

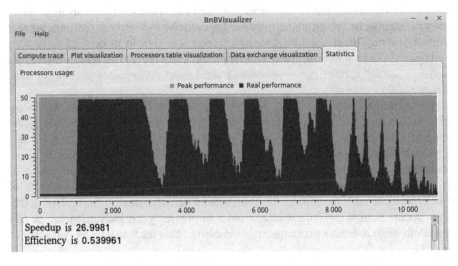

Fig. 5. The performance for small values of T (Color figure online)

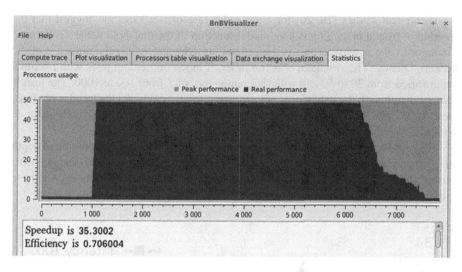

Fig. 6. The performance for moderate values of T

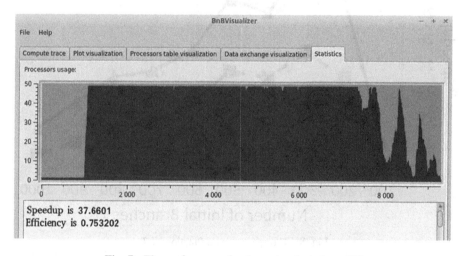

Fig. 7. The performance for dynamic adaptation of T

6.2 Case Study II: Studying Performance of Parallel Frontal Algorithm

In the second case study we simulated the simplest possible load balancing scheme – *frontal branching*. In this approach the master performs T B&B steps thereby producing a number of sub-problems. Then each sub-problem is sent to the respective slave and is solved completely. The results are collected and the best found solution is selected and supplied to the user. The number of available cores is supposed to be larger than the number of subproblems generated by the master.

Theoretical studies [19, 20] for a particular case of B&B method suggest that the speedup of frontal branching is a unimodal function of the threshold value T. We used simulator to check whether this is the general behavior and assess the influence of the network latency. The simulator was run in batch mode on random trees with maximal depth varied from 30 to 50 and with different values of T from 100 to 1000. The results showed that though the behavior is not necessary strictly unimodal the trend is obvious: there is a value of T where the speedup reaches its maximum and then starts to decrease.

Figure 8 shows the plot of the speedup as a function of T for a random tree of depth 40. Two graphs show the speedup as a function of T for zero latency (red) and non-zero latency (blue). We observe quasi-unimodal behavior for both cases. As expected the speedup for non-zero latency is less than for zero latency case.

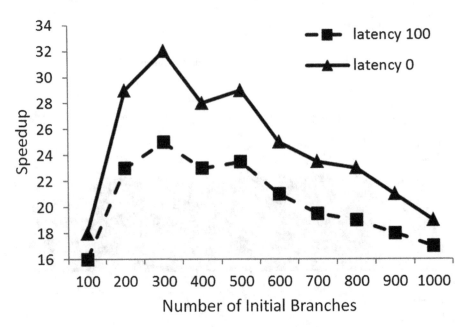

Fig. 8. The speedup as a function of T

7 Conclusions

The paper discussed the simulator of parallel Branch-and-Bound method that can be used for a deep study and comparison of load balancing algorithms. Though the simulation can't completely replace the testing on a real multiprocessor it can significantly reduce the number of expensive runs on a supercomputer. Since the traces produced by the simulator follow the same format as the parallel solver the graphical front-end supports performance visualization for both the simulator and the optimization library. The simulator can run in batch mode to perform large-scale simulation for comprehensive performance analysis, e.g. produce scalability charts [21].

In the future we are going to implement more sophisticated hierarchical interconnect models in our tool and perform a comprehensive analysis and comparison of various load balancing algorithms.

Acknowledgements. This study was supported by Ministry of Science and Education of Republic of Kazakhstan, project 0115PK00554, Russian Fund for Basic Research, project16-07-00458 A, Leading Scientific Schools project NSH-8860.2016.1, Project I.33 of RAS.

References

1. Pardalos, P.M., Romeijn, E., Tuy, H.: Recent developments and trends in global optimization. J. Comput. Appl. Math. **124**(1–2), 209–228 (2000)
2. Scholz, D.: Deterministic Global Optimization: Geometric Branch-and-Bound Methods and Their Applications. Springer, New York (2011)
3. Gendron, B., Crainic, T.G.: Parallel Branch-and-Bound Algorithms: Survey and Synthesis. Oper. Res. **42**(6), 1042–1066 (1994)
4. Lüling, R., Monien, B.: Load balancing for distributed branch & bound algorithms. In: Proceedings of Sixth International Parallel Processing Symposium, pp. 543–548. IEEE (1992)
5. Barkalov, K., Gergel, V., Lebedev, I.: Use of xeon phi coprocessor for solving global optimization problems. In: Malyshkin, V. (ed.) PaCT 2015. LNCS, vol. 9251, pp. 307–318. Springer, Heidelberg (2015). doi:10.1007/978-3-319-21909-7_31
6. Ganglia Monitoring System. http://ganglia.sourceforge.net/. Accessed 12 June 2016
7. Nagios-the industry standard in IT infrastructure monitoring. https://www.nagios.org. Accessed 12 June 2016
8. Stefanov, K., Voevodin, V., Zhumatiy, S., Voevodin, V.: Dynamically reconfigurable distributed modular monitoring system for supercomputers (DiMMon). Procedia Comput. Sci. **66**, 625–634 (2015)
9. Shende, S., Malony, A.D.: The TAU parallel performance system. Int. J. High Perform. Comput. Appl. **20**(2), 287–331 (2006)
10. Adhianto, L., Banerjee, S., Fagan, M., Krentel, M., Marin, G., Mellor-Crummey, J., Tallent, N.R.: HPCToolkit: Tools for performance analysis of optimized parallel programs. Concurrency Comput.: Pract. Exp. **22**(6), 685–701 (2010)
11. Servat, H., Llort, G., Giménez, J., Labarta, J.: Detailed performance analysis using coarse grain sampling. In: Lin, H.-X., Alexander, M., Forsell, M., Knüpfer, A., Prodan, R., Sousa, L., Streit, A. (eds.) Euro-Par 2009. LNCS, vol. 6043, pp. 185–198. Springer, Heidelberg (2010). doi:10.1007/978-3-642-14122-5_23
12. Müller, M.S., Knüpfer, A., Jurenz, M., Lieber, M., Brunst, H., Mix, H., Nagel, W.E.: Developing scalable applications with vampir, vampirserver and vampirtrace. In: Proceedings of ParCo 2007, Jülich, Germany, pp. 637–644 (2007)
13. Mohr, B., Voevodin, V., Giménez, J., Hagersten, E., Knüpfer, A., Nikitenko, D.A., Nilsson, M., Servat, H., Shah, A., Winkler, F., Wolf, F., Zhukov, I.: The HOPSA workflow and tools. In: Proceedings of 6th International Parallel Tools Workshop, pp. 127–146 (2012)
14. Mohr, B.: Scalable parallel performance measurement and analysis tools-state-of-the-art and future challenges. Supercomput. Front. Innov. **1**(2), 108–123 (2014)

15. De Bruin, A., Kan, A.H.R., Trienekens, H.W.: A simulation tool for the performance evaluation of parallel branch and bound algorithms. Math. Program. **42**(1–3), 245–271 (1988)
16. Evtushenko, Y., Posypkin, M., Sigal, I.: A framework for parallel large-scale global optimization. Comput. Sci.-Res. Dev. **23**(3–4), 211–215 (2009)
17. Snir, M., Otto, S.W., Huss-Lederman, S., Walker, D.W., Dongarra, J.: MPI, The Complete Reference. Scientific and Engineering Computation. MIT Press, Cambridge (1996)
18. Lamport, L.: Time, clocks, and the ordering of events in a distributed system. Commun. ACM **21**(7), 558–565 (1978)
19. Kolpakov, R.M., Posypkin, M.A.: Estimating the computational complexity of one variant of parallel realization of the branch-and-bound method for the knapsack problem. J. Comput. Syst. Sci. Int. **50**(5), 756–765 (2011)
20. Posypkin, M.A., Sigal, I.K.: Speedup estimates for some variants of the parallel implementations of the branch-and-bound method. Comput. Math. Math. Phys. **46**(12), 2187–2202 (2006)
21. Voevodin, V., Antonov, A., Dongarra, J.: AlgoWiki: an open encyclopedia of parallel algorithmic features. Supercomput. Front. Innov. **2**(1), 4–18 (2015)

Author Index